Bioengineering of Materials

Books in the Series

Protein-Based Materials
Kevin McGrath and David Kaplan, editors

Biomolecular Electronics: An Introduction via Photosensitive Proteins
Nikolai Vsevolodov

Forthcoming Books in the Series

What Sustains Life?
Dan W. Urry

Materials Inspired by Nature: The Architecture of Functional Supramolecular Structures
Hagan Bayley

Collagen: A Universal Biomaterial for Medical Devices
Richard Berg, editor

Introduction to Microencapsulation
Curt Thies

Nikolai Vsevolodov

Biomolecular Electronics

An Introduction via Photosensitive Proteins

Translated by Marina Georgadze
Edited by David Amiel

Birkhäuser
Boston • Basel • Berlin

Nikolai Vsevolodov
Hyundai Network Systems
A Division of Hyundai Electronics America
Herndon, Virginia, USA
and
Institute of Theoretical and Experimental Biophysics
Russian Adacemy of Sciences
Pushchino, Russia

Marina Georgadze
Jersey City, New Jersey

David Amiel
Cambridge, Massachusetts

Library of Congress Cataloging-in-Publication Data

Vsevolodov, N. N.
Biomolecular electronics : an introduction via photosensitive proteins / Nikolai Vsevolodov ; translated by Marina Georgadze ; edited by David Amiel.
p. cm. -- (Bioengineering of materials)
Includes bibliographical references and index.
ISBN 0-8176-3852-0 (hc)
1. Bioelectronics. 2. Photosynthetic pigments. 3. Protein engineering. 4. Proteins. 5. Rhodopsin. 6. Bacteriorhodopsin. I. Title. II. Series.
QH509.5.V74 1998 98-12277
571.4 ' 5--DC21 CIP

Printed on acid-free paper

Birkhäuser

ISBN 0-8176-3852-0
ISBN 3-7643-3852-0

Typeset by The Alden Group, Limited, Oxford, U.K.
Printed and bound by Hamilton Printing, Renesselear, NY.
Printed in the U.S.A.

9 8 7 6 5 4 3 2 1

To my wife Olga

Contents

Series Preface

The properties of materials depend on the nature of the macromolecules, small molecules and inorganic components and the interfaces and interactions between them. Polymer chemistry and physics, and inorganic phase structure and density are major factors that influence the performance of materials. In addition, molecular recognition, organic-inorganic interfaces and many other types of interactions among components are key issues in determining the properties of materials for a wide range of applications. Materials requirements are becoming more and more specialized to meet increasingly demanding needs, from specific environmental stresses to high performance or biomedical applications such as matrices for controlled release tissue scaffolds. One approach to meet these performance criteria is to achieve better control over the tailoring of the components and their interactions that govern the material properties. This goal is driving a great deal of ongoing research in material science laboratories. In addition, control at the molecular level of interactions between these components is a key in many instances in order to reach this goal since traditional approaches used to glue, stitch or fasten parts together can no longer suffice at these new levels of manipulation to achieve higher performance. In many cases, molecular recognition and self-assembly must begin to drive these processes to achieve the levels of control desired.

This same need for improved performance has driven Nature over millenia to attain higher and higher complexity. For example, the modification of properties of the macromolecules comprising membrane structures was critical to provide partitioning of cellular components and organelles. Tough cellulosic fibers formed that extensively hydrogen bond to allow plants to partially escape the demands of gravity at ground level and occupy new niches. Efficient transduction mechanisms based on tailored material structures are able to interconvert energy as a key to survival for biological systems. Tailored organic-inorganic templates are employed by marine molluscs to toughen ceramics made from available environmental feedstocks. Development of life past the single and oligocellular levels would have been difficult without controls over the synthesis of tailored materials and their assemblies. At the same time, Nature effectively utilizes molecular recognition and self-assembly to drive the formation of complex materials necessary to achieve the performance required for survival. This is essential in biology since all life exists within limited energy budgets. A full understanding of these processes is still needed, since it remains difficult even to predict protein folding, much less the driving forces required to created molecular motors responsible for flagellar motion.

Assembly processes in biology take advantage of phase changes, ionic interactions, hydrophobic interactions, hydrogen bonding, epitaxial growth of inorganic phases, enzyme reactions to alter recognition, and many other forces. All of these processes are enscribed at the primary sequence level but manifest themselves at many levels of organization and hierarchy during the formation of specialized materials. Natural materials also respond and adjust to changes in environmental stimuli, a desirable attribute not well understood or developed in synthetic materials. Finally, natural materials are synthesized, processed and assembled in an aqueous environment, and all of these materials are returned to natural geochemical cycles once their useful lifecycle is completed. An understanding of these processes can only improve the environmental impact of future materials and their handling.

The drive in materials science to control the tailoring of components and interfaces, and the timing of this need with the development of a wide range of tools in biotechnology that can be applied to the same goal, provide the basis for the volumes in this series. Our aim is to address this interface between materials science and biotechnology, and the explore how insights from Nature provide blueprints for materials science. These insights take many forms, from the direct manipulation of biological polymers to tailor structure-function relationships, to the concept of polymer templating to direct and control nucleation and crystal growth in a cooperative fashion. Genetic engineering, nanotechnology, protein chemistry and combinatorial methods, among many other methods, are being used to address this interface between biology and materials science. Progress from these endeavors are already significant, including bioactive gels and polymers, protein-based holographic memories, toughened ceramics, and recombinant spider silks and elastin proteins. The series will include contributions that deal with a wide range of these issues, from books that can serve as useful reference materials, to how-to handbooks, and tutorial volumes. Some examples of volumes planned for the series include protein-based materials, biocomputers, bioelectronics and biotechnology, what sustains life from the point of view of the materials scientist, the architecture of supramolecular materials and devices, and bioceramics.

We thank the editorial board, authors of the various volumes, and our publisher for their efforts and insights into the contributions in this series. One of the biggest problems with this field today is the over-extended breadth of what has been called "biomimetics," "smart," "intelligent" and many other terms that are added on to the word "materials." One of our most difficult tasks in this series is to maintain a focus on core issues, progress, and problems that exist in this field that we are calling "Biotechnology and Materials Science." We believe there is a great deal to be gained by understanding and developing interfaces between biology and materials science that will have a major impact on medical and nonmedical materials. This understanding is already generating new processes to make these materials which will expand in scope in the future.

David Kaplan
Director, Biotechnology Center
Tufts University

Foreword

Evolution has been optimizing proteins for light reception and energy conversion for more than 3.5 billion years, and natural selection has achieved implementations far more sophisticated than is currently evidenced by human engineering. Nature provides both proteins and chromophores that have been highly optimized for converting light into molecular energy with maximal efficiency and reliability. Nature also provides a template for making new molecular systems. The use of genetic engineering and biomimetic organic synthesis has recently provided an array of new materials that have enhanced properties. Thus, we should view nature as both a provider and a teacher.

This book presents a detailed look at light-sensitive proteins, and the use of these proteins in a wide variety of device applications. It is important reading for those interested in exploring the potential use of proteins in device applications. We must examine how nature achieves the level of robustness and efficiency that characterizes those proteins that have been selected for light-transducing functions. Ultimately we may be able to take our knowledge of natural systems and design new molecules that are more suited to energy conversion, data storage and optical processing. This is the goal of biomimetic engineering. We seek to mimic natural systems by using those architectures that have been designed to function in living organisms and transplanting these architectures into devices that require similar structure:function behavior. To accomplish this goal, however, will require the detailed examination of how nature has solved problems of a similar nature. In that regard, both the device engineering and evolutionary optimization of light transducing proteins must select for many variables simultaneously, which include: reproducible synthesis, thermal stability, light-flux stability, cyclicity, energy storage, energy conversion, quantum efficiency and optical cross section. When human engineering is compared to nature, we find that we are readily capable of optimizing selected variables quite well. Nature is still better at optimizing all relevant variables simultaneously. For this reason, the approach presented in this book of coupling biological structure-function discussion with device application is important reading for scientists interested in optimizing either organic molecules or biological materials for device applications.

Dr. Vsevolodov is in a unique position to examine this topic. He and his collaborators were the first to investigate the development of photochromic thin films based on bacteriorhodopsin, the light-transducing proton pump found in the purple membrane of Halobacterium salinarium. The successful development of photochromic and holographic thin films containing this protein, some of which was carried out under Soviet military sponsorship and imposed secrecy during the cold war era, did much to stimulate bioelectronic research in other countries. Some of the details of Project Rhodopsin, the well-funded program that explored the use of light-sensitive proteins in optical computing and image storage, are revealed in this book for the first time. Dr. Vsevolodov has made major contributions to the field of biomolecular electronics, and his perspectives are both important and interesting.

This book is appropriate for a wide audience ranging from upper division undergraduates to active research scientists and engineers. Dr. Vsevolodov satisfies the demands of a general audience by providing detailed background information on each subject to initiate the reader into the issues and controversies relevant to the subsequent, more detailed, discussions. Those scientists and engineers working in the field will appreciate the candid and detailed examination of the topics, and the unique perspective of a pioneering researcher. All readers will find the informal presentation, which is filled with personal commentary, lively and interesting.

Robert R. Birge
Distinguished Professor of Chemistry
Director, W. M. Keck Center
for Molecular Electronics
Syracuse University
Syracuse, New York

Note from the Translation Editor

Concomitant with the creation of a new field comes a new vocabulary. New words like photomedium, photocycle, and photoinduced, have merged with the literature and dialect. This book relies on and supplements the use of such terms. Many of their meanings can be inferred directly while others will be found in the Glossary. Flipping ahead to the extensively cross-referenced items will also aid in defining a term as well as reinforcing and extending an idea or result. The reader is strongly encouraged to make use of these, and other sources, many of which are amply provided within the book.

In a similar way to James Watson's *The Double Helix*, this book tells a story of the climate surrounding scientific discovery and research. In doing so, the paths leading to innovations, for the present generation of scientists and the next, is rendered, if only by a small fraction, smoother and gentler. The cutting-edge is only scary if nobody has been there. For those with a budding interest in a future technology whose groundwork is only now being formed, this book will serve as an introduction and carry them all the way to the edge of research. For those whose interest has already sprouted, this book leads directly to novel experiments, using state-of-the-art techniques, in an area whose boundaries have only recently been limned.

This is an unusual book in many respects. Originally written in a Russian style that is unfamiliar to English readers, its intricately crafted structure is revealed only by some of the same painstaking attention that has gone into the editing of the translation. It is my hope that this effort has provided for the reader a cohesive and reader-friendly text. The book, if read with a mindful eye and a vision of the past, present and future technology, will augment and inspire research as well as provide a scientifically grounded but nevertheless entertaining historical perspective.

David Amiel
Translation Editor

Preface

Until recently, one could encounter terms like nanobiology, molecular engineering, electron transfer through supramolecular wires, molecular computing, artificial neurocomputers, optical memory based on photosensitive proteins, biomolecular optical switching, and so on, only on the realm of science fiction. Today these names are found at scientific syposia, and are now frequently labeled under the umbrella title "Biomolecular Electronics."

My book, published in 1987 in Russia, ended with the words of a Chinese proverb: "The one thousand mile road begins with a single step." Ten years have passed, and headway has already been made towards the establishment of bioelectronics as a valid field of science. Not everyone shares the optimism of the developers of this exotic field. The potential to create a biocomputer (to be defined shortly), which can lead to the construction of artificial intelligence, attracts some, scares others, and confounds the rest. However, as with the discovery of nuclear power, the future of bioelectronics will depend on how society utilizes this new science. The ultimate goal, the creation of an artificial brain, at least as efficient as a human's in size and power, is distant but not unattainable. Before this idea is realized, the electronics industry will see that the microchip has exceeded the limits of miniaturization, and that the era of bioelectronics is around the corner. Once, the transistor replaced the radiolamp, only to be itself replaced by a microscheme. This is how the evolution of scientific applications operate. As for the past adversaries of transistors and microschemes, they have long since become their ardent devotees.

Artificial intelligence is meaningless without a compatible organ of perception, surpassing the human one. Such intelligence also requires a compatible "nutritional" system, which is distributed inside, or in between the biocomputer systems, just as in live organisms. Indeed, the creation of a biocomputer involves an array of scientific tasks which envelop virtually all directions of modern biology, physics, chemistry and electronics. This creates the problem of requiring researches with a truly interdisciplinary outlook. Textbooks on bioelectronics have not been written yet. This book intends to reveal the methodology of what has been done in at least one of the areas of this new science so that a new generation of researchers will be prepared.

This book tells the history and present state of research in one of the most interesting sections of bioelectronics: photosentitive proteins and protein complexes. Thanks to those wonderful creations of Nature, our planet is now full of life and intellect (both, so far, natural). In my opinion, photosentitive proteins and photosynthesizing complexes are the most promising materials for the goals that molecular electronics has set out to accomplish. I intend to prove this opinion within these pages.

Chapter 1 offers a brief survey of natural and artificial molecular complexes, and relates some of the achievements and still unsolved problems of modern molecular biotechnology. This chapter also gives modern definitions of biocomputer and biocomputing. Chapter 2 tells of the previous, and recently discovered rhodopsins (phototransforming proteins of natural and artificial origin). Chapter 3 is dedicated to bacterial rhodopsins as the likeliest candidate for use in bioelectronics. Chapter 4 discusses the history and modern state of photosensitive mediums research (photographic and photochromic). Chapter 4 also compares the optical and holographic characteristics of biological, and traditional photomaterials, and describes their advantages and disadvantages. Chapters 5 and 6 discuss the presently known properties of rhodopsins as biolelectronic and optical elements, and their application to a new prototype of electronic device. The Introduction and Chapter 7 reveal the present and future of bioelectronics and biocomputers, and their connection with the topics of this book.

The book also includes stories about the discovery of bacteriorhodopsin and the creation of the first photochromic films. Some of the episodes, dating back to the cold war era, with plots about highly paid international scientific espionage, read as if borrowed from a James Bond movie.

This book is not a scientific monograph in the traditional sense. Much of the vast scientific material on this subject ranges from physics to microbiology and is, to date, very poorly connected with bioelectronics and biocomputing. One of the tasks of this book is to provide specialists with a general, simple, but sufficiently broad conception about the present state of photosensitive proteins research, and its possible interrelation with prospective technology. Hopefully, the book will serve as an introduction into this new field of science for students, graduate students, and specialists in adjacent fields.

Nikolai Vsevolodov

Acknowledgements

I wish to thank all those who helped me in the process of writing this book, both directly and indirectly. My special thanks goes to Robert Birge who took upon himself a formidable task to review the whole manuscript and made a number of valuable suggestions.

I wish to thank David Amiel, my translation editor, for the huge work on improving the English text of my book. I now read it in English with more pleasure than in Russian. I thank both David Amiel and Alla Margolina-Litvin, my Birkhauser editor, for their relentless efforts and energy without which this book would not be possible.

The multidisciplinary nature of the book has frequently required consulting with specialists in various fields, and I usually received the answers to my questions very quickly for which I am much obliged to Sergei Balashev, Michael Conrad, Lel Drachev, Anna Druzhko, Felix Hong, Evgenii Lukashev, Elena Korchemskaya, Kodji Nakanishi, John Spudich and many others. I thank Vladimir Skulachev who sent me his book on Natural Bioelectric Generators, from which I learned a lot about the history of Bacteriorhodopsin research. I am indebted to Michael Ostrovsky for his help in the process of writing the chapter on visual rhodopsins, to Norbert Hampp who made valuable critical remarks on the general outline of the book, and to Elena Karnaukhova who helped me in creating the Appendix.

1

Introduction

Today it is hardly possible to include in one book all the achievements of molecular and biomolecular research since nearly all sciences are actively involved in research on bioelectronic processes and biomaterials for molecular biotechnology (this term may best describe the entire scope of biotechnological problems at the molecular level, including those of biocomputers). The number of natural and artificial elements and devices of molecular scale, discovered in nature, or created artificially in laboratories, is growing constantly. On this basis, future biocomputing systems may be created or may precede the appearance of entirely new systems of yet unpredictable structure. In Sections 1A to 1E, we offer a general survey of molecular biotechnological problems. However, only one of them, the use of biological molecular light converters, and use of their artificial analogs as biosensors and microelements for a biocomputer was chosen as the subject of this book. Why only biomolecular light converters, we may ask? There are three reasons for this choice.

First, such converters could supply biocomputing systems with the cheapest energy in the world (solar energy). Second, information in such systems is exchanged, not only by using electrons, but also photons; and the latter is to date, the fastest and most effective method of energy and information transport. In addition, as opposed to electrons, overlapping light waves do not interrelate. Thus, no distortion of information or loss of energy occurs. Finally, the application of the well-known effects, bioluminescence and biochromism (and recently discovered bioelectrochromism), will permit the use of biocomputer-compatible displays for information input/output by an optical method, which is the fastest and most effective of all.

Information transport by photons, and information storage by photosensitive molecules, which are far smaller than the nerve cells of the brain, can lead to the creation of a more effective computer than the human brain (I say "more effective" since what would be the point of a computer that is less than or equal to that of a human?); particularly if the dynamic holography method (see Glossary) of associative processing and recording of information is employed.

Optical memory, as stated earlier, is many times more efficient than any other method for information storage. Ten times more information can be stored on one square centimeter of high-resolution photofilm than on magnetic film. A small size hologram, or even a common photograph, can store enormous amounts of information. "Picture logic" computers belong to the new generation of computers with optical memory based on dynamic photosensitive substrates of superhigh resolution. Traditional phototechnology is, however, approaching its limits. To solve the latest phototechnological and optical memory problems, new photomaterials with new properties and characteristics are needed. Systems of optical information processing work in real time, thus requiring reversible (photochromic) photomaterials with diverse speeds of reversion, and with an unlimited number of write-erase-write cycles (see Glossary). Optoelectronics is marching forward, and will also require new devices for the correction, improvement, recognition, and sorting of images.

Optoelectronics has united the developers of electronic and optical systems. Together with traditional integral schemes, elements of optical integral schemes have been developed such as deflectors, modulators, commutators, etc. Hybrid optical-electronic complexes are constantly being developed. To date, optical components fall behind the electronic ones in efficiency. However, they are superior in speed, and in the amount of information processed per unit time, which is frequently more important. Optical memory is more reliable than magnetic and electronic memories since it is not affected by external electromagnetic impacts. In the quest for new photomaterials, developers have been avoiding the organic, and more specifically the biological ones, reasoning that proteins are too unstable for incorporation into phototechnology. Recent discoveries with light-sensitive retinal–protein complexes have proved the contrary. For example, photochromic films and photosensitive three-dimensional elements based on bacteriorhodopsin (a visual rhodopsin analog which will be discussed in detail later) are far more stable and "technology friendly" than many of the traditional photochromes.

It is quite possible that in the near future many artificial elements and devices will be based on natural constructions and schemes; especially devices for light conversion to different forms of energy, in addition to systems of artificial vision and memory. There are numerous possibilities for the application of photosensitive natural proteins in biocomputers and biotechnology. The term "biocomputer" then, defines a system, designed and guided by artificial and natural patterns, and constructed with the aid of natural and/or artificial proteins and bioelements. So far, only the tip of the problem has been explored, but the results are already amazing.

This book will also relate what nature and modern scientists have to offer future generations immediately. The future always appears unattainable until one day it becomes reality, and its roots are embedded into the fabric of everyday life.

1A. Biophotonic Processes

Nearly all biological processes on Earth are triggered by solar energy. Plants, seaweeds, and photosynthesizing bacteria absorb sunlight and convert it into their life energy. In the process of photosynthesis, plants and seaweeds generate structural and nourishing components such as proteins, carbohydrates, and fats. Herbivorous animals store this energy in cells in the form of ATP, or as an electric potential. Energy reserves are expended to move, grow, maintain thermal balance, support brain and nervous system activity, and even to emanate light and electrical discharges. Only 0.1% of the solar energy that reaches the Earth's surface is absorbed by plants, but it is sufficient to provide for all the diversity of life on our planet. A light converter of such effectiveness and productivity has not yet been invented by humans.

Animals, plants, and microorganisms have special organs to absorb photons and to convert these into a form its cells can use. Such structures in plants and seaweeds are called chloroplasts (photosensitive pigment-bearing protein complexes). The most common are chlorophyll and bacteriochlorophyll. Nearly all animals and some bacteria also have special photosensitive structures, usually biomembranes, containing the photosensitive molecules rhodopsin and bacteriorhodopsin (BR).

Chlorophylls and rhodopsins are the most well-known and widespread photosensitive biomolecular solar energy converters on our planet. They are also difficult to classify as either animal or plant. For example, the unicellular seaweed *Euglena gracillis* has a micro-eye like photoreceptor to detect a light energy source, but also has chloroplasts. In the dark, its chloroplasts vanish, and the seaweed survives by digesting food from the external medium while its "eye" "sleeps."

In optical phenomena, photons dispersed by the surrounding medium are refracted by objects, animals, plants and so on, which transmit information about shape and activity. The majority of higher animals perceive visual information through their eyes, or eye-like organs. The retina contains a unique visual protein. The crystalline lens projects images onto the retina, which then transmits them to the brain. Only animals, whose eyes and brain are sufficiently complex can perceive the world in visual images. More primitive animals use their eyes to orient by spectral composition of light and/or by light intensity, and the difference between light and darkness. Until recently, we believed that rhodopsin was present only in animals. During the last decade, however, amazing discoveries have been made which reveal that many bacteria and microorganisms also possess rhodopsins.

In the 1970s, a new group of bacteria was added to the phylum of microorganisms. It was called archaebacteria and could not be classified as prokaryote or eukaryote. Certain features tell us that these are the oldest, and perhaps the first bacteria on Earth. In 1971, a retinal–protein complex was discovered in the surface shell of one of the species of archaebacteria, a salt-lake dweller. It was very similar to the visual rhodopsin of the eye and therefore was called bacteriorhodopsin (BR). It employs sunlight energy for goal-directed hydrogen ion transport across the bacterial cell membrane, generating a potential difference. This potential provides for the synthesis of ATP. This kind of photosynthesis differs from the well-known chlorophyll-mediated photosynthesis, in which light energy induces the separation of electric charges, which also creates an electrochemical potential difference on the membrane.

The discovery of BR (the history of its discovery deserves a separate chapter) opened up the search for new types of visual rhodopsins in bacteria and microorganisms and stimulated a series of unexpected scientific results. For example, in the red eye of *Chlamydimonas*, familiar from high school courses, another type of rhodopsin was found. In the fervor of research, scientists would sometimes observe a rhodopsin in every newly discovered photosensitive protein, however little relation it bore to rhodopsins. However, such occurrences, as we have already mentioned, only inspired the scientists to look for new photoreceptors in well-known organisms.

Two main types of photodetectors of biological origin are known to occur in nature simultaneously: chlorophylls (vegetative and bacterial), and rhodopsins (visual and bacterial). It is strange that to serve the immense variety of photophysical and photochemical processes responsible for the existence and development of almost all life on earth, nature selected, engineered, and dispersed all over the world only two basic types of photoreceptor molecules: retinal-proteins and chlorophylls!

Let us recall some general facts about these two photoreceptors. Chlorophylls are the most common natural light converters on our planet. Chlorophyll was first isolated from the leaves of plants in 1868, and later it was extracted from seaweed. Its analog (bacteriochlorophyll) was found in bacteria. The structure of chlorophyll is dominated by a porphyrine ring, in which nitrogen atoms interrelate with magnesium ions. A carbohydrate tail is linked to the ring (see Appendix). Normally, several hundred chlorophyll molecules are arranged in special photosystems (I and II), which are located in the membranes (thylacoids), while the thylacoids are housed within special cells (chloroplasts).

These chloroplasts constitute the basic organelle, hosting the light and dark dependent reactions of photosynthesis, which lead to the formation and storage of carbohydrates and fats. Figure 1.2 is a close-up view of the photosystems. Chlorophyll is synthesized at the regions labeled light antennas (see Figure 1.1), so called since light is transmitted from this point to reaction centers (RCs). After photon absorption, the RC assumes a different charge, and this potential is transferred to the photosystems and used for the synthesis of ATP from ADP. This is displayed in Figure 1.2.

The Light energy conversion process initiates at the RCs, formed by protein complexes where molecules of chlorophyll, pheophytine, and other

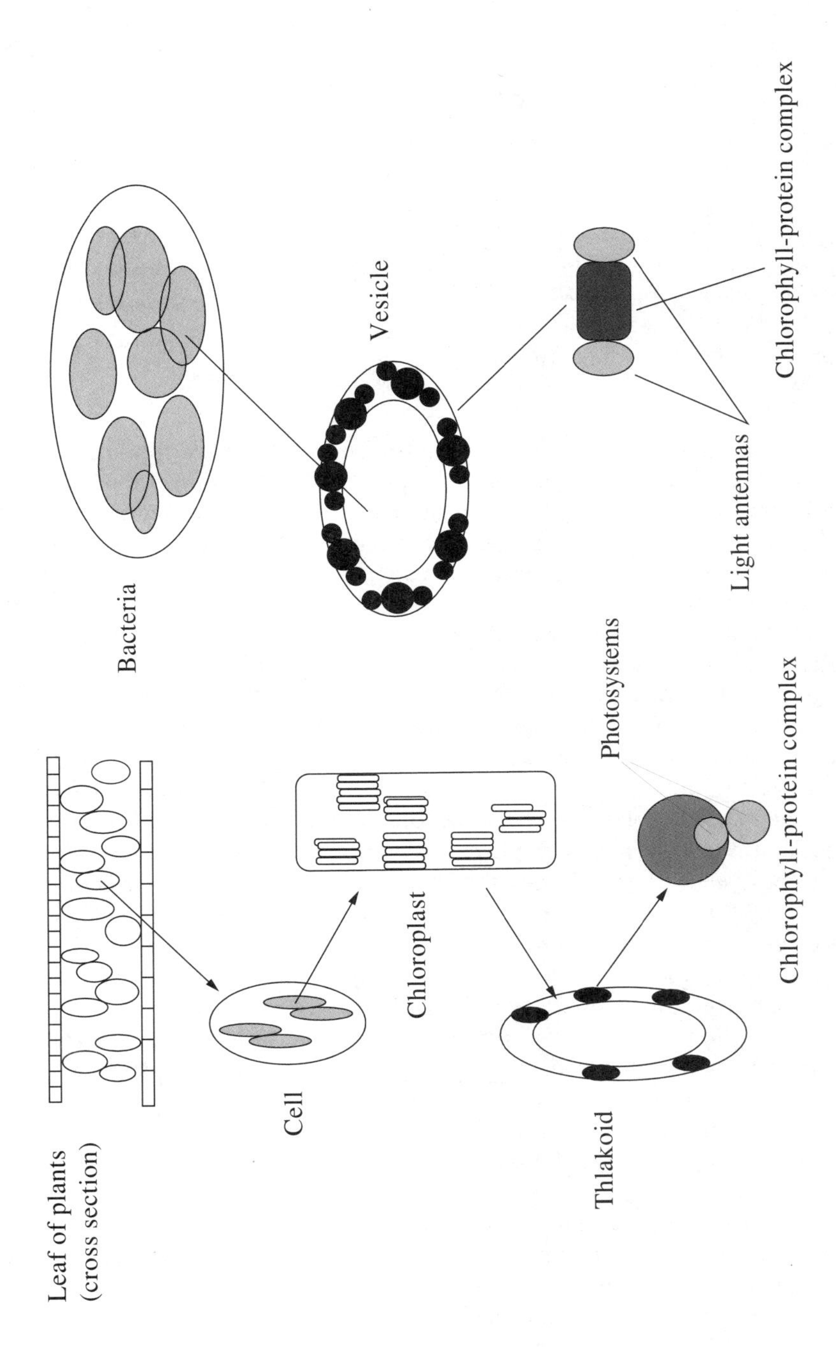

Figure 1.1. A generalized representation of the main structures used in photosynthesis.

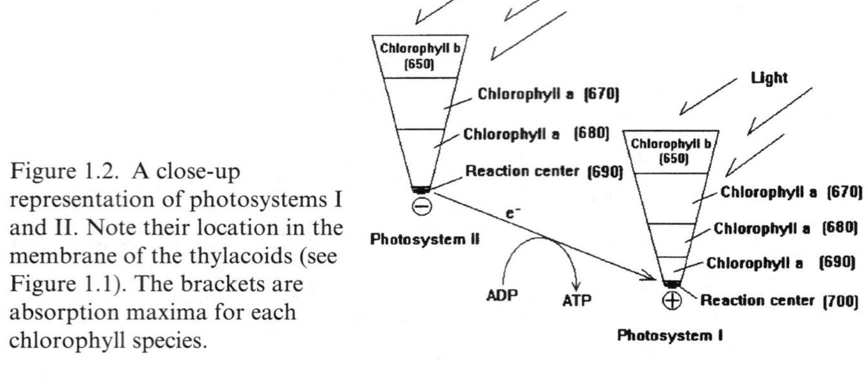

Figure 1.2. A close-up representation of photosystems I and II. Note their location in the membrane of the thylacoids (see Figure 1.1). The brackets are absorption maxima for each chlorophyll species.

photochemically active molecules are spatially arranged (see Section 1B). The relationship between the structure of chlorophyll and its function at the RC of photosystems I and II is well-known, and has been fully covered in the literature [Sybesma 1984]. The efficiency for plants and algae does not exceed 1.5–2%, however, in the RC it may reach 50%. In some artificially modified RCs it goes up to 80%!

Today, 11 chlorophylls have been determined to be similar in their spectral, structural and photochemical properties. Minor structural differences are due to the surrounding medium and to the construction of photosystems, in which a certain type of chlorophyll is functioning (see Section 4F). There is no rigid distinction of chlorophylls between the vegetative and the bacterial groups. Both vegetative and bacterial chlorophylls of Type a were found in *Rhodospirilium rubrum*. Out of 11 chlorophylls, only a and b types have been well studied, and b chlorophyll is employed only in light-collecting systems, but not in photosystems I and II. A relatively recently discovered bacteriochlorophyll g might interest biotechnologists as a photoregistering material for its IR spectrum. Upon IR exposure, the absorption maximum of bacteriochlorophyll g shifts by 100 nm to the blue region. d and e chlorophylls from green and purple bacteria have not been sufficiently researched.

Figure 1.3 shows that the absorption spectrum of natural chlorophylls has two maxima -in the blue and near IR regions. The absorption spectra of natural retinal–protein complexes ranges from UV to red. This fact is important for bioelectronics, since only two common types of phototransporting proteins found on the planet span the complete optical spectrum from UV to IR.

Rhodopsins are the second most abundant photosensitive proteins found in nature. Their basic structure is a retinal–protein complex. Retinal and protein taken separately absorb in the UV region, whereas upon the formation of a complex, the absorption spectrum of a chromophore appears in the visible region. The present definition of "visible region," however, is based on

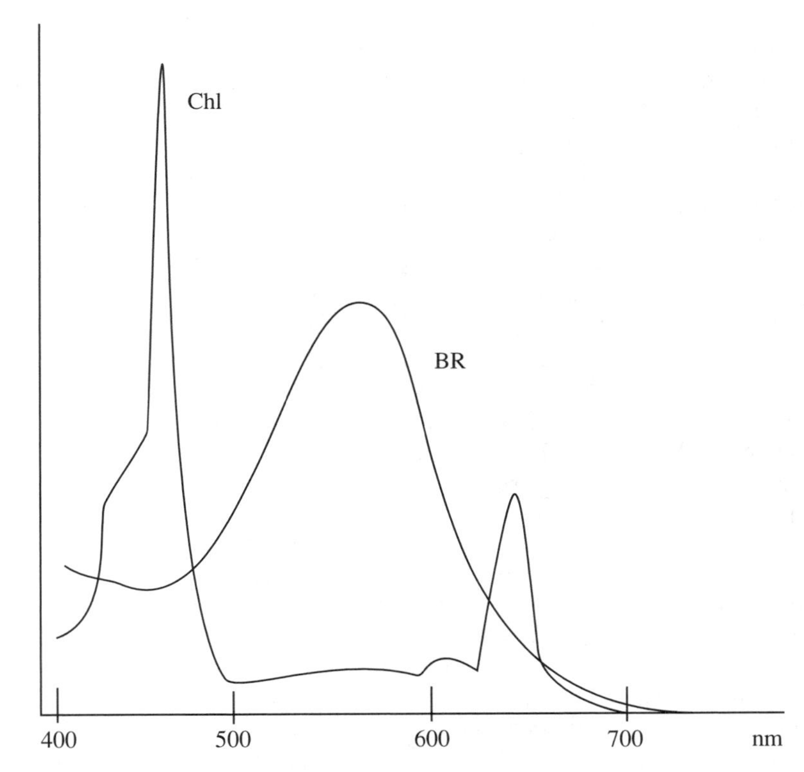

Figure 1.3. The combined absorption spectra of chlorophyll (Chl) and bacterio-rhodopsin (BR) spans the entire visual range of incoming solar radiation, from 300 to 800 nm.

old experiments on human vision, and is not quite accurate. For example, the eyes of some mollusks contain rhodopsin that absorbs in the UV region. It is also known that humans can perceive UV irradiation if an artificial lens transmits in the UV region. Rhodopsins in visual and orientation systems of various organisms may differ slightly in protein structure and/or isomerization of retinal; however, their 3-D structures, and the principles of photophysical processes are similar.

The functioning of all rhodopsins aims at one final goal: the conversion of light energy into an electric potential. The mechanics of electropotential energy utilization by the organism depend on the structure of the photoreceptor apparatus, and is different in bacteria than in the human eye. Upon radiation of the chromophoric center of any rhodopsin, the retinal molecule undergoes certain photoinduced conformational or electrostatic changes which, in turn, lead to conformational changes in the protein. This process is reflected in subsequent changes of the rhodopsin molecules maximum absorption position. At one of the stages of the described photochemical process, depolarization of the rhodopsin molecule occurs, transporting electric charges (ions) across the rhodopsin-containing membrane. The membrane is thus polarized, and the cell can utilize the energy of this polarization for many different purposes.

It is worth remembering that retinal–protein complexes transfer ions across membranes, whereas in chlorophylls, electron charge separation occurs. Both mechanisms are light-dependent. It is not surprising that, in 1971, an attempt was made to create a chlorophyll-based electrochemical element for the conversion of light into electric current.

Curiously, all energetic and photoenergetic processes in cells follow the same pattern, according to which the energies of a charged membrane and a transferred proton are employed to synthesize ATP. One can only admire the simplicity of the idea of a universal energy converter, functioning at the molecular level! Figure 1.4 shows a simplified scheme of the energy conversion processes in biological cells.

The principles at work in the conversion of one form of energy into the other is straightforward. The semipermeable membrane houses two conversion systems. The distance between them is insignificant. One system converts light energy, and the energy of oxidization into a stream of electric charges flowing across the membrane. The second system utilizes the energy of the charges in any other area of the membrane for storage or conversion into working energy. The only difference is in the direction of the charge flow: in chlorophyll-mediated systems it is in the opposite direction.

This mechanism is worth imitating to create an energy supply for future biocomputers.

1B. Bioelectric Phenomena

An impressive example of a natural bioelectric phenomenon is the electric skate. In ancient times, according to reports from ancient historians, skates were successfully employed in medical shock therapy. The electropotential on a skate's skin is a sum of small electropotentials on specifically arranged skate body cells. Today it is considered a common biological occurrence. An electropotential on the surface of a biological cell of any kind happens as the result of intracellular biochemical and biophysical reactions, and is part of its life processes. In photosensitive cells, the potential is created at the expense of absorbed photons. Electric potentials may change at different speeds, sometimes very quickly, resembling an electric discharge in the cells of the mentioned electric skates. The energy of the cell potential is sometimes converted to light (bioluminescence).

Mechanical disturbances, thermal, and even gravitational changes are transformed by special cells into electric signals. Smell, taste, and humidity

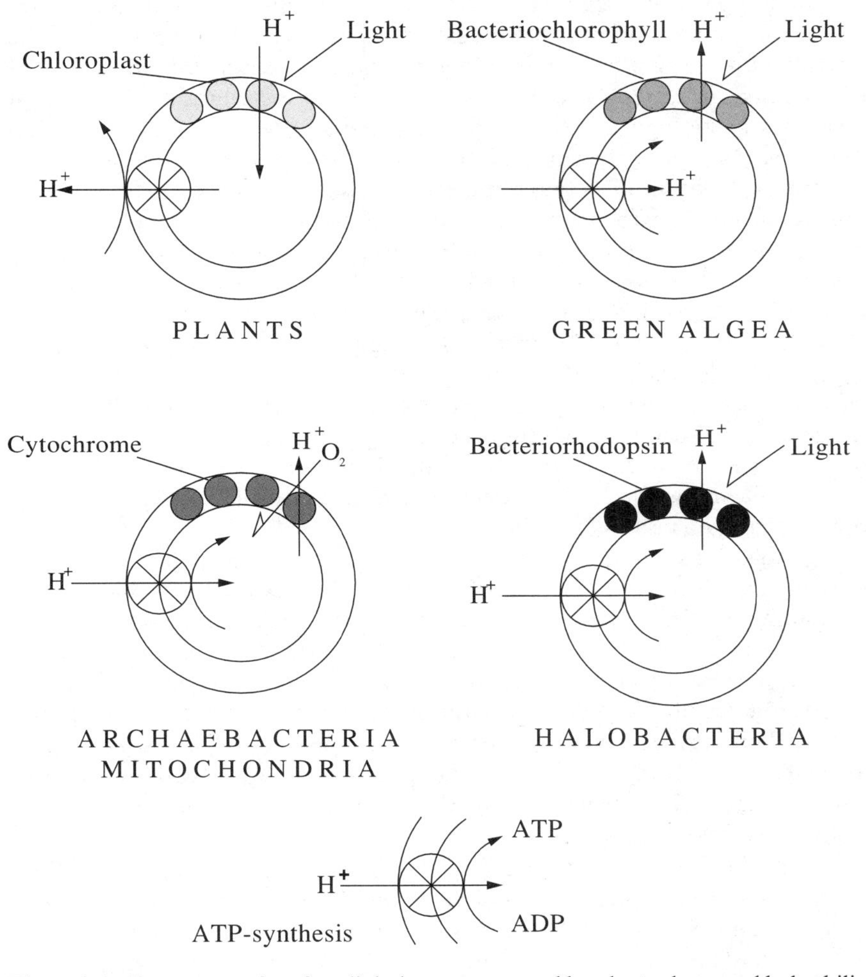

Figure 1.4. Energy transfer of sunlight into energy used by plant, algae, and halophilic cells is based on proton translocation across the cell membrane, as a result of photon absorption by the chlorophyll and retinal proteins. The resultant energy of the membrane potential is used to transform ADP into ATP. In the nonphotosynthesizing cells and mitochondrias, the same process is supported by reduction/oxidation reactions.

receptors also produce electrosignals. At the disturbance, an action potential is created in the receptors. The receptors are connected to axons which act as electrosignal generators. The difference between an action potential and a resting potential enables all receptor cells, neurons, and muscle fibers to produce electric signals of different shapes and magnitudes.

It is well-known that electric potentials on the membrane results from active and passive migration of ions across the membrane. The magnitude of the potential on the membranes of different organelles reflects the prevalence of either anions or cations on either side of the membrane. Many of the chemical processes in biological systems are accompanied by oxidization-reduction reactions based on replacements or transfers of oxygen and hydrogen atoms, or electrons, within or between molecules. This creates a persistent flow of "microcurrents" within live cells, and in biomolecular complexes. The measurement of such currents is a difficult, but possible. After Skulachev (see Section 2.2), fantastic titles like: "proteins as generators of electric current" or "proteins as molecular electric power stations" have entered scientific terminology. Skulachev's colleagues have developed a reliable experimental method, and have shown that long-known enzymes of cytochromoxidase and a newly discovered BR form are, from a simple but useful perspective, molecular generators of electric power. This was shown by direct measurement of electric current in proteoliposome systems (artificial membranes). Their method is now widely employed in measuring ion currents and potentials at the molecular level.

Investigations on the electric current in proteins and biological complexes have a longer history [Krasnovsky 1948]. The photosynthesis of chlorophyll-containing proteins, accompanied by multistage electron transport by a so-called Z-scheme, (See Figure 1.2) has and still is particularly appealing. By this scheme, the electron acts as the main sunlight converter into chemical energy. Electron migration leads to a separation of charges and creates an electric potential on the membrane. This potential, as opposed to the one induced by ion transfer, can bring electrons into motion, thus creating a wire transferable electric current. This provides a theoretical basis for attempts to engineer a chlorophyll-based photoelement.

The problems of sunlight energy conversion in photoreaction centers (PC) of plants and seaweeds is well covered in the literature [Sybesma 1984] [Clayton and Sistrom 1978]. The structure of the RC in photosensitive bacteria has been well studied with the help of various spectral methods. For example, picosecond laser spectroscopy revealed the sequence of electric processes in the RC [Shuvalov and Parson 1981] as a chain of molecular electron transfers (bacteriochlorophylls, pheophytines, chinones, and hem groups) which is strictly conserved in the RC protein. In 15 psec, upon the absorption of light by bacteriochlorophyll, electrons migrate along this chain a considerable distance (30–35 A), accompanied by a loss of energy. These two factors—distance and energy loss—provide for the long-time stability of charges (up to 100 ms) [Shuvalov and Klevanik 1983]. This time period is sufficient to convert the energy of the charge separation into the chemical energy needed to synthesize organic compounds. The quantum yield of sunlight energy conversion to the energy of separated charges in the RC approaches 1. The RC working efficiency, as mentioned above, may be as high

as 80%, and the photosynthesizing process proceeds at room temperature. In other words, the RC functions as an ideal energy converter!

Such characteristics as high molar extinction coefficients and quantum yields of almost 1, allows us to regard RCs as a potential photomaterial for IR light registration. Even simple chlorophyll based memory systems will be able to provide temporary resolution, defined by the time of chlorophyll photoexcitation, up to 10^{-8} sec. These special properties account for the interest in creating artificial pigment complexes based on chlorophyll, and on electron-donor-acceptor proteins. The possibilities of chlorophyll application in biotechnology will be discussed in more detail in Sections 4F and 7A.

1C. Molecular Electronics

Molecular electronics as a term was established in the early 1980s. Naturally, long before that, many scientific groups had been studying electronic processes at the molecular level. However, it was not until the emergence of nanotechnology, and the construction of molecular and atomic resolution devices, that molecular electronics was recognized as a new branch of science. This term is sometimes used to define a new class of molecular resolution devices, such as the STM (Scanning Tunneling Microscope), or a method like the Langmuir-Blodgett.

Those researchers who, in one way or another, stood at the cradle of this new field of science, have only recently been officially assembled. In 1982, the first issue of *Molecular Electronic Devices* came out in print, and 1985 saw the start of *The Journal of Molecular Electronics.* In May 1991, in Hungary, the initial meeting of the Society for Molecular Electronics and Biocomputing (IS MEBC) took place. Since that time, symposiums on MEBC have occurred regularly, and the Newsletter of MEBC has been published. The international book series "Molecular Electronics and Molecular Electronics Devices" has been published by CRC Press (U.S.) since 1992. The number of publications on different branches of research in molecular electronics is increasing, and results of studies look more and more inspiring. Today, molecular electronics labs are found on all continents, excluding only Antarctica.

One of the founders of the IS MEBC, Dr. Felix Hong, wrote: ."...the state-of-the-art progress made in the infrastructure of molecular electronics research, the concerted and concentrated efforts made in a number of countries, and the ever-increasing shift of the scientific public's perception of molecular electronics is shifting from the theoretician's fantasy to the practitioner's new opportunity" [Hong 1994].

The main modern-day strategies of molecular electronics can already be singled out:

* employment of chemosynthesis methods in biology for the synthesis of live system analogs.
* imitation of biological structures and/or their functions, using various scientific methods. This direction is sometimes called "the synthetic design for biomimetics."
* application of natural protein complexes as ready-to-use technological elements.
* the use of photoelectric effects in phototransforming and photosynthesizing proteins for the needs of molecular optoelectronics.
* solution of the so-called "problem of molecular interfacing between biomaterials and the outside world," in other words, elimination of the incompatibility between the biological part of the element and traditional microelectronic elements.
* genetic engineering of new proteins with pre-determined properties and without natural analogs.
* exploration of the molecular recognition mechanism (the "key–lock" principle), which is extremely important for the progress of studies on self-assembling microelements, or even larger systems.

In a 1993 survey, a list of the main achievements in protein applications for molecular electronics was published [Gilmanshin 1993]:

1. A pre-designed protein has been synthesized. Even though the details of its construction may differ from the design, it is, nevertheless, far more stable than any of its natural analogs [Regan and deGrado 1988].
2. Complete synthesis of a gene de novo is also possible for a protein having no natural analogs [Richardson and Richardson 1989].
3. An enzyme for nonaqueous solutions has been "constructed." The importance of this event needs no comment [Arnold 1992].
4. A protein has been created, forming, and supporting a nanometer crystal of a semiconductor [Dameron and Winge 1991].
5. A model ion channel was constructed using peptides. Proteins controlling ion flow across the membrane are the most important elements of a future biocomputer [Lear et al. 1988].

This list is far from complete, and is applicable to both biomolecular and molecular electronics.

Not long ago, yet another new term appeared: "Molecular Monoelectronics." It refers to single-electron processes in inorganic structures. This process correlates with single-electron tunneling, discovered and investigated by Kuzmin and Likharev [1987] and Fulton and Dolan [1987]. The advantages of such systems for use in the computational parts of the computer are obvious, and the possibility of their production has been experimentally confirmed. Devices based on this principle are easily compatible with the elements of modern electronics, and their size is comparable to molecular sizes and membrane thickness. This is all well and good; however, they function only at the temperature of liquid helium, which is not convenient and sometimes is not practical (as, for instance, in compact PCs). On the other hand, it is known that single-electron processes, based on direct electron tunneling also occur in electron-transfer channels of biological systems at normal temperatures. It was proposed to use certain biomolecules (for example, chlorophyll or protoheme) as ready-to-use monoelectronic elements [Gillmanshin and Lazarev 1988]. With this advance yet another term comes into usage : "Biomolecular Monoelectronics." Thus, the modern definition of molecular electronics is expanding [Hong 1994].

1D. Biomolecular Electronics

There is a dual meaning in the definition of the terms biomolecular electronics. Electronics could be defined as the science studying the processes in which electrons and electron currents are involved. Analogously, biomolecular electronics could be defined as the science which studies the identical processes in biomolecules only. For example, charge separation and electron transfer between the photosystems and chlorophyll–protein complexes would be in the domain of biomolecular electronics (see Section 1A). Another example would be electron transfer by cytochromes described in any molecular/biology textbook.

On the other hand, in the literature, when we say biomolecular electronics we imply the use of biomolecules or their complexes for application as independent functional devices capable of interfacing with modern electronic devices. Typical examples would be biosensors, widely used in medicine, or the lesser known photoconverters, both of which are discussed in this book. Let us attempt to explain this duality with examples.

All organisms on our planet, even the most primitive ones, are involved in myriad interactions with the environment throughout their life cycles. Detectors are required to receive information from the environment. Mechanisms to transport this information are also required for processing and reception centers; information which must then be stored for future use. In addition, a mechanism is required to ignore unnecessary or excessive information. From this perspective, organisms may be regarded as biological analogs of the modern digital computer. Furthermore, the transmission of signals in live organisms, and their processing on a microscopic scale, involves electrons and ions. Information storage occurs via the separation of electric charges and electrostatic processes, just as in modern electronic microschemes. However, living systems actively employ molecular conformation and molecular recognition, which together constitute an essential difference from artificial electronic systems.

With all this in mind, we can generate a working definition of biomolecular electronics. It is that scientific field whose goal is to create hybridized instruments and devices consisting of both biological and nonbiological components in which the agents responsible for transport of information and energy are electrons and/or ions.

During the period 1987–1991, several sessions of a Symposium on Molecular Electronics and Biocomputing were organized in Hungary, Russia, and the U.S. The title of the symposium did not contain the word "bioelectronics," but was suggested by "biocomputing."

One of the first real creations of bioelectronics was the biosensor. They are already getting introduced into daily life as a series of minidetectors and indicators that signal mishaps externally or internally (see Section. 5C). The global, long-term vision of bioelectronics is the creation of a biocomputer. The actualization of this idea depends on the answers to two questions. The first is, what is the limit of miniaturization of microelectronic elements? The second is, what are the resolution limits for modern scientific devices? Put simply, how deeply can we observe while constructing the elements of the biocomputer.

Miraculous as it may seem, a single generation of scientists has witnessed the resolution of scientific instrumentation increase from a micro to femto (10^{-15}) scale. The duration of laser impulses has been reduced to femtoseconds, while the resolution of the Scanning Tunneling Microscope has reached the atomic level, permitting direct visualization. In addition, novel computer modeling techniques are able to select which molecular structure may lead to a specific function. Another step on the "thousand mile road" would be the creation of a biomolecular assembly line in which a predesigned biomolecule is assembled and incorporated into a biocomputer. However, this step can not be accomplished until the structure/function relationships are fully elucidated for such biomolecules.

The basic element of every electronic device is a switch. Every calculating device functions as an array of sequentially working switches. Nature widely employs this method of information storage and processing. Many photosensitive proteins may be regarded as single-pole (throw) switches. Several examples will illustrate this point. After photon absorption, a BR molecule switches the potential on the bacterial membrane and activates phosphorylation. Sensory rhodopsins in the same bacteria act as a switch and alter the direction of flagella rotation. In another molecular complex, visual rhodopsin, having absorbed one photon, switches from dark to light processes, triggering a sequence of biochemical transformations, resulting in a nerve impulse (the energy of which is 105 times more than that of a photon). To supply this last process with energy, the internal energy of the organism is expended. A system like this is frequently compared to a phototransistor where a minor change on the base of a transistor leads to a cascade of electrons from the emitter to the collector. This generates an amplifying effect similar to the one in the mechanism of vision (as described above), but supplied from an external energy source (an electric battery).

The processes that occur in the chloroplasts of green plants between photon absorption and the launch of the phosphorylation process can be also regarded as amplifying processes. During photosynthesis, after light absorption, displacement, and separation of charges on both sides of a "photosynthetic" membrane occurs; an event roughly similar to an ordinary charged capacitor. The energy of the separated charges is then utilized on an as-need basis by cell organelles to support vital life processes (similar to the way a capacitor discharge feeds electronic blocks of a scheme). It is known that the discharge of a capacitor may occur at a much slower rate than its charging. This process is also comparable to the work of a rectificator in electronic schemes. Interestingly, recombination of separated charges in biosystems is similar to a leaking capacitor, which diminishes its efficiency. Similar to technologies quest for a more efficient electronic scheme, nature also has to invent ways of inhibiting the recombination processes. The spatial positioning of prosthetic groups within the photoreaction center was shown to be optimal for an efficient forward electron transfer, and for a diminished reverse electron transfer [Kuhn 1986]. This is an example worth following.

From the above examples, it should be clear how nature employs biomolecular electronics. The level of modern experimental science allows the study of electron processes (charge separation in single molecules for example). Any modern research instrument is electronic and thus, by chance, beyond the control of the experimentalist, they are forced to work with a biological and electronic hybrid when peering into electronic processes in biomolecules.

The young biomolecular electronics, as well as the old electronics which we still live by, are endowed with all the essential parts for the assembling of any scheme, such as switches, amplifiers, rectificators, transistors, etc. What we still have to learn from nature are the methods to assemble the building blocks of a biocomputer into a working complex.

1E. Biocomputing: An Overview of What Needs to Be Done

A significant advantage of natural "computers" (biomolecular computers) over the computers of today is that the size of the elements responsible for information processing, storage, and transmission, and the expenditure of energy per unit of information, is much smaller in live organisms. Another distinction is that biomolecular computation in a natural system has neither serial, nor parallel circuitry. Finally, and perhaps most importantly, the brain and systems connected to the brain, are capable of pattern-recognition, self-adaptation to environmental fluxes, process controlling, filtration of information and self-assembly. These distinctions demonstrate that the desired goal (the creation of a biomolecular computer) cannot be attained by simply increasing the number of elements in modern electronic computers, and the simultaneous decreasing of their size. The utilization of specific properties in biological processes, together with the miniaturization of the functional elements down to the molecular level, can lead to the desired results. This is the reason scientists and biophysicists in particular, are engaged in active research with molecules and natural molecular processes responsible for information reception, storage, and processing in live organisms, while mathematicians are trying to discern the laws that describe such processes. Stated simply, it is crucial to understand the interactions between biosystems at the submolecular level.

There is a certain ambiguity to the terms "Biocomputer" and "Biomolecular Computing." Knowledge of structural and functional principles in natural computers might help to develop entirely new design patterns for future computers. The employment of biological organisms would not even be necessary since microelectronics and molecular electronics devices would suffice. In this sense, the term "Biocomputing" stands only for the application of biological principles to brain architecture, its interface with the outside world, and not for the use of biological organisms in Biocomputing. Michael Conrad was one of the first to suggest application of the principles of biological information processing in live organisms to the creation of the computers of the future [Conrad 1972, 1973, 1985]. The fundamental difference between the modern digital computer and the biological one is that the digital computer is fully programmed, whereas the biological (natural) computer is not. According to Conrad and his colleagues, computations within the organism occur at the "mesoscopic scale" [Conrad 1984]. This scale is comparable to the size of macromolecules, and the thickness of an ordinary cell membrane. Ions, enzymes, and macromolecules, participating in diffusion, transport, electrostatic, conformational, and other processes which occur around the membrane, may play the role of active constituents. Each of these processes taken by itself is random and may lead to random events and consequences. However, all of them are interconnected down to the submolecular level via many positive and negative feedback loops. As a result, all biosystems, and the entire organism, adequately react to signal inputs from external and internal receptors. Their collective activity creates what we call "intelligence in life" [Conrad 1990, 1990a].

The creation of a future biocomputer requires a new range of materials. As mentioned before, the first stage of a biocomputer prototype creation need not incorporate materials of a solely biological origin, but it has to be a "smart material." According to the Science and Technology Agency of Japan, smart materials are the "materials with the ability to respond to the environmental conditions intelligently, and to display their functions." Some of the materials that meet these criteria are hemoglobin and bacteriorhodopsin (BR). A conference entitled "Nonlinear electrodynamic and biological systems" took place in 1984 [Adey and Lawerence 1984]. This conference discussed the use of BR for molecular computing elements. Felix Hong studied several models of such elements in 1986 [Hong 1986].

The creation of a biocomputer can be viewed as a two-step process. The first stage advances the problem of compatibility between biological and nonbiological components, and their organization on the submolecular level. This problem has been the topic of many discussions and publications. The second stage brings forward the problems of self-organization, self-assembly and self-repair of the biological system to ensure their survival and progress under changing environmental conditions. Both stages will most likely occur simultaneously.

2

The Distinguished Family of Rhodopsins

Many scientists associate the birth of rhodopsins with the emergence of the first unicellular organisms 3 billion years ago. It is quite possible that BR, discovered by Stoeckenius in halobacteria in 1973 (see Section 2C.1), was the first rhodopsin on Earth, the precursor of all other rhodopsins, including visual rhodopsins. Following the discovery of the first BR (whose function is proton phototransfer across the cell membrane of halobacteria), other BR types with different functions were found. For example, halorhodopsin transfers a chloride ion in a light-dependent manner; phoborhodopsin (PR) and chlamyrhodopsin (ChR) facilitate spatial orientation using light and darkness detectors, while sensory rhodopsin (SR) are determined as the receptors of activation and signal relay, and orient halobacteria by the gradient of light intensity, and by spectral composition.

Rhodopsins (retinal–proteins), are found in the majority of organisms that require the conversion of light into chemical energy. Their functions have diversified over the span of geologic time. In higher organisms, rhodopsins control the perception of light reflected from objects, and their subsequent transformation as nerve impulses sent to the brain. Rhodopsins in higher organisms can even act as enzymes! In lower and unicellular organisms, rhodopsins behave as molecular photosensors for light-dependent, spatial orientation in the surrounding medium. In some bacteria, rhodopsins convert light energy into the energy of ion translocation across the cell membrane to conduct ATP synthesis, and/or to maintain physiological osmotic pressure.

Over the last two decades, new rhodopsins or rhodopsin-like complexes were discovered or synthesized. Some have been used both in theoretical and applied research. The "construction" of all rhodopsin molecules is relatively simple: retinals or their analogs are bonded to proteins (opsins) of a specific structure via a Schiff base. The photoactivity of such complexes generally depends on the stoichiometry of the retinal. It is the retinal portion of the retinal–protein complex that serves as the chromophore in nearly all natural rhodopsins.

The choice of retinal seems remarkable considering the tremendous impact it has had on evolution. Nature's reasons for choosing retinal are perhaps the following:

* a high molar extinction coefficient,
* the extremely high quantum yield of retinal as a chromophore,
* low isomerization time (upon photon absorption this does not exceed a few picoseconds).

These properties of retinal make rhodopsins highly efficient molecular light converters.

The initial advances in rhodopsin research were concerned with BR but, as is often the case in science, advances in BR research propelled the study of VR.

The question inevitably arises, is it possible to construct a visual photoreceptor on a molecular scale which would be more effective than rhodopsins? To create an artificial eye using new rhodopsins as the main photoreceptor? Nobody has yet approached these problems. However, any contemporary biochemist can synthesize new rhodopsins and model its photoreceptive mechanism...

As previously stated, the discovery of BR launched a chain reaction of interest in retinal–protein complexes in laboratories worldwide, attracting renewed attention to VR, the study of which had started more than a century before. This chapter will survey all natural and artificial rhodopsins currently catalogued—and attempt to unravel some of the mystery about their extraordinary properties: properties whose existence could not have been predicted just a few years ago.

2A. Visual Rhodopsin (VR) in Vertebrates

Most vertebrates have eyes allowing visualization of the external world. To avoid a lengthy discussion of the eye's structure, we need only remind the reader that the visual system of vertebrates resembles an autofocus camera, complete with a lens, muscles to focus, and a photosensitive retina upon which the image is projected from the lens. The retina contains a layer of visual cells (rods and cones) that host organelles containing the molecules with retinal-protein (RP). It is in the retina that RP functions as the photosensitive agent converting light energy into an electric potential. This potential triggers a series of biochemical reactions whose end result is a nerve impulse. Wald [1968] offered a classification of VR according to the types, retinal-1 and retinal-2, and the type of opsin found on rods and cones. By his classification, rod opsin complexed with retinal (11-*cis*-retinal, or commonly called vitamin A1 aldehyde) got the name of rhodopsin while the cone opsin complex with retinal-1 is called iodopsin. The rod opsin complexed with retinal-2 (11-*cis*-3-dihydroretinal or simply vitamin A2 aldehyde) is called porphyrhodopsin while the cone opsin complex with retinal-2 is called cyanopsin. This classification is old but still used in the literature. Today, scientists prefer to divide retinal–protein complexes into two classes according to the types of retinals:

1. Rhodopsins—proteins (opsins) chemically linked to retinal-2 which serves as a chromophore.
2. Porphyrhodopsins which have retinal-1 as a chromophore.

These two retinals are the main chromophores in all visual pigments of vertebrates. Porphyrhodopsins are predominantly found in the eyes of fresh-water fish and some amphibians. The proteins may display structural differences which affect the absorption spectra of VR, but have virtually no effect on its function.

The importance of understanding how the functional properties of rhodopsins are connected with their structure is obvious. As a first step in understanding these relationships, the mechanism for the conversion of light energy into nerve impulses needed to be elucidated. Photopotential measuring techniques (on BR films) have been used to measure the photo-dependent potential in VR. This led to the discovery of the electrochromic effect (see Section 2A.8) which is well-known to physicists. Today, a large variety of physical and biochemical methods are employed to study this problem. Mutagenesis (see Sections 3B.1 and 3B.2) has shed some light on the role played by amino-acids, and amino-acid groups in the protein, and is among the most useful methods [Khorana 1992].

2A.1. General Structure of Rhodopsins

VR molecules in vertebrates possess a relatively simple structure. It consists of a protein molecule of modest size (molecular weight comparable to human and bovine opsins–39–40 kDa) and retinal. Recently, amino-acid sequences in human and bovine opsins have been identified [Ovchinnikov 1987]. This revealed the secondary and tertiary structures of VR [Hargrave et al. 1983] [Adamus et al. 1987]. An amino-acid sequence comparison in human and bovine opsins reveals, with the exception of a few residues, their homological coincidence (see Figures 2.1 and 3.1). An entire protein is composed of seven trans-membrane α-helices. The seven helices are labeled A–G. The exact arrangement of helices within VR has not been determined yet. Opsin is highly hydrophobic and doesn't crystallize, excluding the use of X-ray analysis to determine tertiary structure. Therefore, the arrangement and depiction of helices in Figure 2.2 repeats the precisely known tertiary structure of BR (see Section 3A.1).

Before light absorption, retinal in VR is 11-*cis* isomerized, and is covalently linked to Lys296 protein in the G-helix through the protonated Schiff base.

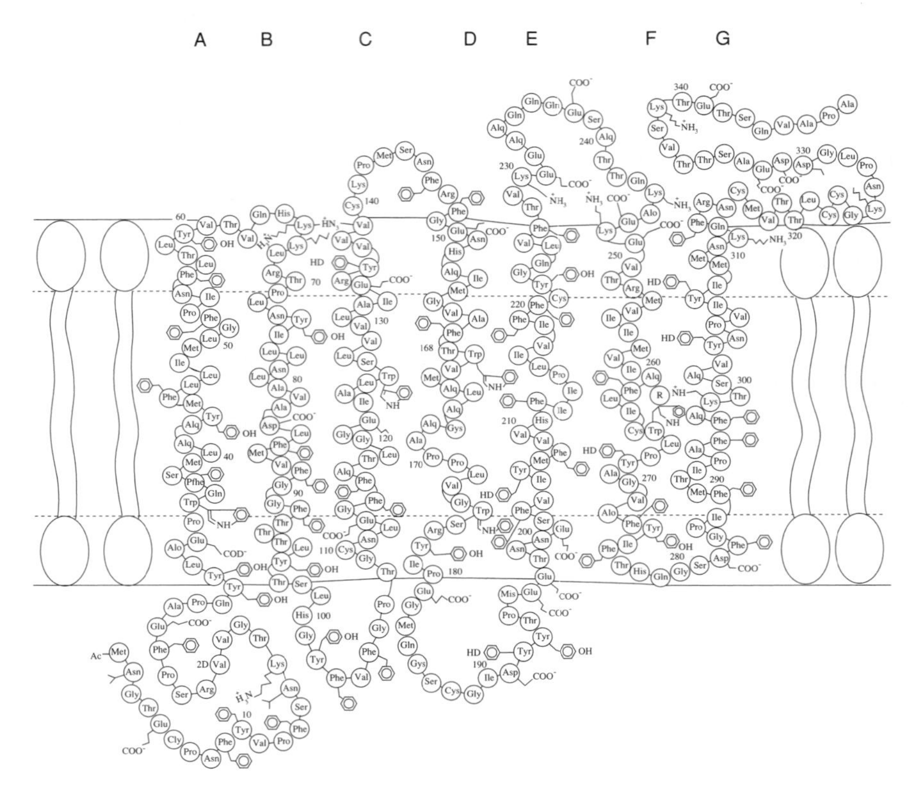

Figure 2.1. Model of the secondary structure of visual rhodopsin, modification of Ovchinnikov (a) and Hargrave [from Mirzadega and Liu 1990] (b). The attachment site of retinal (R) is Lys296. The numbering from A to G refers to the seven α-helices extending across the protein.

This linkage shifts the absorption maximum of free retinal from approximately 370 to 440 nm. The reasons for its further shift (to 500 nm) are still under discussion. There exist a few models for the spatial binding of retinal to opsin in VR, and for its behavior after light absorption.

As early as 1958, it was suggested that retinal interacts with some charged opsin groups, located sufficiently close to the Schiff base and to the atoms of retinals polyene chain [Kropf and Hubbard 1958]. Twenty years later, based on this hypothesis, a model for the external point charge was proposed [Honig et al. 1979]. According to this model, two negative charges (counter-ions) are located close to the chromophore, and one of them stabilizes the protonated state of the Schiff's base. The second counter-ion, sitting close to C-13, creates a local electrostatic field, causing a further shift in the spectrum from 440 to 500 nm (see Figure 2.3).

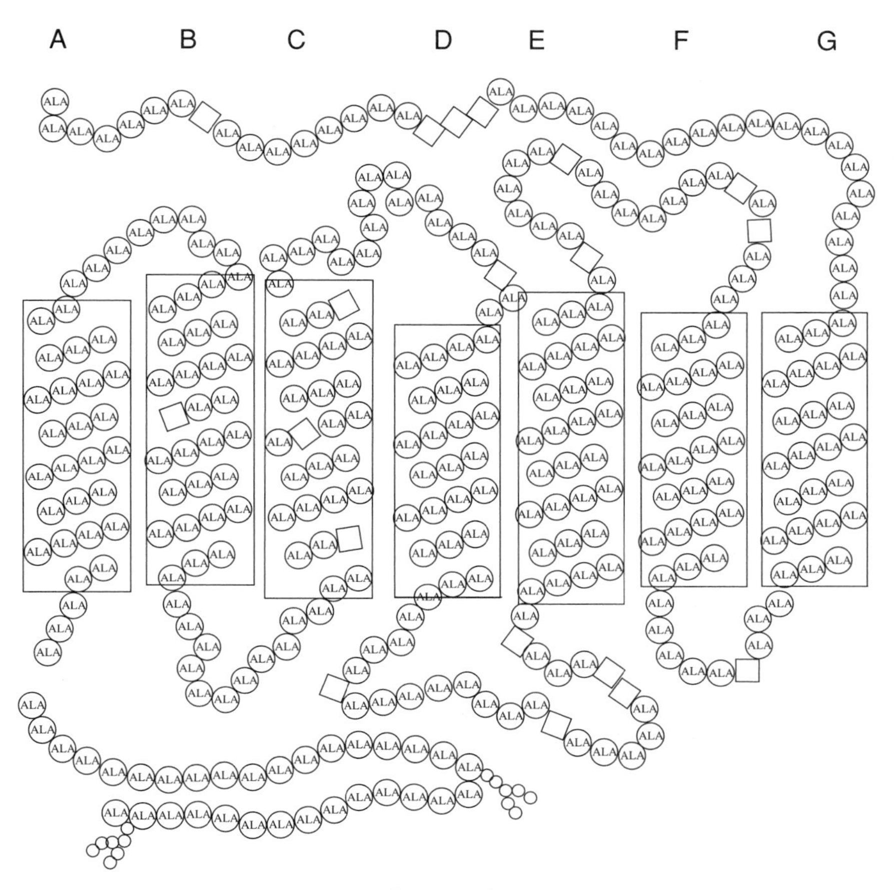

Figure 2.1(b).

Another model depicts which particular amino-acid residues participate in the formation of counter-ions. The protonated state is stabilized by the counter-ion formed by the amino-acid cluster Lys, Glu, Arg. This protonated state then reacts with the negatively charged residue, Glu-113. The second counter-ion is formed by the group of opposite charged amino residues Asp, Glu, located by the atom of C-13 retinal, which reacts with the negatively charged Asp [Nathans 1987]. Studies on the replacement of Glu-113 by glycine (Gly) prove that Glu-113 stabilizes the protonated state since such a replacement shifts the absorption maximum of VR from 500 to 380 nm [Zhykovsky and Oprian 1989].

Interestingly, in the blue-sensitive VR cones, absorbing at 420 nm, the second counter-ion Asp, is totally lacking. This points to the strong influence of the counter-ions on the absorption maxima in VR and possibly determines

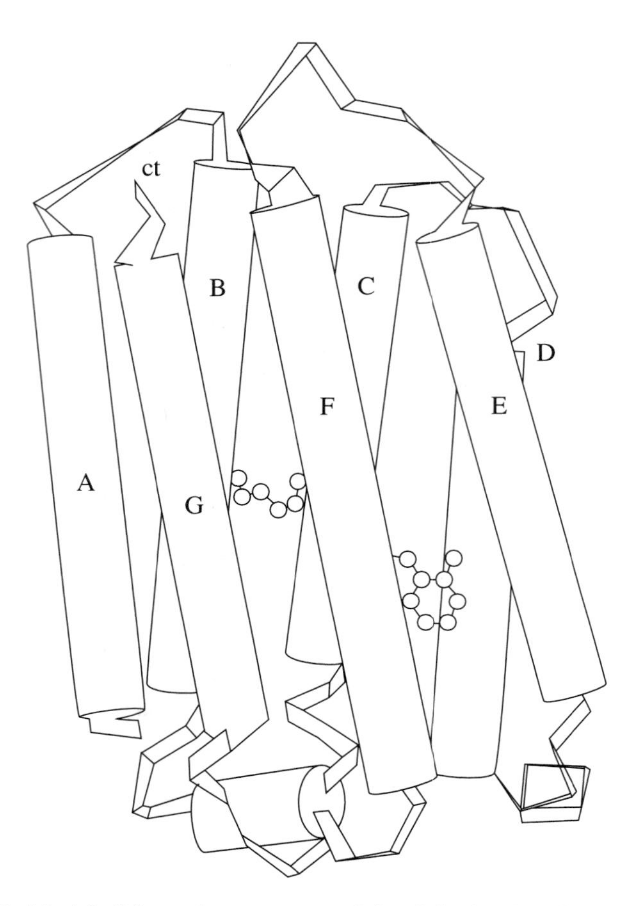

Figure. 2.2. Model of the tertiary structure of visual rhodopsin. The Seven α n-helices extend across the segment membrane roughly perpendicular to its plane. [Henderson et al. 1990].

the differences between the blue-, green-, and red-sensitive VR molecules. However, researchers have not come to full agreement on the problem of the opsin shift. One thing is clear: the shift is connected with a change in the electrostatic surrounding of retinal, and this change is itself dependent on the separate charges on the helices of the polypeptide chain, and on the general tertiary structure of opsin. Electrochromic properties of VR molecules indirectly support this conclusion (see Section 2A.8).

As for the post-absorption structural changes in the VR molecule, there's no universal model either. Using two-photon spectroscopy and site-directed mutagenesis, it was shown that Glu-113 of the C-helix is involved at the first stage of the photochemical process [Birge 1989] [Sakmar et al. 1989]. A model

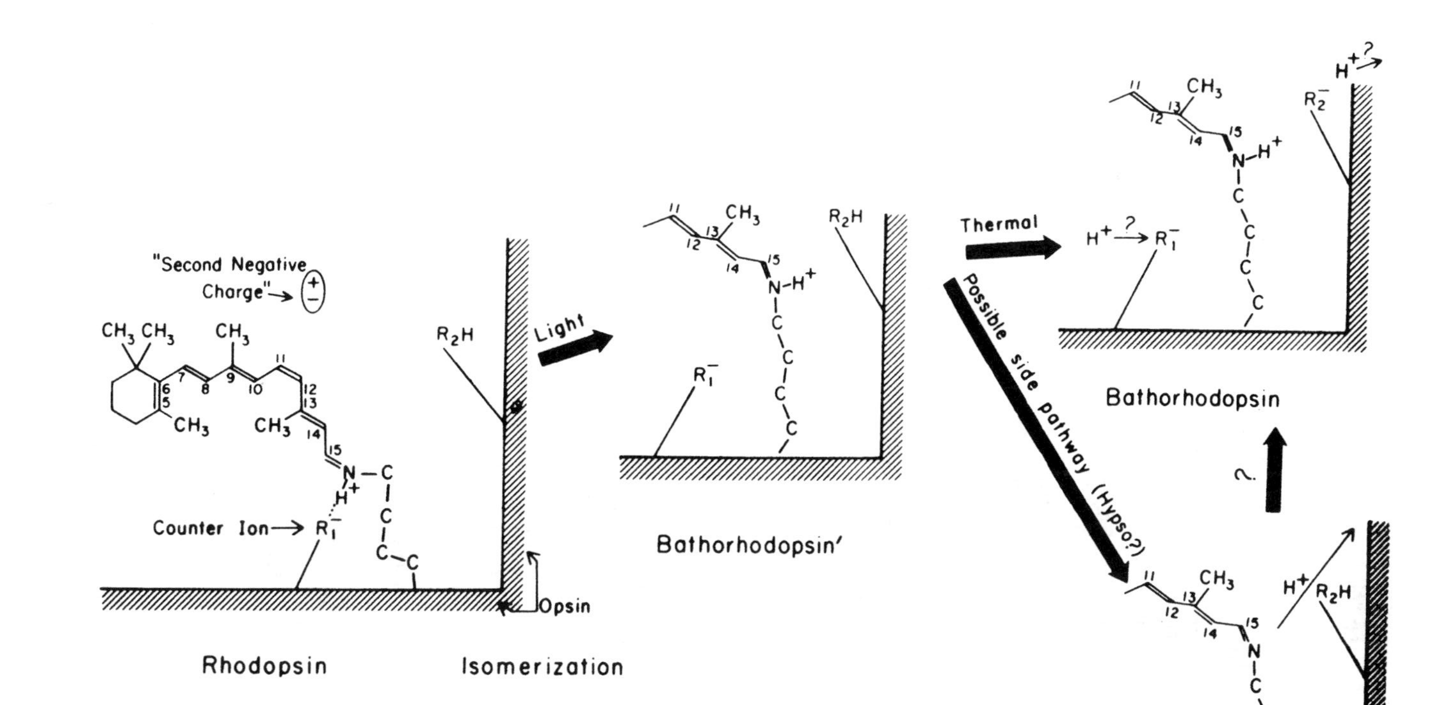

Figure 2.3. Model of the proton translocation mechanism across the chromophore center of visual rhodopsin. 11-*cis* retinal is depicted with its Schiff base forming a salt bridge with a negative counter-ion. The additional charge pair near the 11-12 double bond represents the group (or groups) that regulate the absorption maxima of VR. Any isomerization will produce charge separation as shown in the first step in the figure (for chemical formulas and isomers, refer to the Appendix). Possible proton transfer resulting from charge separation is also shown [Honig et al. 1979].

was proposed which accounts for the reorganization of positively charged atoms by the C-11-C-13 retinal chain during the process of photoexcitation [Birge et al. 1987]. Birge and colleagues suggested that charge reorganization is the principal factor in the creation of a barrierless excited state potential surface for a single bond isomerization [Birge and Hubbard 1981, 1990]. Participation of bound water molecules in the photochemical process must be also taken into account, for one or more water molecules in VR do occur in the bound state [Rafferty and Shichi 1981].

It is generally accepted that the coincidence of rhodopsin absorption maximum with solar radiation provides the necessary conditions for the most effective use of light that reaches the eye. On the other hand, it would intuitively seem that the more quanta that hit the retina, the more effectively rhodopsin must react. The number of quanta, coming from the sun, is greater in the red (for instance, at 600 nm) than in the green spectral region. However, rhodopsins, absorbing at 600 nm, are a lot less common than those absorbing at 500-510 nm. As early as 1972, Govardovsky suggested that most rhodopsins absorb at about 500 nm because at this wavelength the relation between the instructive signal and thermal noise is more efficient than at 600 nm [Govardovsky 1972].

2A.2. The Photochemical Process in VR is Irreversible

At normal temperature and neutral pH, VR molecules irreversibly bleach upon exposure to white light. In the process of photolysis, the molecule passes through a sequence of spectral and conformational transformations. At the final stage of photolysis, the molecule disintegrates into free retinal and protein. The irreversibility of the photochemical process is the fundamental difference of VR from bacterial rhodopsins. In BR, the photochemical process is fully reversible (see Section 2C).

In a few femtoseconds, upon photon absorption, the VR chromophore enters a state of electronic excitation which leads to the isomerization of retinal, and to the formation of the first intermediate—bathorhodopsin [Yoshizawa and Wald 1963]. The formation of bathorhodopsin is a photophysical (a light-induced conformational change) process during which retinal isomerizes from 11-*cis* to the all-*trans* state, storing about 32–35 kcal/mol of energy [Green et al. 1977] [Rosenfeld et al. 1977] [Cooper 1979]. A further process is photochemical, and is called the dark process. It proceeds at the expense of stored thermal energy. At this stage, VR undergoes a "dark-relaxation" (de-excitation) period, in which several intermediates, with different absorption spectra and lifetimes are sequentially formed. All stages of the photochemical process are described in detail in Section 2A.3.

At very large (not physiological) light intensities, an additional bathorhodopsin, preceding the photophysical intermediate appears. It is called hypsorhodopsin and is unlikely to play any important role in the physiological process of vision [Kobayashi 1979] [Pande et al. 1984]. Still another photophysical intermediate, also preceding bathorhodopsin, is photorhodopsin or P-Batho [Yoshizawa et al. 1984]. Some properties of photorhodopsin are still not defined (for instance, its exact absorption maximum), and it is often not included in the scheme of VR photochemical process.

At the final stage of the photochemical process, before the VR molecule disintegrates into retinal and opsin, deprotonation of the chromophoric Schiff base occurs, and the biochemical reactions responsible for optical nerve transmission are triggered. The Schiff base is protonated at every intermediate step excluding the final one. This important fact has been confirmed in many different ways [Oseroff and Callender 1974] [Palings et al. 1987] [Schroder and Nakanishi 1987] [Birge et al. 1988] [Covindjee et al. 1988].

The key difference between VR and BR lies in the fact that the photochemical process in BR (both *in vivo* and *in vitro*) proceeds according to a cyclic scheme (Figure 3.5), whereas in VR it is cyclic only *in vitro*. As Figure

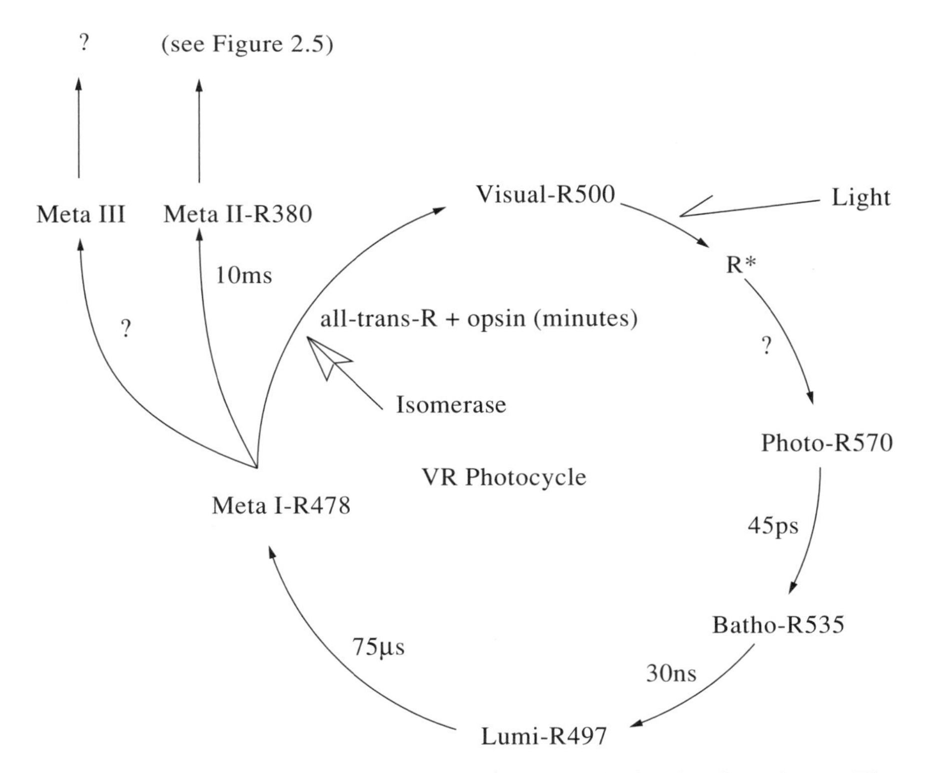

Figure 2.4. Photochemical reaction cycle of the visual rhodopsin of vertebrates. The R-numbers are absorption maxima (nm) for each intermediate. Times before and after intermediates are ascension and descension times of each species.

2.4 shows, the photochemical process *in vitro* ends at the stage of Meta II and/ or Meta III intermediates.

At these stages, retinal separates from opsin and remains in the all-*trans* state. To isomerize back to the ground state (11-*cis*-retinal), the organism uses a special enzyme called retinal isomerase. Only upon the action of this enzyme can a photoactive VR complex form again and re-initiate the photochemical process [Law et al. 1989].

2A.3. Photoinduced VR Intermediates

Under normal conditions, a VR molecule in the ground state absorbs at 498 nm. Upon photon absorption, the first (photophysical) intermediate—bathorhodopsin—is formed. This is the first step in a sequence of dark-adapted intermediates from lumirhodopsin to metarhodopsin II (Meta II) (see Figure 2.4). Some experimental data suggests the existence of two more candidates for the roles played by these photophysical intermediates. They are photorhodopsin, (absorption maximum at 560 nm [Shichida 1986]) and hypsorhodopsin (absorbing at 435 nm [Yoshizawa 1972]. The formation of hypsorhodopsin is possibly connected with multiphotonic processes [Yoshizawa et al. 1984]. However, the existence of photorhodopsin, which in some experiments precedes bathorhodopsin, is still hotly contested [Kandori et al. 1989]. Photorhodopsin is not observed at helium temperatures [Shichida 1986]. Figure 2.4, therefore, shows bathorhodopsin as the only product of the primary photophysical event. All other intermediates result from dark processes which occur at the expense of bathorhodopsin.

In 1963, it was already shown that isomerization of retinal from 11-*cis* to 11-*trans* occurs in the VR chromophore upon photon absorption [Yoshizawa and Wald, 1963]. Retinal remains in the 11-*trans* state until lumirhodopsin is formed, and then transforms to all-*trans*. This classic work was followed by a great number of models to explain retinal's post-illumination behavior. Beginning in 1987, researchers agreed that VR retinal occurs in two initial states: 9-*cis* (isorhodopsin) and 11-*cis* (rhodopsin). After photoisomerization to 9-*trans* and 11-*trans*, the corresponding primary photoproduct (bathorhodopsin) with protonated Shiffs base is formed [Pande et al. 1984] [Palings et al. 1987].

As for the place of isomerization in the 11-*cis* retinal polyene chain, the early models with imaginative names like "Bicycle Pedal" [Warshel 1976] and "Hula Twist" [Liu et al. 1985] did not include the possibility of isomerization due to simultaneous rotation around the retinal double (C11–C12) and single bonds. However, Resonance Raman spectroscopy [Palings et al. 1987], FTIR [Genter et al. 1988], retinal replacement with analogs [Derguni and Nakanishi, 1986], and some other methods, clearly showed that rotation occurs only around the C11-C12 double bond.

Experiments revealed that rhodopsin forms bathorhodopsin in a few picoseconds with a quantum effectiveness of 0.67 [Hurley et al. 1977]. These measurements perfectly fit theoretical calculations [Birge and Hubbard 1981] which provide the values of 2.3 psec and 0.61, respectively. The quantum effectiveness of isorhodopsin conversion to bathorhodopsin is three times less [Hurley et al. 1977]. Bathorhodopsin stores 35 kcal—the energy of one light quantum at 500 nm [Cooper 1979]. This energy is enough to support all the dark stages of VR photolysis, and to inhibit thermal molecular processes responsible for thermal noise in the photoreceptor.

The intermediate that follows bathorhodopsin is called lumirhodopsin. Its formation takes a few dozen picoseconds, during which time retinal relaxes from 11-*cis* to the all-*trans* state. At the formation of lumirhodopsin, retinal rotates around the C-6-C7 single bond, and changes its ion ring position relative to its protein surroundings [De-Grip 1988]. During the post-lumirhodopsin period, until the conclusion of photolysis, retinal remains in the all-*trans* state. Thus, the formation of other intermediates is determined only by conformational processes in opsin, and by the electrostatic interaction of charged groups with retinal [Protasova et al.1991].

The next intermediate—metarhodopsin (Meta I) is formed in dozens of microseconds and absorbs at 475 nm. Low-temperature spectroscopy data points to the significant conformational reconstruction occurring in the VR protein molecule at the Meta I and Meta II stages of formation. These reconstructions envelop even the lipid surrounding of the protein molecule [De-Grip 1988]. Meta I to Meta II conversion ends the dark-adapted part of the photochemical process in VR. An enormous spectral shift (from 475 to 380 nm) happens at this stage, and is due to the deprotonation of the Schiff base.

Meta II formation promotes new molecular conformations in Opsin's tertiary structure. Hydrophilic protein areas on the plasma surface of the photoreceptor membrane react separately with transducine, rhodopsinkinase and arestine. As a result, a cascade of biochemical reactions, responsible for adaptation, transduction, ion channel blocking, and receptor potential generation is launched (see Section 2A.5).

Hydrolysis and formation of free all-*trans* retinal and opsin occurs in concert with Schiff base deprotonation. Opsin returns to its initial conformational state, whereas all-*trans* retinal, aided by an isomeric enzyme, transforms to 11-*cis*. The process of VR resynthesis has a few stages [Dowling 1987], and results in the formation of a new VR molecule which concludes the photocycle.

The re-protonation of the Schiff base may cause the formation of yet another intermediate Meta III, with an absorption maximum at 460-465 nm. The role of this intermediate and the reason for its formation are not quite clear since it does not participate in the generation of the cascade of biochemical reactions [Liebman et al. 1987].

2A.4. Electrogenic VR Intermediates

Between 1964–1967, a receptor photopotential was registered on isolated eye retina illuminated with a pulse of light [Brown and Murakati 1964]. This had never been observed before. This new potential was called Early Receptor Potential (ERP) [Cone 1967] and consists of two components with different polarity and differently charged components: the fast and the slow. A plausible reason for its appearance is the charge displacement occurring in VR during photolysis. The fast negative charge is registered as early as bathorhodopsin formation, while the slower positive charge appears at the Meta I to Meta II conversion. ERP can be detected *in vitro* even if the retina is frozen. The formation of the slow potential is not connected with the protonation of the Schiff base [Trissl 1979].

At physiological wavelengths, ERP becomes increasingly small and has no direct effect on the mechanism of transduction. However, its detection under artificial conditions provides insight into the mechanism of the electronic and conformational reconstructions occurring in VR photolysis. The overall data shows that registration of the fast component is connected with the formation of a dipole moment in the VR molecule which is itself caused by a change in the electrostatic interaction between retinal and opsin. The slow component reflects conformational reconstructions in the entire VR molecule at the stage of Meta II formation [Ostrovsky 1990]. The duration of membrane discharge depends on the resistance of the membrane, and enables computation of the difference in channel conductivity between the rod disks and the plasma membrane. For example, it has been shown that the photoreceptor membrane of rod disks, as opposed to the plasma membrane, contains no channel structures [Rebrik et al. 1986].

A so-called Late Receptor Potential (LRP) is created on the plasma membrane of the outer rod segment (on the retina) using hyperpolarization caused by the blocking of photoinduced conductivity, and is directly related to the biochemical mechanism of the amplification process.

Amplification is achieved through a cyclic biochemical reaction with the participation of transducin (a peripheral membrane protein belonging to the G-Protein family [Bennett et al. 1982]. Upon the completion of the dark-adapted photochemical process and the formation of Meta II, the Schiff base is deprotonated. The Meta II intermediate formation stage is the key step in triggering the cascade of amplifying reactions as shown in Figure 2.5.

G-protein transducin (GT) plays an important role in this reaction. In the dark, transducin binds to GDP and is inactive. During VR photohydrolysis, Meta II reacts with GT-GDP, forming a GT-GTP + GDP complex. This complex then dissociates into subunits Gt(α) and Gt($\beta\gamma$). Meta II is then released for the new stage of interaction with transducin.

At this stage, while unbound to the photoreceptor disk membrane, and with the Meta II lifetime far exceeding the time of its interaction with the GT-

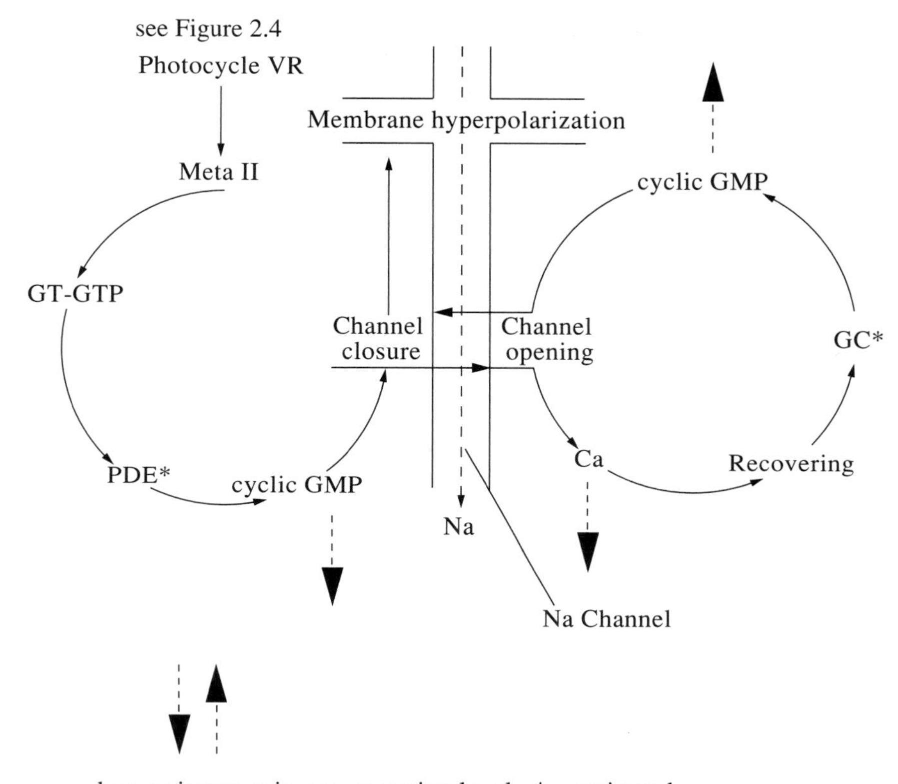

Figure 2.5. Information cycles involved in visual excitation and recovery. Photoisomerization of visual rhodopsin (VR) to the Meta II intermediate triggers a cycle leading to cyclic GMP hydrolysis and closure of the cytoplasmic membrane channels which generate nerve impulses. Channel closure also induces a drop in the calcium ion concentration, which leads to the activation of guanylate cyclase (GC) and a re-opening of channels.

GDP molecule, VR activates hundreds of GT-GDP molecules. At the next stage, GT(α) activates phosphodiesterase enzyme (PDE). This exceptionally active enzyme hydrolyzes up to 3000 cyclic GMP molecules at the rate 5' GMP/sec. In model systems *in vitro*, the total amplification from the two stages may reach a million! Since every photoactive VR molecule hydrolyzes up to a million cyclic GMP molecules, their concentration decreases. This decreased concentration causes the sodium channels to close, and hyperpolarizes the outer plasma membrane of a photoreceptor cell. The ability of cyclic GMP to regulate sodium conductivity was first described by Fesenko and colleagues [Fesenko et al. 1985]. The Modern view of a transducin cycle is well presented in a minireview [Stryer, 1991]. The amplification gain is approximately 100 times greater *in vitro* than *in vivo*.

As for the mechanism of adaptation to high intensity light, it has been suggested that Meta II, when exposed to a large photon flow, partially binds to a specific protein called arrestin [Wilden et al. 1986]. Hence, GT activation is limited and the electric signal never reaches the critical limit.

2A.5. Electrical Signals from the Retina

The retina of vertebrates hosts millions of photoreceptor cells (generators of photoelectric signals). Figure 2.6 shows the structure of two types of such cells. The plasma membrane of a photoreceptor cell has special channels for sodium ions. In the dark, these channels are open to a stream of ions. Upon light absorption, the VR molecule triggers the channel-blocking biochemical

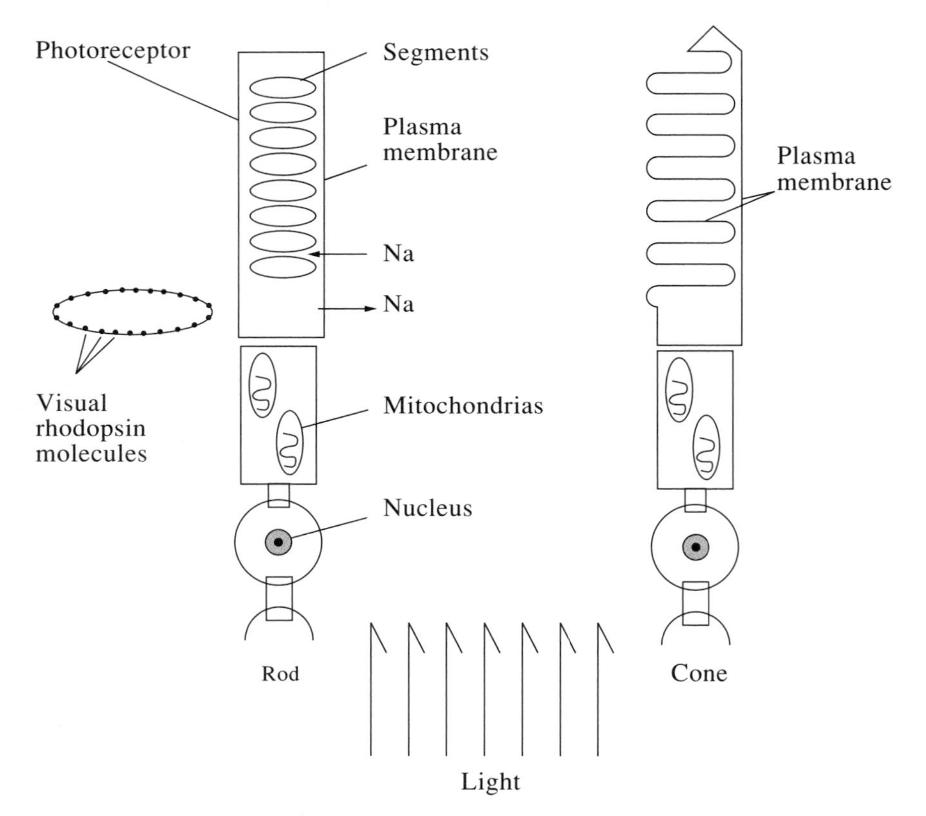

Figure 2.6. A simplified model of vertebrate eye rods and cones which act as photoreceptor cells.

process that leads to hyperpolarization of the membranes, and generates a negative potential inside the cell (see Section 2A.4). The energy of one photon can block up to a million ions, and generate a potential of up to 1 mV which then generates a single nerve impulse. However, the photoreceptor background noise level is very high. Therefore, it takes about a hundred photons to generate one nerve impulse and to simultaneously activate 10-20 rhodopsin molecules [Barlow 1988]. It is still not clear what causes background noise. Nevertheless, there are plenty of theories to which the reader can refer [Birge 1990].

The retina is a specialized photosensitive nerve center which not only fixates visual information, but also performs its initial processing. Studies of morphology and functions of the eye retina have more than a century of history behind them [Ramon and Cajal 1892].

Figure 2.7 shows a simplified structural representation of the visual system.

Visual neurons differ from common nerve cells in some specific features. For example, cone and rod neurons, horizontal cells, and bipolar cells, generate slowly changing oscillations of the potential and therefore cannot generate impulses which are recognized by nerve cells and fibers. These three neuron types comprise the first (entrance) layer of the retina in which information perception, processing, and transduction operate by an analog

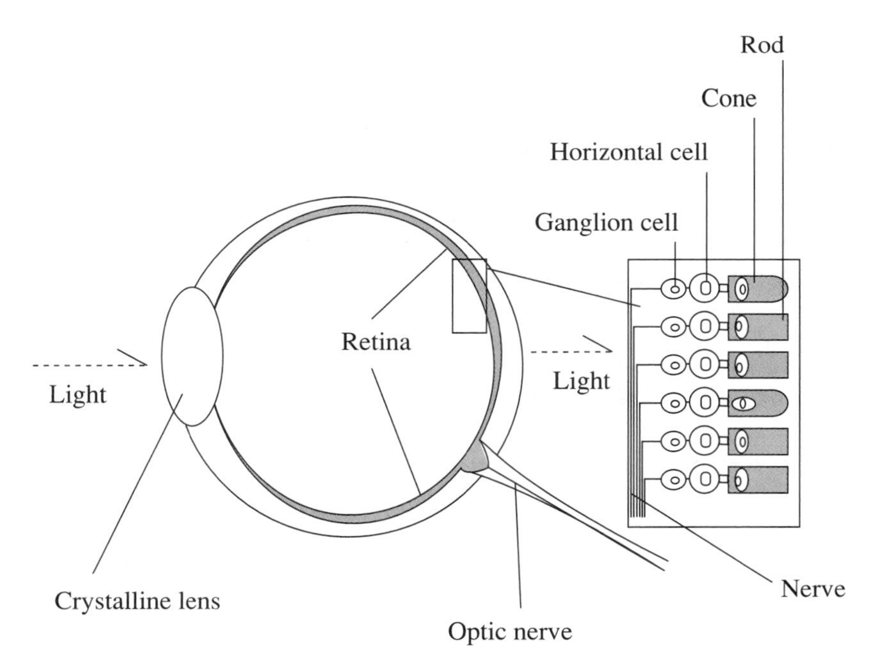

Figure 2.7. A close-up view of the retina. Surprisingly, the light has to pass through the nerve cell layers before it reaches the cones and rods.

type of mechanism. The second (exit) layer of the retina incorporates bipolar axons, amacrine, and ganglion cells, that work in a pulsed operation mode, and send discrete signals to the brain about partially processed information. Thus, the retina resembles a digital analog computer with analogous input and digital output (see Figure 2.8).

The retina contains clusters of simple nerve networks, organized into more complex ones. The properties of the latter are so interesting that cyberneticists, mathematicians, and, of course, computer scientists are drawn to studying them. We admit, however, that our understanding of the retina as the primary element for visual information processing is very superficial.

What we know of the sequence of events proceeding from the initial absorption of a photon at the retina, to the subsequent generation and transmission of a nerve impulse to the brain may be summarized as follows:

* ion current flows in the dark through photoreceptors from the inner to the outer segment (see Figures 2.6 and 7). The electric potential, generated on the outer membrane segment, blocks the ion current and induces a wave of photocurrent. This photocurrent spreads along the receptor from the outer segment to the inner one, thus modulating the release of a chemical mediator during the synaptic event.
* simultaneously, an electric potential passes to the adjacent photoreceptors of the retina by means of electrical contacts, i.e. one photon can excite a few dozen retinal rods since their energetic barrier is much lower than that of a single photon. In retinal cones, this barrier is greater by two orders of magnitude. However, electrical contacts between them exist nevertheless.
* due to the electric contacts between retinal rods, and a temporary delay in signal response, the retina acts as a spatial filter for high-frequency signals [Detwiler et al. 1980] which may be the very first step of visual information processing.
* a mediator depolarizes different types of horizontal cells which are electrically linked into independent networks. These networks have separate input ports for different photoreceptors (rods or cones each sensitive to a different color) and do not interact.
* simultaneously, the same mediator affects bipolar cells. Numerous dendrites are connected to many photoreceptor synapses, forming triads with the dendrites of horizontal cells of several synapses at a time. It is agreed that the bipolars control the generation of high spatial light intensity frequencies in the retina.
* parallel information processing, occurring in the input layer around the entire surface of retina, dramatically enhances the speed of light signal translation in an analog fashion.
* bipolars connect the input and output layers of the retina and are its input elements. The functional organization of the output layer is far more complex. Further information processing occurs in the output layer by

means of the digital method. A detailed description of this mechanism of its neuron networks has not been attained.

* we know that ganglionic cells fall into several types. Each of these selects and transmits into the brain one strictly defined parameter of a videosignal; –X-cell transmits the constant component of a videosignal; –Y-cell transmits the information on temporal changes in the videosignal; –W-cells of several types transmit the information on motion, direction, speed, etc.
* information from the different ganglion cells is received by distinct sections of the visual cerebral cortex.

The organization of the visual cerebral cortex, its sequential and parallel methods of information processing by the different types and classes of cells is an object of speculation, and is beyond the scope of this book.

Even our simplified scheme of videosignal processing in the retina reveals how much future biocomputer developers have to learn from nature. For the history and more detailed treatment on the neurophysiology of the retina, the reader is referred to the following monographs: Polyak [1941] and Dowling [1987].

2A.6. A Brief Look at Visual Pathways and Mechanisms

Section 2A.5 described the basic principles governing light conversion into electric and nerve signals in the retina. However, the visual process in general, is much more complex than the previous treatment implied, and, in many aspects, more unknown than the process of photon conversion into nerve impulses. In all likelihood, it is not a single, but a concert of visual events, proceeding in parallel or in tandem. For example, adaptation to bright light which supports vision in the illuminance range, from units up to millions of lux, initiates an expansion of the pupils diameter. Subsequently (or simultaneously?), retinomotor mechanisms turn on: retinal cones are drawn outward, while retinal rods are screened by light-absorbing melanin which moves between the outer segments of photoreceptors. In higher vertebrates, the retinomotor effect is only faintly expressed, and there must be some other mechanism for protecting rods against overexposure.

Amazingly, during the process of light adaptation, the contrast sensitivity of the eye (i.e. its ability to distinguish small gradations of illuminance) doesn't change. The contrast range is proportional to the slope of the curve of the dependance of the retinal electrical response upon light intensity. The steeper the slope, the greater the contrast range. On the other hand, the functional illuminance range of the retina diminishes with the increase in steepness. This contradiction has a natural solution. When the illumination changes, the curve shifts along the illuminance scale without altering its

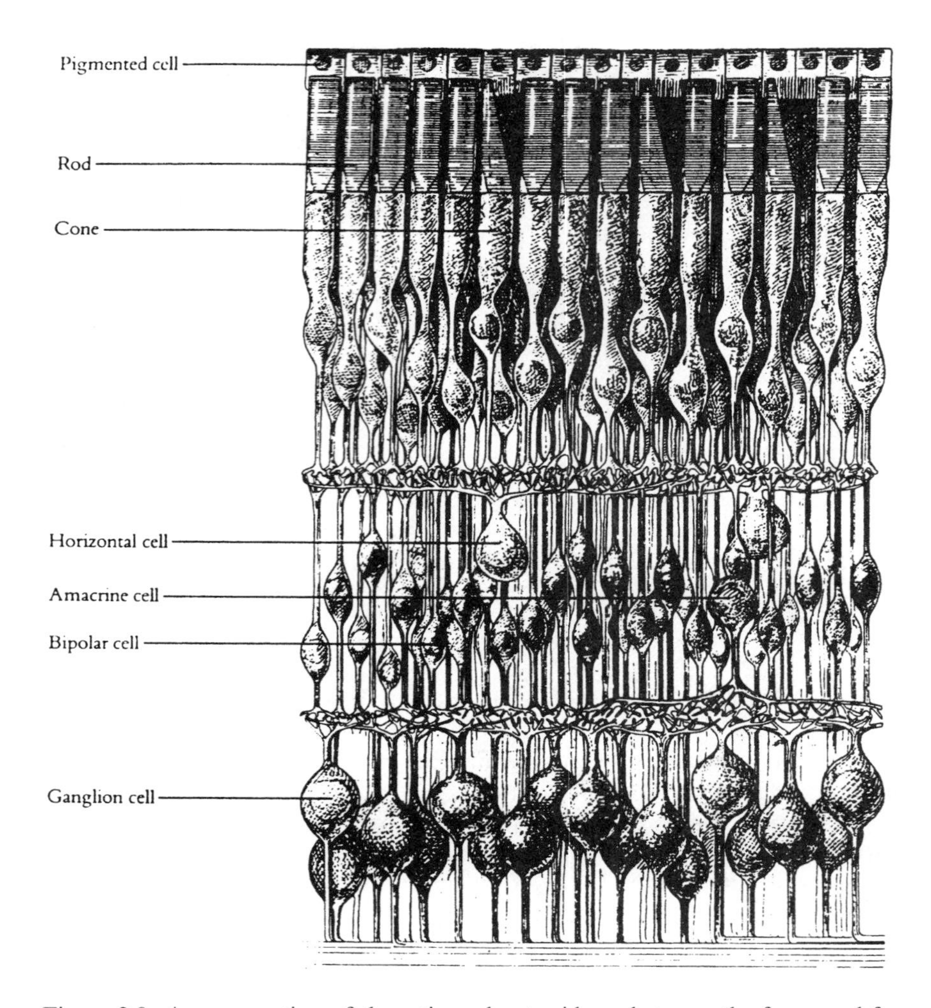

Figure 2.8. A cross-section of the retina, about midway between the fovea and far periphery, is about one-quarter mm [Hubel 1988]. Note the resemblance to the microscheme section of a modern computer.

steepness. Since this shift is not instantaneous, it takes the eye a few seconds to recover detailed vision. Ca^{2+} ions play a significant role in the mechanism which accelerates visual adaptation to reduced light. As we said in Section 2A.5, decreased cyclic GMP concentrations induce a partial blocking of ion channels in the membrane, including the calcium-dependent channels. The concentration of Ca^{2+} ions inside the cell then decreases, causing a reverse opening of the channels, and bringing the cell back into a dark-adapted state. This process functions as a negative feedback loop which accelerates the restoration of light sensitivity in the retina.

In the post-receptor adaptation mechanisms of some vertebrates, an essential role is played by transitional processes in horizontal cells [Byzov and Kuznetsova 1971]. In such cells, amplitudinal features of the response to illuminance changes are very steep and lie in a narrow range. However, if the light intensity change is greater than this range, it shifts to a new illuminance region in a few dozens of seconds. There exist other mechanisms of light adaptation about which little is known.

Figure 2.8 shows a cross-section of a typical vertebrate retina. Note the resemblance to a microchip.

After the preliminary videosignal processing event occurring in the retina, electrical impulses from ganglion cells are projected to the midbrain. According to the presently accepted view, ganglion cells of different types are arranged in the retina as mosaic structures called rasters [Wassle 1982]. A similar order appears in the arrangement of fibers in visual nerve cells which connect specific elements in the retina to certain brain structures. It is quite possible that the raster-like structure of the retina are spatially reflected in the functional areas and structures of the brain [Torrealba et al. 1982]. This implies a parallel information type of mechanism which, as mentioned previously, offers significant advantages.

As far as color vision, 99.9% of authors agree that the reproduction of colors in the brain is supported by three types of cones absorbing in the blue, green, and red spectral regions. This theory is appropriately named the three-component theory. It is of interest that in all three cone types, it is iodopsin that functions as the chromophore. This is a sensible observation since iodopsin has different absorption maxima due to conformational changes in the opsin molecule.

However, some scholars are skeptical about the three-component theory, and not without cause. A major reason is that the very existence of iodopsin has not been sufficiently proved! Other evidence against the three-component theory is provided by alterations of vision in alimentary hemeralopia (see Glossary) patients. Vitamin A levels in such patients drastically decrease, which must cause the distortions in vision. However, color vision remains undistorted, while night vision fails [Novokhvatsky and Alyi 1983].

2A.7. Measurement of Early and Late Receptor Potentials

In Section 2A.4, we described how back in the 1960's a photopotential called Early Receptor Potential (ERP) was registered on isolated retinas. A powerful light flash of 1.0 msec bleached 20–40% of VR. The VR photopotential has two components—the fast (R1) and the slow (R2) [Brown and Murakati 1964] [Cone 1967]. The origin of ERP is now studied using different techniques, in particular, with direct trans-membrane potential registration [Bolshakov et al. 1979] [Drachev et al. 1981] [Trissl 1979]. Photoreceptor disks

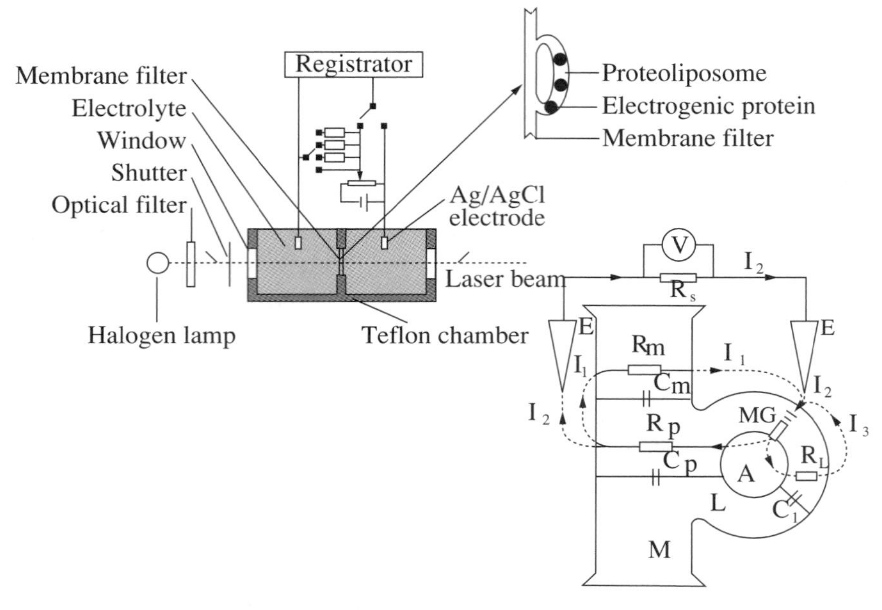

Figure 2.9. The typical device for measurement of the electrogenic function of membrane photosensitive proteins, and the path of currents (I) in the proteoliposome-planar membrane system. MG—molecular electric generator (e.g. BR); C and R are the equivalent electrical resistance and capacity of membranes [Drachev et al. 1979].

or outer segments are built into a flat artificial membrane which divides two electrolyte containing compartments. A photopotential is measured between the electrodes. Figure 2.9 shows a typical arrangement of how these potentials are measured. Experiments show that a membrane filter, saturated with lipids, works best. Photoreceptor disks or segments are easily built into such a membrane, forming a stable system for multiple photopotential measuring [Drachev et al. 1981].

Figure 2.10 shows a typical photoelectric response of ERP from the disks. The first phase is negative with a magnitude around 0.3 mV. A positive phase of 20–30 mV consists of the fast (2nd) component and the slow (3rd) component. The fast negative phase corresponds to the R1 component; the fast positive phase corresponds to the R2, and the third (slow) phase refers to the self-discharge of the membrane itself. R1 lasts a few nanoseconds and is related to photophysical processes in the VR molecule. Most likely, it reflects a charge shift in the chromophoric center resulting from isomerization of retinal at the stage of bathorhodopsin formation. R2 forms in dozens of msec, and reflects conformational reconstitutions in the opsin molecule and the stage of proton uptake. Proton uptake occurs at the stage of Meta II

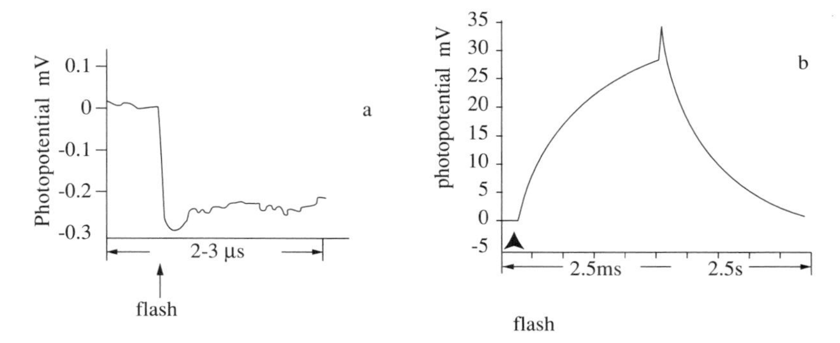

Figure 2.10. The kinetics of the photopotential on photoreceptor disks from cattle retinas. Laser pulse 15 ns (530 nm): (a)—the faster negative phase (R1); (b)—the slower positive phases (R2 and R3) [Drachev et al. 1981].

formation. However, proton translocation across the disk membrane doesn't seem to occur, and R2 is not related to Schiff base deprotonation during Meta I to Meta II conversion. Proton uptake, most likely, is related to the release of a proton-binding group in the VR cytoplasmic domain [McConnell 1968] [Trissl 1979]. Such proton behavior in VR deviates significantly from the behavior and role of a BR proton where it acts as a photosynthesizer (see Section 2A.1).

Processes related to photopotential generation in VR and BR will be discussed later [Hong and Okajima 1987] [Ostrovsky 1989] [Hong 1995, 1995a].

Direct measuring of photopotential in VR allows comparison between the electric characteristics of disk membranes, and other segment membranes. For example, the disk membrane discharge (3rd phase) happens in seconds, whereas the outer segment membrane discharges in milliseconds. This time gap allows estimation of the membrane resistance (or conductivity). The resistance of disk membranes equals 1–2 Mohm/cm^2 which is 3 orders of magnitude greater than the resistance of plasma membrane segments [Rebrik et al. 1986]. Disk membrane resistance is comparable to the conductivity of many membranes undergoing energetic transactions (mitochondrias, chloroplasts, BR etc.) without channel complexes. On the other hand, at saturating light intensities, with all channels closed in the plasma membrane, its resistance increases, and coincides in value with the resistance of the disk membrane.

2A.8. VR Response to an Electric Field (Electrochromism)

It is known (see Section 2A.1) that the light quantum absorbed by the VR molecule causes shifts in the absorption spectra and a redistribution of electron density in aldimine retinal. This results in changes in the electrostatic interaction in the adjacent regions of the protein [Rosenfeld et al. 1977] [Birge and Hubbard 1980]. As a result, charges separate and the membrane polarizes. This is how light energy transforms into electric energy. The magnitude of this effect depends on the degree of cooperativeness of a sufficiently large number of the membranes [Bolshakov et al. 1979].

The action of an external electric field must change the light-induced dielectric polarization. Electroinduced shifts of rhodopsin absorption spectra can then be expected. Such an effect was observed in base-dried photoreceptor membrane films from the bovine retina [Borisevich et al.1979 and reference therein]. Later, a comparative study between photo- and electro-induced spectral shifts in the photoreceptor membrane film was conducted [Protasova et al. 1987]. The application of an external electric field to the photoreceptor membrane changes the electron-conformational state of opsin. The action of the field causes a substantial shift of the VR molecule absorption spectra, which is not accompanied by retinal isomerization [Protasova et al. 1989]. These and other results provide significant proof that static charges, retinal atoms, and the electrostatic field of opsin influence spectral shifts in VR; both in the ground state, and in the process of photochemical and photophysical transformations. A description of one of the methods of film preparation from visual photoreceptors is provided since such films can serve as models for optical converters.

1. A photoreceptor membrane suspension was pasted on a glass base in the proportion of approximately 0.8 mg protein weight per cm^2 surface area.
2. It was then dried at room temperature in an exiccator with selicogel.
3. Measurements were made at room temperature with 65–70% humidity.

It is noteworthy that at this level of humidity, the water content in the membrane layer reaches 45–50%, and the films are fairly wet. The water content in the protein and its surroundings significantly affects its functioning. This fact must be remembered when experimenting with protein film preparations.

Photoexcitation of the potential and/or discoloration of the films was induced by 1 ms duration white light impulses. At the discoloration of rhodopsin, the film displayed the sequence of intermediates similar to those in the photoreceptor membrane suspension; however, their lifetimes grew dramatically longer. Upon subjection to an external electric field, the absorption band shifted to the red spectral region, and its new position was stabilized by an electric field. Removal of the field started slow discoloration of the rhodopsin, resulting in the formation of the intermediates whose spectra resembled those of Meta I and Meta II. If, at this stage, the field is repeatedly applied to the film, it causes the formation of an unstable product whose absorption spectrum resembles the initial spectrum of rhodopsin. Interestingly, this product also forms at the illumination of a previously discolored

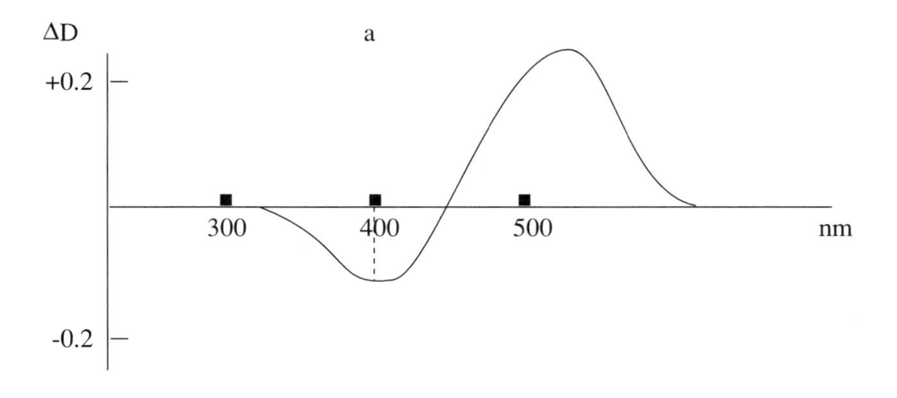

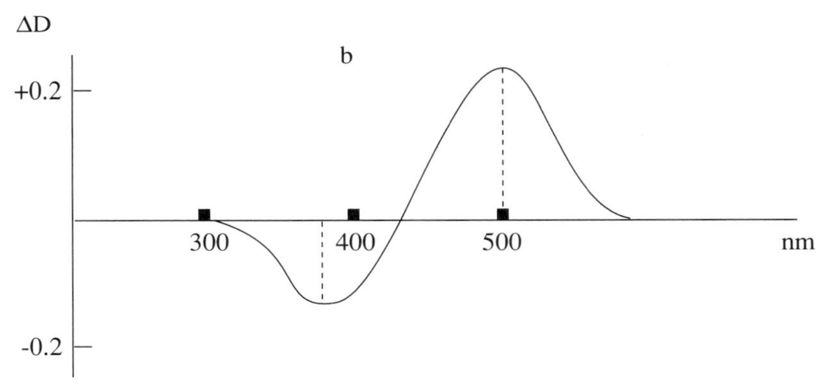

Figure 2.11. Electroinduced regeneration of rhodopsin on the dry film of a photoreceptor membrane by an electric field: (a) the field was applied to electrobleached rhodopsin; (b) the field was applied to photobleached rhodopsin.

film. An electric field, when applied to bleached film, causes the formation of the unstable product absorbing at 475 nm. The above-described results are partially presented in Figure 2.11

It is important to understand the mechanism of electric field influence on VR photoprocesses. Recent data does provide single answers to the many questions. Some results, if taken seriously, make the problem more confusing. For example, electrical regeneration of VR photoproducts in films can be observed only two days after light or electric field application. During this time, some unknown dark-adapted process occurs in the film, apparently, bringing the protein into a state of equilibrium. Electroregeneration may occur only after the equilibrium has been achieved. Experiments in titration of SH-groups suggest that opsin plays the dominant role in this process. The electric field action gradually reduces the number of such groups from 6–7 to

1–2. However, this effect is achieved only by periodic and multiple field application. A stationary field induces no such effect. Contemporary knowledge does not suffice to explain these phenomena.

Electroinduced processes are also difficult to explain from the viewpoint of photoinduced retinal isomerization. The electric field applied to the film has the power of about 1500 V. Accounting for the width of one membrane, it generates about 10–100 mV. Every VR molecule requires about 0.1 eV of energy. However, this energy is obviously not sufficient to isomerize retinal [Protasova et al. 1991]. An important caveat must be issued, which is that the results of all experiments with high water content in the preparations may be distorted, and be entirely dependent on water electrolysis. In this event, the pH must change which in its turn may produce effects unrelated to photochemistry, but resembling ones that are.

The list of unanswered questions and ambiguous data can be continued; however, the electrochromic effect is real, and has interesting applications to bioelectronics.

2A.9. The Boundary Between Photochromism and Photoisomerism in Rhodopsins

Ignoring the details of the photochemical and biochemical mechanisms of vision, a global perspective on rhodopsin behavior in rods and cones supports the view that rhodopsin is the principal agent in the photochromic process.

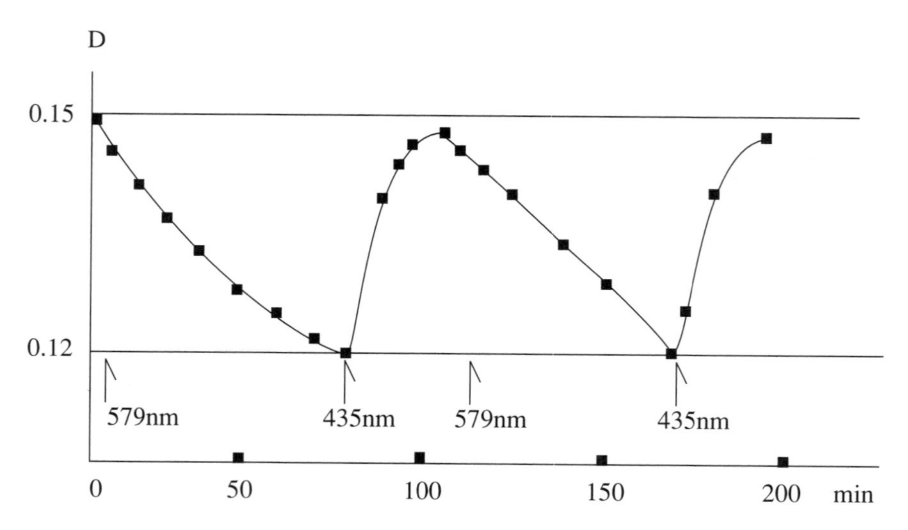

Figure 2.12. Photochromic (reversible) changes around 510 nm in the optical density of a frog visual rhodopsin suspension upon exposure to red (579 nm) and blue (435 nm) light [Krongauz V et al. 1975].

Figure 2.12 presents a graph of optical density changes vs. wavelength action. Note the reversibility depending on the wavelength [Krongauz et al. 1975]. This photochromic process involves the energy created by the organism, and is irreversible *in vitro*. Under physiological conditions, there exists at least two ways to restore VR after its dissociation: photoisomerization and dark-adapted enzyme isomerization of *trans*-retinal, and 11-*cis*-retinal. Photoisomerization may occur in response to bright light flashes, or prolonged illumination. Dark-adapted isomerization at the expense of retinisomerase (esterisomeraze) enzyme has long been doubted since it was thought that it was also light-induced, and ought to be labeled a photoisomeric process. It should be possible to create a full photochromic scheme of VR *in vitro*, however, we lack reliable scientific knowledge on this subject. BR, on the other hand, has its reversibility problem solved in a simple and exquisite way (see Section 3A.2). Why didn't the same solution work for VR? That is still a puzzle.

2A.10. The Influence of Ions (Ionochromism) on VR Spectra

The influence of ions on the positions of VR optical spectra has not been sufficiently studied enough to draw general conclusions [Crescitelli, 1978]. However, the ionochromic process in itself is interesting both for the working mechanics of retinal–protein complexes and for its application to biosensors. The works of Ostrovsky and colleagues proved the existence of ionochromic processes *in vitro* [Slobodyanskaya et al. 1980]. They showed that chloride additions to a digitonin extract of outer segments of chicken photoreceptors leads to a bathochromic shift in the iodopsin absorption spectrum. We have to remember that iodopsin is a visual pigment of the cones with retinal-2 as the chromophore. It was then revealed that Cl^- ions do not affect the rhodopsin spectrum of the chicken eye. The author suggests that iodopsin is an ionochromic anion-binding pigment, interrelated with chromophoric molecular centers.

2A.11. Rhodopsin in Fish Eyes

Studies of VR in fish have long been separated from the main pathways of rhodopsin studies in other vertebrates. It was held that VR structures in fish are formed in an aqueous environment, and bear little resemblance to VR found in other vertebrates. In 1937, it was shown that in saltwater fish, the VR pigment contains only retinal-1, while in fresh water species it has only retinal-2 [Wald 1937]. However, later studies proved that this rule is not general. Approximately half of all freshwater fish have a mixture of retinal-1 and retinal-2 [Knowles and Dartnall 1977]. This proportion depends not only on the type of fish and its age, but it also has a seasonal dependence [Bridges and Yoshikami 1970]. In species of minoga and sturgeon, 2–3 weeks after migration from the sea to the river, rhodopsin is fully replaced by porphyrhodopsin. The reverse replacement takes place after the reverse migration [Munz and Beathy 1965] [Crescitelli et al. 1985]. Light intensity changes may also change the proportion of rhodopsin and porhyrhodopsin [Bridges and Yoshikami 1970].

In fish visual pigments, absorption maxima vary much more those of other vertebrates. Some fish have color vision and differing pigments in their cones. For example, in the retina of *Caprinus carpio*, three types of pigments, absorbing at 462, 529, and 621 nm are found. This is different to the pigment on the rods absorbing at 523 nm [Tomita et al. 1967]. Most land vertebrates have rod pigments absorbing in the range of 490–500 nm, whereas in aquatic vertebrates or amphibians, this range is extended to 470–545 nm. This fact can be explained by the change in spectral composition of light hitting the water.

All of the information above permits a general conclusion to be drawn: *in vivo*, fish rhodopsins modify with uncommon ease in response to environmental changes. This process is facilitated by easily modified protein and retinal types.

2B. Rhodopsins of Invertebrates and Microorganisms

Startling as it may seem, the eyes of many invertebrates, including insects, are much more developed than the human eye. Insects, like humans, recognize the colors of the visible spectrum and are no less light-sensitive than humans. Furthermore, they are able to see in the UV range and the direction of light-wave polarization, and can distinguish the shapes of rapidly moving objects. Such are the attributes of a millimeter sized biological tool! Specialists in modern optoelectronics can, as yet, only dream of constructing such an instrument. Invertebrates convert light quanta into nerve impulses based on the same principles, and, like vertebrates, use rhodopsin as a photoreceptor. The simple organization of the insect nervous system, coupled with a high reproduction rate, make them ideal models for gaining insight into the mechanics and evolution of visual systems.

The discovery of rhodopsins in bacteria, insects, and unicells has created new research possibilities. The mystery surrounding eye origin reminds us of the renowned question: what came first the chicken or the egg? Rhodopsins simply do not fit into natural selection theory. Why would different groups of animals, distanced from each other by time and morphological differences, with no common ancestor, possess identically structured eyes? One modern theory finds it quite possible that in the process of evolution, photoreceptors emerged no less than 40 times, each time independently [Salvini-Plawen 1982]. The opposing theory implies that rhodopsins and photoreceptors emerged from one source [Vanfleteren 1982] (see also Section 2C.5).

To support different theories, their creators put forth arguments that have less in common than a bull and a fruitfly. On the other hand, it was recently shown that bull, fruitfly, and octopus have similar primary, secondary, and tertiary rhodopsin structures. This adds to the confusion surrounding the origin of rhodopsin. The only possible explanation is by postulating the existence, at an early evolutionary stage, of a general scheme for the structure of the visual system which was optimal for survival [Zavarzin 1941] [Gribakin 1987]. This may explain why visual systems differ only in their structural properties, and not in the details of their construction. Figure 2.13 shows the structure of a squid photoreceptor, while Figure 2.14 depicts a model of the primary and secondary structure of octopus photoreceptors. It is clear that, in spite of all the diversity in appearances, the structural principles being implemented, and the basic chemical components being used, are identical in all photoreceptors.

In all invertebrates, the plasma membrane is the fundamental component of the photoreceptor cell, and contains rhodopsin as the photoconverter. The membrane itself assumes a certain configuration depending on the environmental conditions, and on the type of organism. These forms may be quite exotic. All receptors are evolutionarily separated into two types: ciliaric and

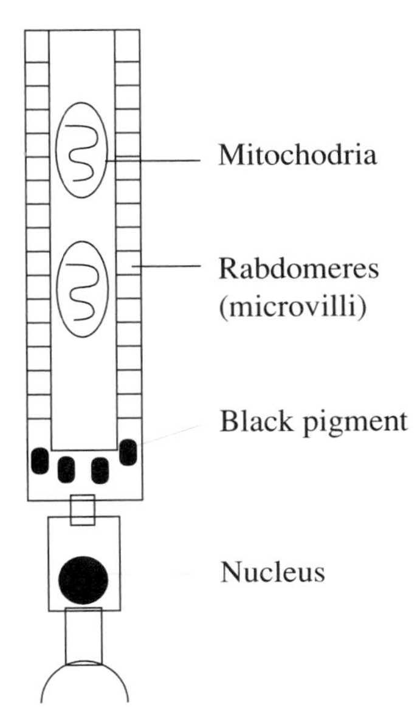

Figure 2.13. Simplified depiction of a squid photoreceptor.

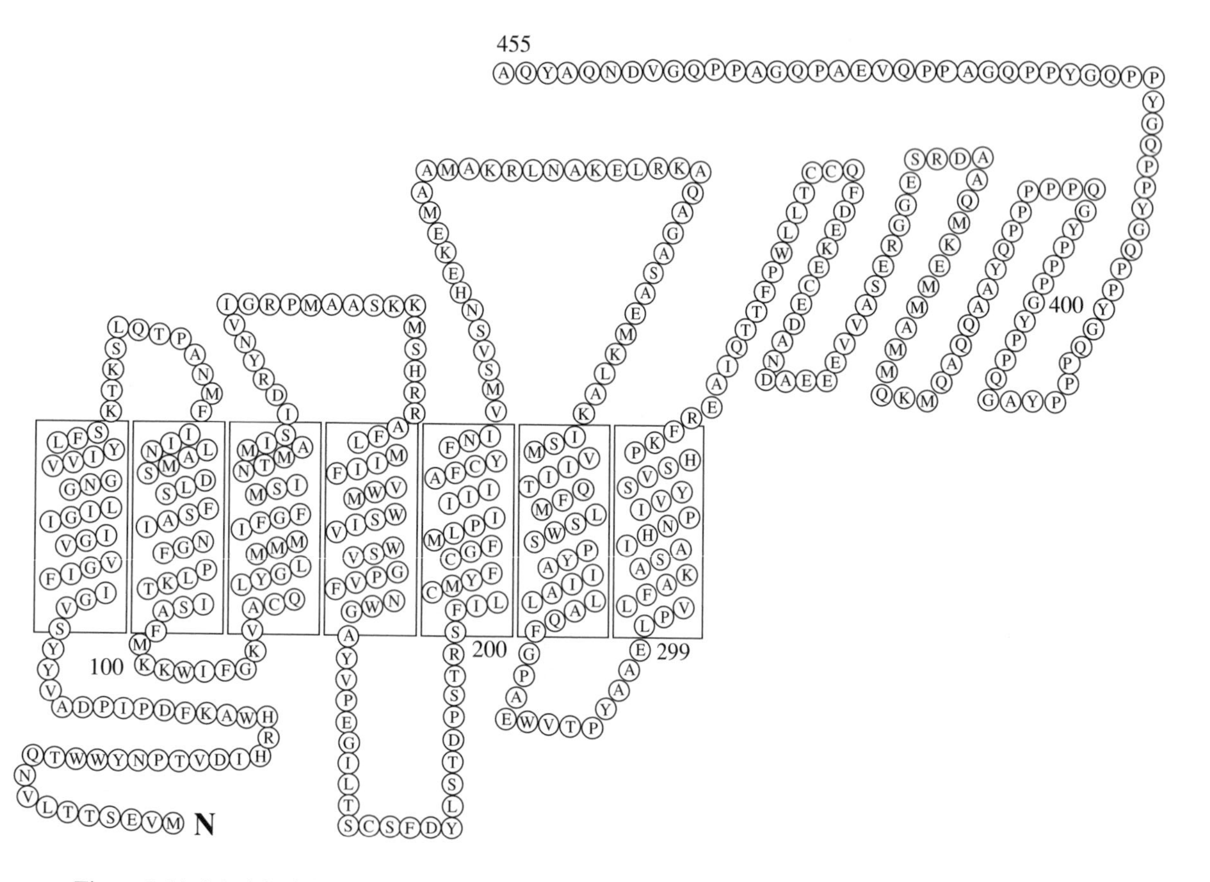

Figure 2.14. Model of the secondary structure of octopus retinal-protein [Ovchinnikov et al. 1988].

rhabdomeric. According to modern theories, the former originated from the plasma membranes of cilia or flagella, while the latter developed from the fingershaped prominences of the plasma membrane. However, this book relates to those visual structures that are naturally analogous to optoelectronic systems. To learn more about the morphological classification of photoreceptors, the reader should refer to a classical work of Salvini-Plawen and Mayer [1977]. From the optoelectronics standpoint, the eyes of some insects and mollusks (whose diameter rarely exceeds 10 μ), have interesting optical components. For example, the photoreceptor in the eye is normally shielded from excess light by special cells, or granules which contain light-absorbing pigments. In the eyes of *Nerilla* though, instead of granules, there are specifically arranged sets of reflecting plates. A light-focusing lens in *Dinophilus* consists of one pigment cell where two photoreceptors are sealed in. *Nerilla* has two special light focusing corneal lenses which is itself a complete optical system! [Eaken et al. 1977].

There are many astonishing examples of eye similarity in morphologically distant animal phyla. For example, there is a similarity in the organization of retinas in the octopus and in *Polycheta vanadis* [Elkin 1982]. Conversely, members of one animal group can exhibit a combination of eye components coming from different groups. The eye of the sea-mollusk *Pecten irradians* has two retinas, one of which is ciliaric, and the other rhabdomeric [McReynolds and Gorman 1970]. These facts suggest the existence of a single generalized scheme of vision.

Special mechanisms regulate the process of vision, optimizing it to every environmental fluctuation. Invertebrates, like vertebrates, have a photoreceptor membrane with a day-night regeneration cycle. In addition, the photoreceptor of an invertebrate enlarges in the dark, and its light sensitivity increases, attaining a one quantum level of sensitivity. For example, the crab *Crapus*, has a photoreceptor with a square surface area 20 times greater at night than in the daytime. Normally, such cyclic mechanisms are autoregulated by a photoreceptor, without intervention from another system (brain, kidneys, etc.). Amazing exceptions do occur. It has been shown that the photoreceptor cycle in the horseshoe crab *Limulus* is governed by electro-impulses emanating from a circadian rhythm inside the brain, and is totally independent from its photorecepting apparatus [Kaplan and Barlow, 1984]. Exceptions such as these are increasing in number. If this trend continues, many of the presently held ideas about the vision process will have to be abandoned in favor of new ones which can explain some of these anomalies.

2B.1. Some Natural Examples of Rhodopsin-based Photopigments

Invertebrates which have eyes or some sort of light gathering system are numerous. Scholars researching them are less numerous. Only a few visual pigments have been discovered, and even fewer have been thoroughly studied. However, the results of these studies can fill many volumes. In this section, a general survey on the photo-pigments of some representative specimens of mollusks, crayfish, insects, and unicells is presented. Such creatures were selected for their connection with optoelectronics and other emerging fields. Other organisms of interest for biomicroelectronics will be described in following chapters.

Photopigments of mollusks, arthropods and microorganisms have the following common properties:

* high photochemical quantum yield,
* photocycle reversibility, not involving the internal energy of the organism,
* chemical linkage between retinal and protein via the Schiff base,
* stabilization of the tertiary structure of biomolecules in the membrane at the expense of the lipid surrounding,
* insolubility in water (excluding bee rhodopsin),
* absence of rotational and lateral diffusion.

The VR photocycle of cephalopods is different from vertebrate rhodopsin. It has two Meta I forms, an acid and alkaline form, to conclude its photocycle. The VR photocycle of squid and octopus is shown in Figure 2.15a and b.

The chromophoric structure of rhodopsin apparently depends on the type of photoreceptor. For example, the squid *Watasenia scintillans* has three types of photoreceptors absorbing at 484, 500, and 470 nm. The bases of their visual pigments are 11-*cis*, 3-dihydro, and 4-hydroxyretinals, respectively [Matsui et al. 1988]. This squid may have a uniformity of color vision since blue light prevails on the sea bottom, and green light prevails on the waters surface. The

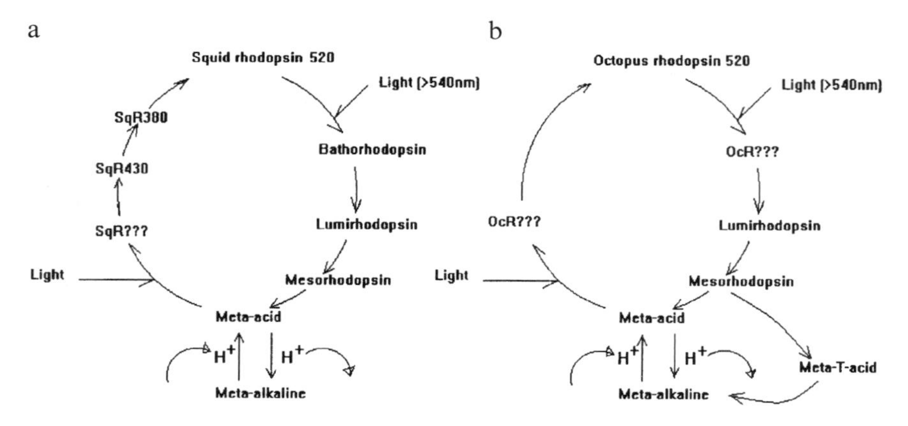

Figure 2.15. Photochemical reaction cycles of retinal-proteins in the (a) (*Todarodes pacificus*) squid and (b) (*Paroctopus defleini*) octopus. [Adapted from Tsuda 1978, 1979].

exact molecular weight of opsin and its amino acid sequence in *Watesenia* [Nashima et al. 1979] and octopus [Ovchinnikov et al. 1979] has been deduced. Opsin structures in these two creatures are virtually identical to their vertebral counterparts, with molecular weights in the range of 46–51 kDa. In some cephalopods, apart from the usual visual receptors, extraocular photoreceptors have been found. These control exclusively the local illuminance level [Hara and Hara 1987]. The absorption maximum of rhodopsin extraocular photoreceptors is 480 nm, and is similar to eye rhodopsin in its structure and main characteristics. From the retina of the gastropod *Conomulex luhuanus*, a rhodopsin has been isolated, absorbing at 474 nm, and fulfilling a similar function as rhodopsin found in cephalopods [Ozaki et al. 1987]. Nerve impulse sequences are also generated in a similar fashion; hyperpolarization (to create a potential drop) of membrane cells residing in the retina [Liebman et al. 1987].

The eyes of arthropods and mollusks are color sensitive. Most sea crayfish do not recognize colors since they have only one type of VR. Freshwater crayfish who have red-green, yellow, and blue-violet visual pigments are likely to have color vision [Eguchi et al. 1973]. Rhodopsin concentration in the crayfish rhabdomere is 2-3 times smaller than in the vertebrates. 11-*cis*-retinal is the principal rhodopsin form, however, 11-*cis*-3-dihydro-retinal may also synthesize in the eye of the crayfish *Procambarus*, forming porphyrhodopsin [Suzuki et al. 1984]. Furthermore, quantitative proportions of rhodopsin and porphyrhodopsin of this crayfish are seasonally dependent. Little is known about the photochemical properties of crayfish rhodopsins. Photoregeneration, for example, can proceed only at extremely low illuminances. The faintest light damages the retina. This mechanism of dark regeneration of rhodopsin is totally unexplained. Another riddle is that the dark-stage formation of rhodopsin *in vivo* is not temperature dependent while regeneration *in vitro* from metarhodopsin is thermally regulated. Furthermore, the regenerative effect *in vitro* is observed only at 25°C, significantly higher than the accepted physiological range. Such uncommon behavior of rhabdomeric rhodopsins hinders their study.

The majority of arthropods and some cephalopods distinguish light polarization, dichroism reaching up to 14° [Mote 1974]. Photoreceptor cells have special dichroic analyzers of various structures. A dual system of microvilli (rolled-up into tubular photoreceptor membranes) arranged perpendicular to each other, is one such example (See Figure 2.16).

The mobility of VR molecules in microvilli membranes must be limited or very inhibited to prevent damaging cellular sensitivity to the selected direction of the incidence polarization vector.

Insect VR studies began in the late 1970s, however, we still do not have precise data on their photocycles. Various species of flies and bees have been the central focus of research. Different from vertebrate VR, insect VR has a low tolerance to detergents, which encumbers many biochemical techniques.

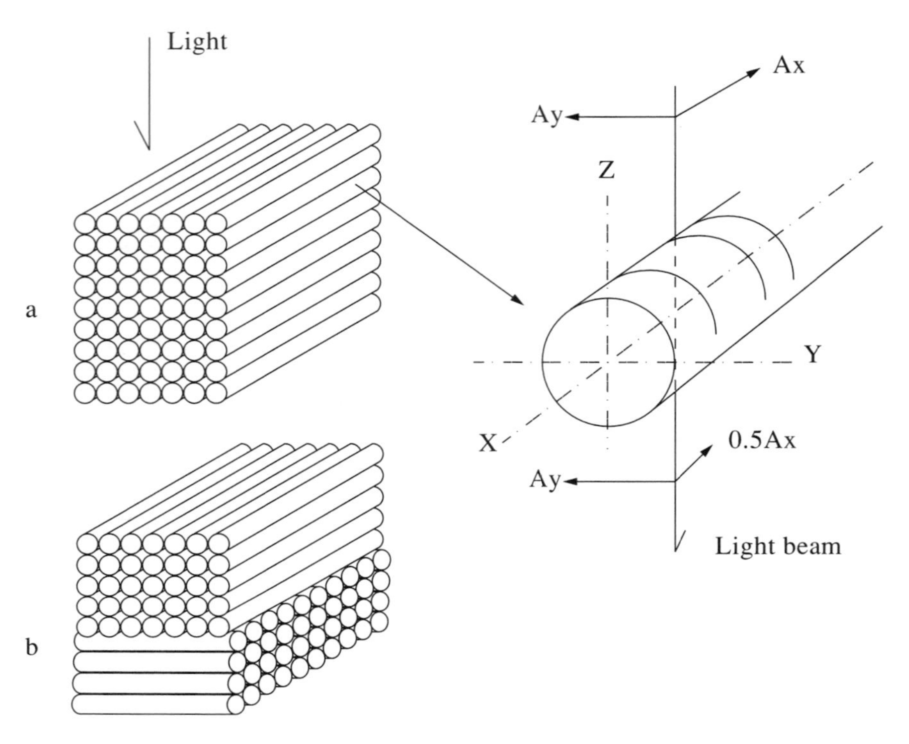

Figure 2.16. Depiction of a dichroic analyzer based on photoreceptor membranes rolled into microtubes (microvilli). Amplitude of the light wave (A); polarization along X (A_x) absorbed more than perpendicular A_y: (a) the rhabdomeres with parallel microvilli: (b) the rhabdomeres with perpendicular microvilli.

In addition, insect receptors contain large quantities of screening pigments (ommochromes) which are difficult to separate during experimental trials. To avoid this, researchers often use so-called white-eyed insects (moths) in which the synthesis of ommachromes is blocked by site-directed mutagenesis.

Similar to cephalopod VR, insect VR transforms into stable metarhodopsin upon illumination. A unique property of the Meta stage is that it has both bathochromic and hypsochromic shift spectra [Bennett and Brown 1985] [Langer et al. 1986]. Rhodopsin regeneration from metarhodopsin depends on the type of photoreceptor and may be accomplished photochemically or biochemically. In the first case, the photocycle is fully reversible, in the other case, VR regeneration requires the expenditure of biological energy from the insect [Schwemer 1983].

The chromophore of most insect visual pigments is 11-*cis*-retinal, and for a few others, 3-hydroxyretinal [Vodt 1987]. The latter may serve to adjust visual sensitivity to UV radiation. According to one theory, this occurs by the complexing of 3-oxyretinal with opsin, forming a pigment which absorbs in the range of 330–370 nm. Now being itself a visual pigment, it transmits UV-light quanta into the VR molecule. Energy transfer is accomplished through the use of a dipole–dipole mechanism [Kirschfeld et al. 1977].

A system of rhabdomeres (structurally and optically similar to modern optical fibers), often serves as a detector of polarized light in insects. For example, bee photoreceptors have two sequentially arranged coaxial rhabdomeres, whose transmission depends on the vector of polarization of the rhabdomere with the incident light [Gribakin 1973].

Besides the differences, vertebrate and insect VR also have common features. The molecular weight of the insect opsin approaches the vertebrate opsin in mass and lies in the range of 35–40 Da. In the process of VR photolysis in insects, phosphorylation and dephosphorylation have been observed. The extinction coefficients in vertebrate and insect VR coincide. However, the current state of knowledge about the photochemical and biochemical visual processes in insects is not at the level at which we understand the corresponding processes in vertebrates.

2B.2. Retinochrome: A Photoisomerizing Enzyme

Retinochrome was first discovered in 1965 in the photoreceptors of the *Ommastrephes* eye, and in its extraocular photoreceptors [Hara and Hara 1965]. This photosensitive pigment has all-*trans*-retinal as a chromophore. Retinochrome was later found in many cephalopods [Hara and Hara 1972]. In the visual process retinchrome performs two important functions: it isomerizes (photodependently) any retinal isomers into 11-*cis*-retinal, and also serves as a VR donor in the squid eye [Hara and Hara 1973], [Sperling and Hubbard 1975]. Thus, it maintains, at a constant rate, elevated 11-*cis*-retinal concentrations in the photoreceptor. Retinochrome was also found in photosensitive bubbles in the visual cells of certain mollusks [Hara and Hara,1987].

The absorption maxima of retinochromes depends on the type of mollusk and oscillates between 490–503 nm. Their molecular weight is modest, around 24 kDa, which is 2 × less than the weight of rhodopsin in mollusks. However, its molar extinction coefficient is 1.5 × higher, making its light sensitivity increase as compared to rhodopsin. Retinochrome photolysis undergoes a cyclic stage. The photocycle, and the dark-adapted reactions occurring in the squid *Todarodes* is presented in Figure 2.17 [Hara et al. 1981]. Different from common rhodopsin, the retinal molecule of retinochrome chromophore transforms from the all-*trans* to 11-*cis* state, while its first intermediate is lumiretinochrome (absorption maximum at 475 nm). Metaretinochrome occurs in two states: part of the retinal molecule binds with protein through the Schiff base and absorbs at 470 nm, the other part occurs in a free, uncomplexed state. Both intermediates are photoactive and can do two things; interconvert in light, or return to the ground state, i.e. transform back

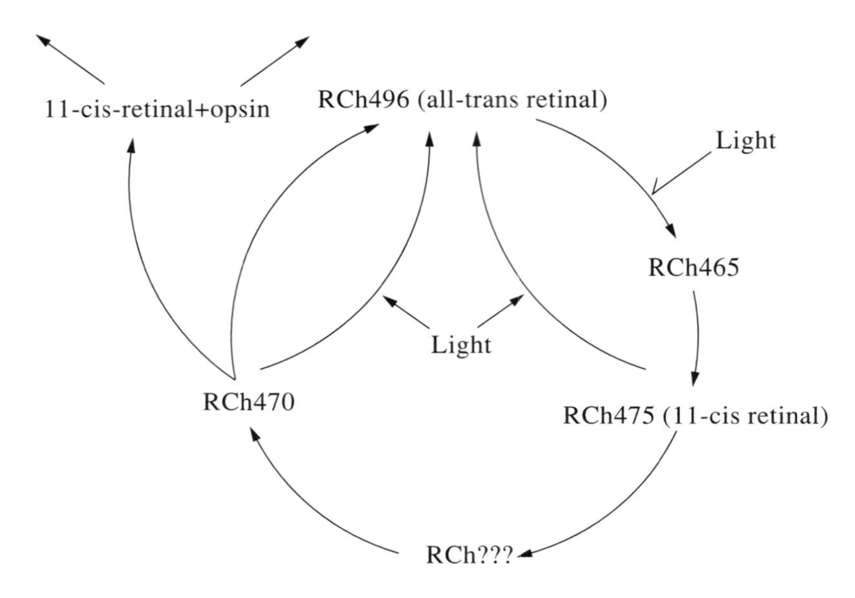

Figure 2.17. Photochemical reaction cycle of the squid (*Todarodes pacificus*) retinochrome [Hara T and Hara R 1965] [Tokunaga et al. 1990].

into retinochrome. The prevalence of one or the other depends on the wavelength and light intensity.

The part of metaretinochrome molecules in which retinal separates from protein (aporetinochrome) has the properties of the enzyme class called isomerases. This was confirmed by the following experiment: aporetinochrome was incubated with free retinal in three different isomeric states, all-*trans*, 13-*cis*, and 9-*cis*. In the dark, they reacted with protein, forming three different pigments with different absorption maxima. Upon exposure to orange light, all three retinals convert into one isomeric 11-*cis* state! It is concluded that the retinochrome protein acts as an isomerase. It is worth remembering that 3-dehydroretinal easily binds to aporetinochrome, forming a stable pigment. Retinochrome research is forging ahead with ongoing success [Tokunaga et al. 1990] [Kinumi et al. 1993]. Chapter 7 will discuss the prospects of retinochrome application as the photochromic agent in opto- and bioelectronics.

2B.3. Retinal-binding Proteins

Retinal-binding proteins (RBP) form a special class of rhodopsins. Different from previously described VR, they are water-soluble, and have low photosensitivity. The first RBP was isolated from the honeybee eye in 1958 [Goldsmith 1958]. Of all presently known rhodopsins, RBPs are the only ones in which the chromophore is formed not by binding across the Schiff base, but by utilizing stable, noncovalent bonds [Ozaki et al. 1987].

The RBPs of squids and cattle are functionally similar. Squid RBP participates in the transport of 11-*cis*-retinal, and enlists the use of the isomerase properties of retinochrome (see Section 2B.2). Being less light-sensitive, RBP transports this retinal isomer to opsin without losing it along the path. The chromophoric spectrum of squid RBP has two maxima at 330 and 400 nm. As opposed to retinochrome, this pigment does not have a photochemical cycle. At a light intensity greater than physiological, the isomeric composition of the pigment changes. In addition, the number of 11-*cis* isomers decreases, while the 13-*cis* and all-*trans* forms increase. This result shows that under artificial conditions, PBP similar to retinochrome, has photoisomerase properties. Despite 30 years of research, honeybee RBP largely remains a mystery [Pepe et al. 1984].

2B.4. Rhodopsins for Use in Detecting UV Light

A photoactive retinal–protein complex with an absorption spectrum of the ground state occurring in the UV range is a fairly rare natural occurrence. It may be that such rhodopsins are not essential in nature, or possibly, researchers are not persistent enough in their quest. It is known for certain that there exist three types of photoactive rhodopsins with an active spectrum in the UV range of optical range.

Rhodopsin, isolated from the eye of *Ascalaphus marcaronius* (owlfly) absorbs at 345 nm. It has a fully reversible photocycle (Figure 2.18) with three intermediates, two of which are photoactive [Hamdorf et al. 1973]. In the *Limulus poliphemus* median eye, little-known photorecepting cells absorb in the UV range [Lisman 1985].

The mollusk *Tridacna maxima*, dwelling in Indo-Pacific coral reefs, has several thousand eye dots hosting three types of photoreceptors which absorb at 540, 480, and 360 nm. The photoreceptor with an absorption maximum at 360 nm is found only in this particular genus. Its occurrence is apparently due to the uniqueness of the living environment. These creatures dwell on the highest surface of the corals, at the precise point where UV radiation is not absorbed by water [Wilkens 1984]. An eye dot with three photoreceptors of different spectra, spanning the entire visible spectrum, theoretically must be able to recognize colors. But why would a mollusk need color vision?

2B.5. Chlamyrhodopsin: A Recently Discovered Photopigment

In freshwater pools and reservoirs the unicellular mobile green algae, *Chlamydomonas reinhardtii* dwells. This algae branched off the evolutionary

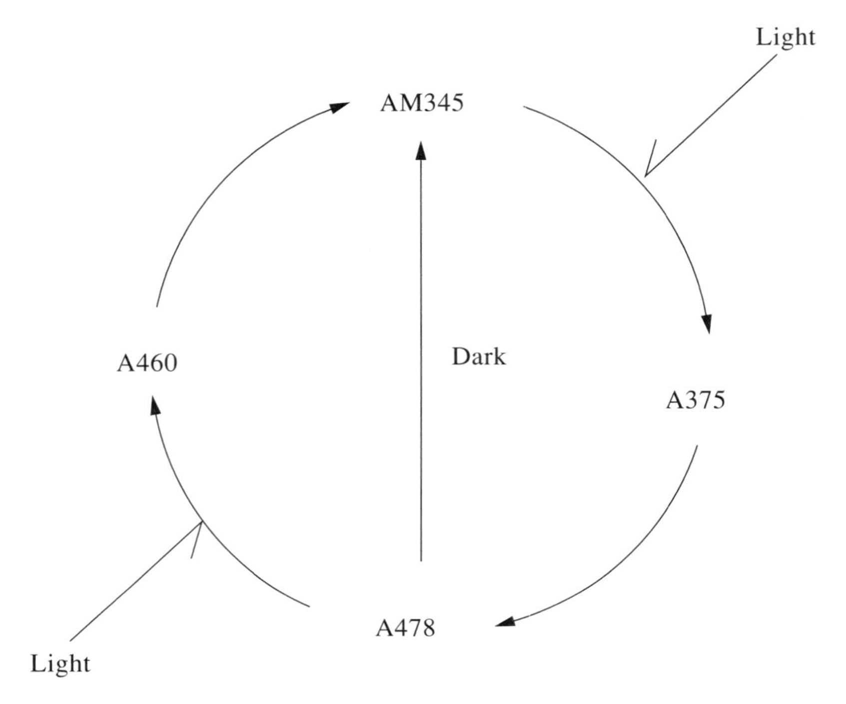

Figure 2.18. UV-photochemical reaction cycle of *Ascalaphus marcaronius* (owlfly).

tree with plants about a billion years ago, and evolved distinct from the fungi and animals that separated at a later junction. Its cell structure is typical for eukaryotes. It has a chloroplast and a strange red eyespot which helps the cell to orient and move relative to the direction of a light source. The structure of this eye consists of membranes which enclose carotenoids and are separated by cytoplasmic membranes. The carotenoids have a high refractive index, whereas the refractive index of the cytoplasmic membranes is low. As a result, we have a wonderful imitation of a dielectric laser mirror! As a consequence, the eyespot has a relatively narrow directional beam pattern without the use of a lens [Foster and Saranak 1989].

The eyespot is responsible for optimizing the negative and positive phototaxis in the cell. In the plasma membrane, overlying the eye structure is the pigment [Foster and Smith 1980]. Recently this pigment has been isolated and called chlamyrhodopsin [Deininiger et al. 1995].

The FN68 mutant contains only a little photopigment, and only at high light intensities is phototaxis observed. However, after a 10 minute incubation with retinal or retinal analogs, negative phototaxis appear with a 1000 times lower threshold [Foster et al. 1984]. This fact points to the existence of an opsin-like protein in the mutant. Many retinal analogs form photoactive pigments in *Chlamydomonas*, excluding a few analogs with a ring structure at C9–C10 of the polyene chain.

By 1990, more than a hundred chlamyrhodopsin analogs had been studied [Derguini and Nakanishi 1986] [Nakanishi et al. 1989] [Foster et al. 1989] [Lawson et al. 1991]. Some of them are listed in Table 2.1. A big opsin shift into the blue spectral region, upon reconstitution with the analogs of acyclic aldehydes of hexanal and saturated n-hexanal types, leads to the appearance of photoreceptors with active spectra in the UV range [Foster et al. 1988] [Nakanishi 1991].

More precise knowledge of the mechanisms of retinal–protein complexes in live organisms can be achieved by studying the work of their analogs under

Table 2.1. Analogs of chlamyrhodopsin

No.	Name and structure (in brackets)* of retinal analogs	Absorption maximum (nm)	Opsin shift (1/cm)	Maximum sensitivity**	Ref.
1.	(3)	505	1750	10	1,4
2.	(2)	483	1670	10	1,4
3.	(17)	490		10	4
4.	(18)	474		25	4
5.	(11)	472		35	4
6.	all-*trans* (1)	501	2510	95	1,4
7.	tetraenal (13)	461		120	2
8.	(16)	480		150	4
9.	dienal (4)	446	9030	220	1, 2
10.	hexanal (5)	354		660	1,2,4
11.	(6)	340	7630	270	1
12.	(19)	469		300	4
13.	(7)	505	2930	300	1,4
14.	(8)	476	5310	350	1,4
15.	n-hexanal (9)	339		400	1, 2
16.	(15)	470		430	4
17.	(10)	467	1520	550	1,4
18.	trienal (14)	459		860	2
19.	(12)	448	1210	2500	1,4
20.	(20)	450		2800	4

* Structure (see Appendix).

** Phototaxis action defined as the reciprocal of the threshold for phototaxis at a peak of each analog. The units are 10^{-18} m^2s/photon.

References: 1. Nakanishi 1991.
2. Nakanishi et al. 1989.
3. Nakanishi and Crouch 1995.
4. Foster et al. 1989.

in vivo systems. *Chlamydomonas* has turned out to be the ideal living system for such experiments. This system generated an unexpected result; the isomerization of retinal is not necessary to start photactivation processes, suggesting the importance of the dipole moment, and charge changes to activation [Foster et al. 1991]. We know a lot about photochemical processes in chlamydorhodopsin, however, the mechanism of its phototaxis is still not clear. Difficulties with the isolation of a photoreceptor protein do not allow experiments to be conducted *in vitro*. There is hope that such obstacles can be overcome with the enlistment of modern methods in physical chemistry. By using a suction electrode technique, it has been demonstrated that chlamydorhodopsin activates an electrical signal and a photoreceptor channel within 80 ms after exposure to a flash of light [Hegemann et al. 1995]. The results of the above demonstrate a direct connection between chlamydorhodopsin and protein complexes controlling cell flagella. Articulating this mechanism is only a matter of time.

2C. Bacterial Rhodopsins Found in Halobacteria

Salt-loving bacteria are called halobacteria. In 1971, a new retinal–protein complex, called bacteriorhodopsin (BR) was discovered in one of the halobacteria named *Halobacterium halobium*. More halorhodopsin discoveries followed including the discovery of two sensory rhodopsins. Since 1990, the halobacteria have also been called by a new name, *Halobacterium salinarium* with both appearing in the literature.

The isolation of a rhodopsin-like protein from bacterial membranes in 1973 stirred first the biologists, then the biophysicists joined in, and eventually all inquisitive scientists, despite their areas of expertise, became enamored. Although structurally similar to visual rhodopsin, BR differs from the latter in its extraordinary stability against external conditions, and an ability to function where ordinary proteins denature. Its photocycle, from the uptake, to proton translocation, is fully reversible, and requires no extra energy above the energy of the absorbed proton. Compared to the huge photoreaction centers in plants and algae, or to the unstable molecule of visual rhodopsin, BR stands unique, more like an extraterrestrial creation than an object of earthly origin.

2C.1. Discovery of a New Proton Pump

The stories of nearly all scientific discoveries contain some enigmatic aspect. A random event, inexplicable at first, occurs in the work routine, and leads unexpectedly to a discovery, or drastically changes the orientation of the research. Recently Walter Stoeckenius, a scientist with half-a-century of experience, published an article called "From Membrane Structure to Bacteriorhodopsin" [Stoeckenius 1994] where he gave a detailed narrative of how chance brought him to the discovery of a natural, molecular, light-driven proton pump.

By the 1960s, it had become evident that cell membranes are not merely passive films, separating cells and organelles not only from each other, but from the extracellular medium, allowing ions of different species to diffuse through them. The diffusional permeability of membranes is selective, i.e., is different for different ions. Furthermore, a few factors indicated that membranes contained molecules and complexes responsible for the active transport of not only ions, but other objects in different directions across the membrane. The Davson–Danielli model postulated the membrane as a three-layer structure (lipid layer included between two layers of protein). With the appearance of the electron microscope, the three-layer hypothesis was confirmed. Many scientists in the 1960s were studying membranes, hoping to find and explain the source of active transport mechanisms inside the cell. However, the Davson–Danielli model failed to explain the mechanics of many cellular processes.

In those days, Stoeckenius was studying the structure of mitochondrial membranes. He showed that proteins are found within the lipid layer, as well as on its surface. To prove his own membrane model, he needed a convenient research object. Simultaneously, Brown reported strange properties belonging to *Halobacterium halobium* which did not have a cell wall, and inhabited saltwater reservoirs [Brown 1963]. As the concentration of salt dropped, the cell membrane disintegrated into fairly large fragments. Stoeckenius considered these fragments suitable for his research. He also discovered in Brown's work certain contradictions between results and conclusions, which further stimulated his interest toward the new membrane. Soon he discovered two properties of the fragments: first, the membrane sediment, after centrifuging, acquired a purple coloring; second, the X-ray diffraction of this sediment revealed a series of sharp reflections, indicating the presence of some well-ordered crystalline lattice.

For the time being, the purple coloring remained unexplained, but the crystalline lattice inspired Stoeckenius to contemplate a new membrane model. In 1968, he obtained a relatively pure residue of the purple fragments. The work was stepped up in 1969 when Allen Blaurock and Dieter Oesterhelt joined the team. Blaurock, an X-ray analyst, was a perfect candidate to study the structures of new membranes. Before joining Stoeckenius, he researched the structure of the rod cell outer segments of the retina, and was very familiar with the properties of visual rhodopsin and with VR experimental testing methods.

Oesterhelt was a biochemist and had researched electron transport chain components on membranes of different types. For the purpose of revealing the mystery surrounding the purple membrane (PM as it was named by Kunau and Stoeckenius in 1968 after Kunau had cleaned the membrane off the main fractional additions) [Stoeckenius and Kunau 1968] an extremely lucky combination of specialists were culled together to become teammates. The label "Purple Membrane" had its shortcomings. Halobacteria tended to be confused with purple bacteria, and the color of pure PM, when purged of carotinoid additions, appeared... violet. Unfortunately, these facts were not uncovered until some time had elapsed, and all attempts to change the term failed.

During this period, Blaurock confirmed the crystalline structure of PM, and suggested that proteins are arranged in a lipid layer as two-dimensional hexagonal lattices. In 1970, a structural model of PM was proposed with an implication that proteins are incorporated in a lipid layer, i.e., they form an integral membrane protein [Blaurock and Stoeckenius 1971]. The type of the protein was still a mystery. A hint came from two directions. Stoeckenius discovered a phototaxis in halobacterial cells. At that time it was accepted that phototaxis cannot occur in anything but photosynthetic bacteria (which Stoeckenius fortunately did not know). Simultaneously, Oesterhelt, based on his past experience with animal rhodopsin, suggested the existence of its analog in PM. The idea was denounced as ludicrous, but Oesterhelt confirmed his suspicion by extracting retinal from PM suspension [Oesterhelt and Stoeckenius 1971]. It became evident that PM contains some retinal–protein complex. Authors called it bacteriorhodopsin. The presence of a photosensitive protein in halobacteria seemed to explain its phototaxis (see Glossary).

The most important riddle was still unresolved. What other use besides light orientation could halobacteria have for PM which often took up as much as half of the membrane surface area? Both Stoeckenius and Oesterhelt struggled with this problem and soon obtained amazing results. The concentration of hydrogen ions in the outside medium increased upon light exposure. That is how BR was first recognized as a light-driven ion pump [Oesterhelt and Stoeckenius 1973.]. This term, despite possessing more of an engineering than a biological slant, has nevertheless become familiar to all researchers of BR and its analogs. Later experiments on vesicles fully confirmed the preceding results by demonstrating a photophosphorylating ability of vesicles that had BR i.e. their ability to activate ATPase, using the energy supplied by the light-induced photon transfer across vesicle membranes [Racker and Stoeckenius 1974]. These results were a brilliant confirmation of Mitchell's theory on chemosmosis. In the same year, it was shown that the cell spectrum and the PM absorption spectrum coincide. The conclusion (or, better, suggestion) was that phosphorylation occurs in bacterial cells with PM-containing cell surfaces upon exposure to light [Danon and Stoeckenius 1974].

The next step was to determine PM's role in bacterial metabolism, and if possible to reveal the mechanism of light energy conversion into proton transport. Roberto Bogomolni and Rich Lozien, at the time, actively researched the photoreaction process in BR using a flash spectrophotometer which they had assembled on their own. Stoeckenius possessed the required amount of PM suspension for these experiments. In a short time, the main intermediate spectral products were found and studied, and a scheme of a BR photochemical cycle was created [Lozier et al. 1975]. It had been known before that a Schiff base (which binds retinal to protein) protonates and deprotonates in the duration of a photocycle [Lewis et al. 1974].

In 1975, together with Koji Nakanishi's group, they received preliminary results, pointing to retinal isomerization from the all-*trans* to 13-*cis* form in the duration of a BR molecule photocycle [Pettei et al.1977]. In less than five years the main questions were generally answered. In another year Stoeckenius proposed the first model of a transport mechanism [Stoeckenius 1976].

Toward the end of his story, Stoeckenius emphasized the chain of remarkable circumstances and happy coincidences that had led him to his discovery. By chance, his early works on myelinated nerve fibers inspired him to study membrane structures. His subsequent choice of a mitochondrial membrane as an object for structural research made him turn to bioenergetics. Finally, Brown's self-contradictory report influenced his decision to make halobacterial membranes a new research object. The result was the discovery of PM. Even though PM proved an ideal object for structural research, scientific curiosity drew Stoeckenius outside the boundaries of his own competence as an electron microscopist.

Thereafter, his scientific interests and research profile shifted directions drastically. Unburdened by the superstitions of specialized knowledge, he could stage experiments which no narrow specialist could ever conduct. Happy encounters and coincidences helped him to create a team able to solve an entirely new scientific problem in record time. Not everybody immediately understood that a new rhodopsin had been discovered. Stoeckenius recalls how his theses, offered at the 6th International Photobiology Conference in 1972 was mistakenly included in the "DNA-Repair Mechanisms Session, instead of the "Vision and Photosynthesis" session. Similar situations are frequent when big discoveries are involved: comprehension arrives later—if ever. Fortunately, BR was not only accepted as a new scientific artifact, but it also liberated the mentality of many scientists, enabling them to pursue further research. A series of entirely new natural rhodopsins and their derivatives were discovered in this atmosphere. As a whole, research results afforded a vision on the mechanism of sunlight energy conversion, and biological evolution from a wholly new perspective.

The BR mechanism is still argued about, and different theories are being put forth. These theories will be reviewed in Chapter 3. To close the story of the discovery of a light-driven proton pump, we'd like to quote its author/discoverer, speaking on its functional mechanisms: "it is postulated that the

Schiff base proton is transported, driven by PK changes of the Schiff base, and moves along chains of proton-exchanging groups, and that conformational changes of the molecule switch the connection of Schiff base from the external to the internal surface thus rendering the process vectorial." He finishes with the following words: "This first sketch of a model has withstood the test of time quite well" [Stoeckenius 1994].

2C.2. The Story of Discovery Unfolds

In Russia in 1973 (the year of publication of what is now known as a historical paper by Oesterhelt and Stoeckenius [1973], a project called "Rhodopsin" was started by the academician Yuri Ovchinnikov. The project's goal was to conduct complex research on visual and bacterial rhodopsins, and also to obtain federal funding for the research. During my first visit to the U.S. in 1989, I understood that a passion for big projects with loud names is not accidental. Obtaining funds for research from the government or, more accurately, from the pocket of the taxpayer, is possible only if a scientific project is nicely presented. There is a parallel to be drawn to American cafes: no food more bland and better designed is to be found on the planet. There is the another side to this blandness, since American food is much healthier than the tasty, but cholesterol and fat-laden dishes of Russian and European cuisine. The same is true in science: if a dry scientific idea is neatly packaged, it is likely to obtain support, possibly to the benefit of the taxpayer, and on rare occasions, to the benefit of all humankind. On the other hand, if a scientific project is related in a precise, conservative manner, gold mines may be concealed, but government officials will remain unmoved, and a brilliant idea may be lost forever.

The "Rhodopsin" project was presented nicely enough, and was launched by an appropriate organization nearly simultaneously with the onset of BR research in the U.S. Not long before that, the Institute of Biophysical Research (where I started working after graduating from Moscow University) had been moved from noisy, dusty Moscow to an idyllic suburb on the beautiful Oka river. Construction of an academic biological center went on at full speed. Ideal conditions for fruitful work by graduate students and science specialists, from biologists to physicists, was soon achieved. To work on the Rhodopsin project, the efforts of two research institutes (ours, and the Institute of Bio-organic Chemistry in Moscow, directed by the project initiator academician Ovchinnikov) and another large research group at Moscow University led by Vladimir Skulachev, were united. The first results appeared as soon as 1974. In an international journal, a paper by Skulachev and co-authors was published under the intriguing title: "Electrogenesis by Bacteriorhodopsin Incorporated in a Planar Phospholipid Membrane"

[Drachev et al. 1974]. Simultaneously, the author of this book constructed the first flash spectrophotometer equipped specifically for BR studies. Both myself and Bogomolni had assembled flash spectrophotometers on our own [Kaushin et al. 1974]. The idea of deciphering the BR amino-acid sequence, proposed by academician Ovchinnikov, apparently dates back to this period. From 1973, when his lab acquired the first milligrams of the PM suspension, a young and talented biochemist, Nadik Abdulaev, took to working on it. In 1976, they published a report on the influence of proteolytic enzymes on the purple membrane [Abdulaev et al. 1976].

Similar research (deciphering of the BR structure) was started in the U.S. by the group from a Nobel Prize laureate, Khorana. The race was afoot. "Who will be the first to decipher BR structure?" The Russians won this scientific "marathon," preceding Khorana by almost a year [Ovchinnikov et al. 1979] [Khorana et al. 1979]. It was an important step toward understanding the functional mechanisms of BR and its analogs. The "marathon" story is in itself interesting (see Section 3A), as well as the story of the production of the first photochromic films on BR base (see Section 4D).

BR was the first membrane protein whose amino-acid chain was fully deciphered. The Stoeckenius stage of BR research, by that time, was completed. Stoeckenius having determined the function of the BR molecule in halobacteria as a molecular light-driven pump for active hydrogen ion transport across the cytoplasmic membrane (see Section 2C.1). This definition is commonly accepted, as is the Skulachev definition of BR as an electrical energy generator [Skulachev 1976].

Skulachev is a biochemist by education; a bioenergeticist by calling. There is a magical cadence to these words: bioenergetics, bioelectronics (biocomputer for me). Skulachev, being a strict follower of Mitchell, and a devotee of his Chemo-osmotic theory, had long been interested in the role of proton potential in cell energetics. At their meeting in February 1973, Stoeckenius shared with him his guess that BR might act as proton potential generator. Skulachev immediately understood that a new type of photosynthesis had been discovered. In the same year, his biochemical lab was joined by a physicist Lel' Drachev, who started experimenting with BR, trying to measure the ion flow across purple membrane using well-known methods.

The first experiments demonstrated the ability of BR to generate photo current [Drachev et al. 1974]. Proteoliposomes from BR incorporated in a flat phospholipid membrane and separating two electrolyte-containing compartments were exposed to light. The potential difference between the light-exposed electrolytes reached 0.3 V. This was more than enough to provide energy for cell ATP synthesis. The works of Drachev and colleagues are nowadays accepted as classical, and his methods with slight modifications are applied in many labs around the world. That is how BR helped to supplement a substantial aspect of Mitchell's hemo-osmotic theory, i.e., its electric part. Mitchell's theory, ever since, has appeared perfect.

Today, BR as a biochemical product, can be ordered by catalog. In the 1970s, it had to be "homemade," or begged for from Stoeckenius. Russian scientists, however, had more luck than the others: in the basement of our Institute, Lyna Chekulayeva, a microbiologist, quickly assembled a gigantic photocultivator, in which the mud-color culture of halophilic microorganisms was bubbling day and night, brightly illumined by fluorescent lamps. Aided by only one apprentice, for many years she provided for the needs of the Rhodopsin project participants, and for the constantly increasing number of those who wished to join this exotic protein research. Of all the BR participants, she was the only microbiologist who named a cell culture of halophils after her native town, and not herself.

BR also proved to be the first membrane protein with a determined spatial structure. Richard Henderson, an American scientist, brilliantly combined X-ray structural analysis, electron microscopy, and statistical analysis, which allowed a description of the tertiary structure of BR at 7A precision. He proposed the first 3-D model of BR 4 years before its amino-acid chain was deciphered [Henderson and Unwin 1975]. Even today, Henderson continues to research different aspects of the structure and functioning of retinal–protein complexes.

By 1979, in 5 years of research, more was learned of the BR molecule than had been learned about animal rhodopsin in a century. This accounts for the extraordinary stability of the BR protein molecules incorporated in PM. They are resistant to high temperature, to acids, alkali, and to many bio-organic solvents. The PM suspension can be kept in a refrigerator for years. PM phospholipids, simple ethers of phosphoglycerine and fatty acids, are much more stable compared to those of a plain membrane. A sample of PM suspension in water can remain at work for months, and films based on polymer and PM solution for years. PM extraction from bacteria is a very simple: the bacterial biomass is placed into pure water and subjected to osmotic shock. Of the large cell particles, only PM remains, for the crystalline lattice keeps them from deteriorating. Centrifuging leaves pure PM sediment, containing nothing but BR and lipids. Thus, without any chemical filtration, a 100% pure experimental product is obtained!

Due to PM capacity for self-assembling, the substitution of retinal in a BR chromophore by its natural and artificial analogs presents no difficulty either. Today, the menu of photoactive BR analogs is counted by the dozens. The BR "family" is growing, and not only by synthesizing analogs. A few more BR-type natural retinal–protein complexes have been discovered. Even in a halophilic cell, BR is not the only photosensor. In 1979, Spudich and Stoeckenius thoroughly investigated the effect of halophilic phototaxis (see Glossary) observed by Stoeckenius as early as 1973. The study revealed the presence of photo- and hemosensory taxis controlling systems [Spudish and Stoeckenius 1979]. Two years later, Spudich reported the discovery, in halophils, of a photoreceptor which can transport Cl^- ions upon exposure to light. In other words, he discovered a new molecular pump [Harurkar and Spudich 1981]. However, this is not the end of the BR story. Further explorations revealed it to be a "multifaceted child" or a "black box'" with more than two dozen input–output channels.

2C.3. Biosynthesis of Purple Membranes

In the summer of 1980, a 20–25 cm surface water layer in the Dead Sea turned a bit purple. This layer had a high concentration of *Halobacteria halobium* (up to 2×10^3 cells/ml). A study showed that the membrane surfaces of the cells contained BR [Oren and Shilo, 1981]. In 1988, the waters of salt lakes in the Krasnovodsk vicinity were changing in color ranging from yellow to bright violet due to the presence of halophils with different concentrations of carotenoids and PM. Such events are rare, and depend on many natural factors. The normal halophil concentration in natural reservoirs is low, and does not go beyond 10^5. High concentration solutions are synthesized in lab conditions.

In lab conditions, halobacteria are cultivated in lumostats, similar to those used for the synthesis of monocellular green algae [Rogers and Morris 1979]. The cell division period is 7–8 hours at a temperature of 39–40°C; the overall cultivation time depends on the researcher and varies from 3 to 7 days. The BR concentration in cells depends on the conditions of cultivation. Temperature, pH, light intensity, spectral composition, and the chemistry of the medium all play significant roles [Danon 1977]. Many years of observation show that natural cell cultures under artificial conditions grow better in winter than summer [Chekulaeva, unpublished data].

Under anaerobic conditions, or when suppressing the activity of the cell respiratory chain with biochemical reagents, the content of ATP in the cell becomes exponentially lower. When exposed to light, cells start synthesizing PM, and the ATP content sharply increases which points to the phosphorylating property of PM [Danon and Stoeckenius 1974]. One widely accepted method of anaerobic halobacterial growth in light was described by Oesterhelt [Oesterhelt 1982].

When a cell biomass is obtained, it gets resuspended in a big volume of water with the addition of an enzyme called DNAasa. After osmotic shock, and the splitting of the DNA, the suspension is centrifuged several times. The sediment is 100% pure PM, which is either resuspended in distilled water to required concentrations, or is dried in a leophyl dryer for storage. The centrifuging method can be replaced with the method of filtration. On a 10 liter sample of culture extract to yield 0.4–0.6 g of PM (dry weight) [Neuman and Leigeber 1989].

It is commonly accepted that BR and PM biosynthesis in halophils occurs according to the following scheme [Sumper et al. 1976]:

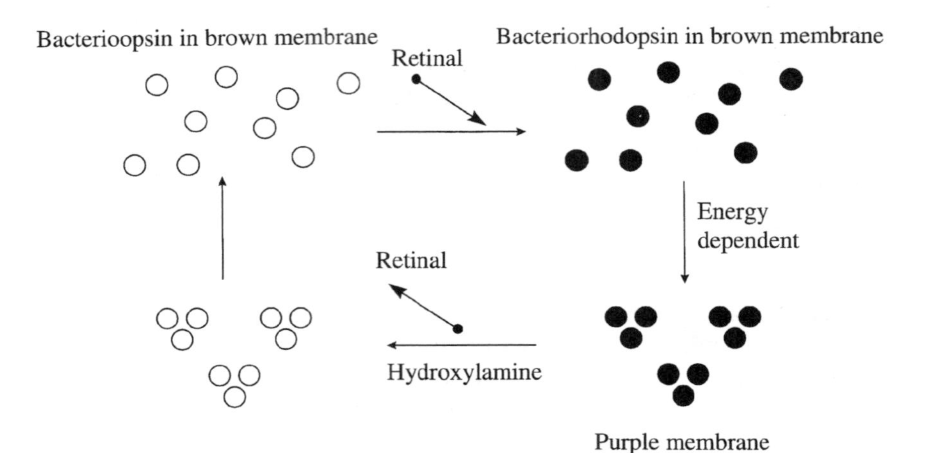

Diagram 2.1

Under optimal cultivation conditions, the square surface of the cell membrane occupied by PM may reach up to 80%.

2C.4. Purple Membranes: Two-dimensional Crystals

The main property of PM (which may also be its function) is the formation of a hexagonal crystalline array of trimers. Lipids fill the gaps at the base of the lipids, and BR molecules. The crystalline lattice is hexagonal with a unit cell size of 63 A [Blaurock and Stoeckenius 1971]. In this kind of structure, BR does not denature below 80°C (in some cases up to 160°C!), and retains its chromophoric properties in a pH range from 00.0–11.0.

PM consists of 25% lipids and 75% BR molecules by weight, and is 4.5–5 nm thick. A cross-section of PM fragments varies up to 1 micron, and depends on the type of cell culture and on the conditions of growth. BR molecules penetrate all through PM, i.e., BR is a trans-membrane protein. Exploration of the crystalline structure by electron and neutron diffraction methods provides a full picture of the α-helix polypeptide string segment arrangement in BR molecules (and of protein molecules arrangement in PM) [Henderson et al. 1990]. Figure. 2.19 shows electron-density maps of BR with a distinctly visible array of seven polypeptide columns in a BR molecule, and a

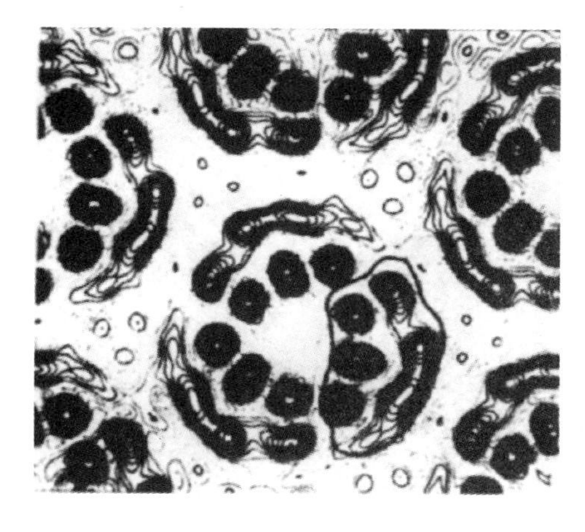

Figure 2.19. Hexagonal trimer packing of BR molecules from purple membranes. One of the three trimer molecules is circled by a line within which a projection of seven α helices can be seen [Henderson and Unwin 1975].

3-D configuration of three BR molecules. The shape of PM can be artificially modified from crystal to orthorhombic [Michel et al. 1980] without altering BR's properties.

From the colorless strains with suppressed retinal synthesis, the so-called "white membranes" (WM) are extracted. In these membranes, apo-protein (bacterio-opsin) is packed as a two-dimensional crystal monolayer of hexagonal structure. The addition of retinal into bacterioopsin suspension, or into the medium of colorless cell culture growth, leads to immediate recrystallization of normal PM [Neugebauer et al. 1978].

One of methods of retinal suppression in a wild-cell culture is to add nicotine to the cultivation medium. This causes the formation of so-called brown membranes (BM) instead of PM, the former containing no more than 5% of BR molecules and cytochrome (which adds brown coloration to the membrane) [Sumper 1982)]. The BM protein does not form a crystalline lattice, and BR molecules exist in a monomeric state. All-*trans* retinal additions into the BM suspension lead to the formation of new BR molecules and their instant crystallization in PM [Sumper and Herman 1978]. The crystalline lattice of such PM is not uniform since the 5% monomeric BR molecules are not built into it [Hiraki et al. 1981]. Figure 2.20 shows the spectra of BM, WM and PM [Mukohata et al. 1981].

2C.5. Halobacteria: A Riddle for Biological Evolution

Halobacterium halobium is an archaebacteria whose habitat is concentrated salt lakes, ponds, and flats. The strangeness in behavior and structure of this microorganism was discovered back in early 1970s when it became obvious that two energetic systems of phosphorylation exist within one bacteria; the

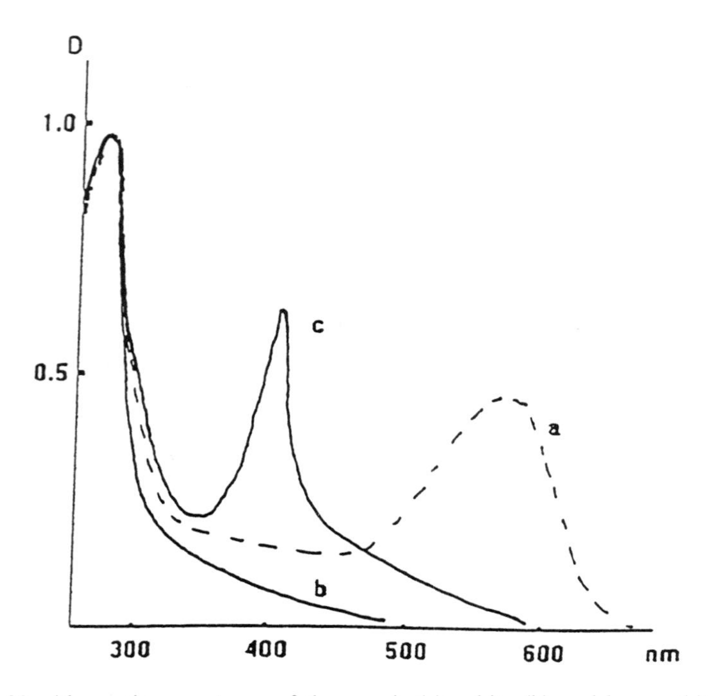

Figure 2.20. Absorption spectrum of the purple (a), white (b) and brown (c) membranes of *Halobacterium halobium* [Mukohata et al. 1981] [Hiraki et al. 1981].

system of oxidization, and the photosynthetic system [Oesterhelt and Stoeckenius 1971] [Danon and Stoeckenius 1974] [Stoeckenius 1976]. The halobacteria had always occupied a special place among unicellular prokaryotes [Werber 1980]. They live in high-concentration saltwater, where no other microorganism survives. It is enough to mention that the concentration of salt in the Dead Sea reaches 34%! High concentration of carotinoids in bacterial cells is what makes the water acquire a purplish color. The town of Krasnovodsk (red water) in Russia got its name for its proximity to red water lakes with high concentrations of carotenoids in halobacteria. The purplish coloring of lake water appears only during the periods when bacteria synthesizes PM. Such periods are irregular and depend on changes in water chemistry, oscillations of temperature, solar activity, seasonal changes etc. In 1980 and 1988, the environmental conditions in the regions of the Dead Sea and Krasnovodsk must have been favorable for active PM synthesis. The bacteria can be found in salt crystals on the littoral area, or on salted fish a few days after preparation.

Halobacterium halobium is shaped as sticks of approximately 0.5 mm in diameter, and 3–10 mm in length. The shape of cells depends on the outside medium and can be oval, or even spherical (for example, at low magnesium

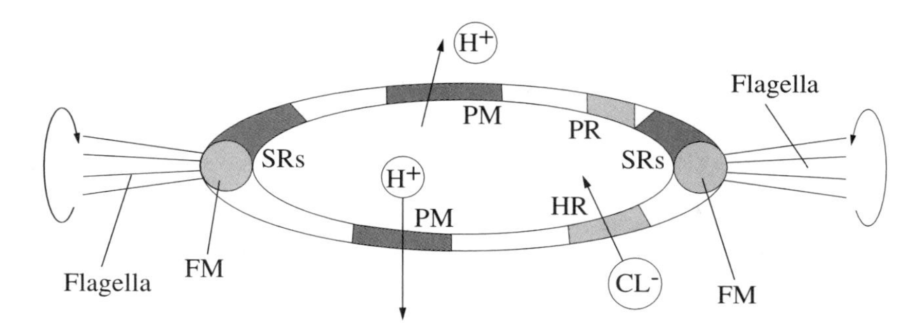

Figure 2.21. Representation of a sectioned *Halobacterium halobium* cell. FM—flagella motor. PM—BR molecules-containing purple membranes. HR—section of halorhodopsin-containing cell membrane. SRs—sensory rhodopsins SR-I and SR-II-containing sections of the cell membrane. PR—phoborhodopsin (or SR-II). H^+ and Cl^- ions of hydrogen and chlorine. The scheme displays a random arrangement of photoactive membranes.

concentrations). On the opposite ends of the cell, the moving devices, flagella, are located. In Figure 2.21, the membrane structure is shown on a schematic cell cross-sectional view. The halobacteria belong to prokaryotes, but morphologically differ from them. The cellular wall has none of the regular "construction parts," and consists mainly of glycoprotein. The bacterial membrane consists of a protein-lipid layer with specific areas colored purple and red. PM contains only one photosensory protein—BR, whereas in red membrane, three new types of rhodopsin were found: halorhodopsin (HR) [Mukohata and Kaji 1981], sensory (or slow) rhodopsin (SR), and phoborhodopsin (PR) [Stoeckenius and Bogomolni 1982] [Spudich and Bogomolni 1983]. The red membrane also contains the electron-transport chain and complex ATP-synthezase/ATPase. Nowadays, the functions of all four rhodopsins are well-known: BR is a light-driven proton pump; HR is a light chloride pump; SR and PR (the modern name are SR-I and SR-II) are also light-driven and control the phototaxis of halobacteria.

Halobacterium halobium is not the only member in a vast array of new rhodopsins, however surprising this already seems. As mentioned above, one cell unites two phosphorylation systems, a normal one, and a photosynthetic one. A brief highlighting of some of the differences between systems having rhodopsin and those without follows.

* the photosynthetic system is extremely simple, with proton transport carried out by one single molecule. This is contrasted with photosynthesizing systems in green plants (without Rhodopsin) which is fairly complex, and involves several spatially and functionally interdependent molecules,
* photosynthesizing halobacteria employ visual rhodopsin for its photosynthesizing purposes, instead of common chlorophyll,

* due to the specific construction of the cell wall and the crystalline packing of BR in PM, bacteria survive under extreme environmental conditions (high temperatures, sun radiation, high salt concentration, redox reactions) without losing its photosynthetic apparatus.

All of these oddities are still the object of scientific interest, and are actively discussed [Oesterhelt 1989] [Bogomolni and Spudich 1992].

Let us imagine that halobacteria were the first unicellular predecessors of life on Earth. What impelled nature to invent another photosynthetic protein—chlorophyll? Clearly BR and HR are light-driven ion transport systems, and SR and PR are signal systems of orientation and movement. It would seem that these four rhodopsins would be enough to supply every organism on the planet with photosensors and photoconverters, however, nature invented chlorophyll not man. Could the visual rhodopsin in the eyes of humans and (and many other creatures), be of bacterial origin, echoing back to an earlier phase in evolution a billion years ago? These questions have no answers yet, and halobacteria remains a problem in theories about evolution [Bayley 1979].

2D. Modification of Rhodopsins

Modern science has a wide variety of methods to modify rhodopsins. We understand the modification of rhodopsin molecules as the structural change of the protein or amino acid sequence and/or chromophore substitution by its analog. Modification always leads to a cardinal change of optical and photochemical characteristics of rhodopsins. Certain characteristics of photochemical processes in any rhodopsin molecule can be, without a doubt, altered by changing the parameters of the surrounding medium (temperature, pH, etc.). However, under the definition of a "modification of the rhodopsin molecule," only the following methods are understood:

1. Specific change to retinals (retinal by retinal analog substitution)

The substitution of a rhodopsin molecule for its analog in any rhodopsin leads to partial or complete alteration of the previous properties and/or to the appearance of new ones. The initial stage of a goal-directed substitution of retinals in a rhodopsin molecules requires apoprotein (i.e. retinal-freed rhodopsin protein molecule) formation from the initial rhodopsin molecule. This task can be approached in two ways: by using the photoinduced hydroxylaminolysis of the rhodopsin molecules when retinal molecules bind with hydroxylamine molecules, vacating the Schiff base place. The second is using a microbiological method where artificially created bacterial cultures do not form retinal; thus, the Schiff base place is initially vacant. In both cases apoprotein (i.e. retinal-free protein) is formed. Naturally, the microbiological method is inapplicable for vertebrate VR.

The methodology of creation of retinal-modified rhodopsin, i.e., of a rhodopsin analog, is relatively simple. Once the selected analog of an artificial or natural retinal is added to apoprotein suspension, the self-assemblage of a new rhodopsin molecule occurs spontaneously. The effectiveness and speed of the self-assemblage depends on many factors (light, temperature, protein and retinal analog types, etc.). Reconstruction is impossible for only a few retinal analogs (see Sections 3C.1 and 3C.2).

2. Random mutagenesis

Random mutagenesis may always occur during bacterial cloning. At a large number of random mutations, a rhodopsin with an altered opsin amino acid content may form in the mutant bacteria. As a result, rhodopsin characteristics may also change. If these changes interest the experimenter, he may start cultivating a new strain from the new clone [Oesterhelt and Krippahl 1983]. A few presently widely used strains are the result of such experiments (see Table 3.3)

3. Site-specific mutagenesis

The methods of modern gene engineering allow programming the replacement of certain pre-assigned amino acids in opsin [Soppa and Oesterhelt 1989]. In this case, photochemical properties of rhodopsin change, but unlike method 2, they can be controlled. This permits the direct study of the processes taking place in rhodopsins and their analogs (see Section 3B.2)

4. Chemical modification

Modification of amino acid residues of opsins may be assisted with ferments or chemical reagents, removing or replacing entire fragments of a polypeptide chain. It is known that the proteolysis of amino acid tail groups (C- and N-terminal tails) of proteins and polypeptide chain loops between the helices changes a few characteristics of a photocycle [Abdulaev et al. 1976] [Rosenheek et al. 1978] [Hristova et al. 1986]. It was found that the modification of BR with carbodiamide and several amines which neutralize the –COO tails or convert them into positively charged ones [Ovchinnikov et al. 1982]. The modification of BR, isolated from wild type and mutant halobacterial cultures, by cyclohexadione retards the kinetics of the formation and decay of the M intermediate (See Section 3A.3). In such a suspension, in partially dehydrated films and polyacrylamide gels, M lifetimes reached 10 minutes [Lukashev et al. 1993]. Interestingly, detachment of a few fragments of a polypeptide chain does not interrupt the photocycle, and, furthermore, the splitting of the BR molecule in two uneven parts does not affect generation of a membrane photopotential in the smaller part which has a chromophore [Abdulaev et al. 1978].

5. Mixed method

In the rhodopsins produced as described in 2 and 3, retinal may be substituted by its analogs using the methods described in 1. Combination of all the methods opens new possibilities for the research of photochemical processes in retinal–protein complexes of natural origin, and the perspectives for their application in future biotechnology. Several examples of these new possibilities are discussed in Chapters 5 and 6.

3

The Unique Properties of Bacteriorhodopsins as Energy Converters

A quarter of a century has passed since the discovery of PM (purple membrane) and the first bacterial rhodopsin (BR) was isolated in a wild cell culture of *Halobacterium halobium*. Since that time, microbiologists and geneticists have synthesized dozens of artificial halobacteria cell cultures, have reconstructed hundreds of BR analogs (genetic and mutant invariants), and have even discovered new bacterial rhodopsinoid molecules. Such analogs differ in their ground state absorption spectra, the lifetimes of intermediates, the ability to transport protons, etc. Only a few BR analogs have no photochemical activity (no photocycle), and are not capable of creating a potential drop. Nearly all retinal-protein complexes, to a greater or lesser extent, are capable of energy conversion. However, despite a knowledge of BR protein structure (primary, secondary, and tertiary), the retinal binding location, and despite the multitude of theories put forth, and the behemoth experimental databank, the mechanism of BR functioning is still largely a mystery!

In this part of the book the reader will find a description of all presently known bacterial rhodopsins, and the majority of their artificial analogs. My aim in doing this is to alter the view of Rhodopsins as simply ion pumps or electrocurrent generators. In fact, as has been stated many times, Rhodopsins function as universal energy converters, which as yet, has proved to be beyond the reach of the human mind.

3A. Wild-type Bacteriorhodopsin

Wild type bacteriorhodopsin is the name of any BR extracted from natural bacteria. If the cell culture was artificially synthesized, or the BR was modified, the adjective "wild" is omitted. In spite of the availability of many BR modifications, many scientists prefer to work with BR extracted from wild cell cultures. The rigidity of PM in wild bacteria is higher, and their experimental behavior is, in many cases, more stable than their artificial counterparts. A concentrated PM water suspension from a wild cell culture can be stored for weeks at room temperature, and for years when refrigerated at 5–8°C. Freezing for longer periods may cause the formation of aggregates from PM; however, after ultrasonic sonication the aggregates disappear. A leophyl-dried PM preparation probably has no storage limit. BR molecules in films, prepared by the author on a base of polyvinyl alcohol and PM suspension in 1978, have not lost their photochemistry until the present day. When heated to 100°C, the chromophoric center of the suspension disintegrates, and the water suspension becomes colorless. Upon lowering the temperature, self-assemblage of PM occurs, with full or partial restoration of the photochemical properties of BR.

The PM extracted from wild cell cultures of bacteria from salt lakes in Russia and the U.S. are similar in photochemistry, but differ in their mechanical rigidity, and in their PM lipid content [Chekulaeva, unpublished data]. It is worth recalling that monomolecular BR is not stable, and, when speaking about experiments with BR, the authors imply PM, or a BR molecule in a lipid medium.

3A.1. Chemical and Physical Properties of BR

The BR molecule is a retinal-protein complex, consisting of a protein molecule (BR) and a retinal molecule (oxidized A vitamin form also called Vitamin A aldehyde) bound by a Schiff base. The protein has a molecular weight of 26.534 DA, and 248 amino-acid residues in a polypeptide chain. The chain has seven α-helices in dimeric structure, and exists in a globule-shaped 3-D structure (See Figure 3.1).

BR belongs to the class of transmembrane proteins, and penetrates the entire thickness of the membrane contacting both the cytoplasmic and the external surfaces. Retinal is linked to the 216th lysine residue located approximately two-thirds the distance from the cytoplasmic surface of the membrane. In the dark, retinal occurs in all-*trans* and 13-*cis* isomeric configurations. In the light, all 13-*cis* retinals isomerize to the all-*trans* configuration. The full sequence of amino acid residues of a BR molecule was first deciphered in Ovchinnikov's lab [Ovchinnikov et al. 1979], and later, but independently, in Khorana's lab [Khorana et al. 1979].

Investigations on 3-D protein structure have been going on for nearly 25 years. The first model for the general spatial structure of BR was proposed as early as 1975 by Henderson [Henderson and Unwin 1975] [Engelmann et al. 1980] (See Figure 3.2).

The seven α-helix polypeptide columns are traditionally abbreviated from

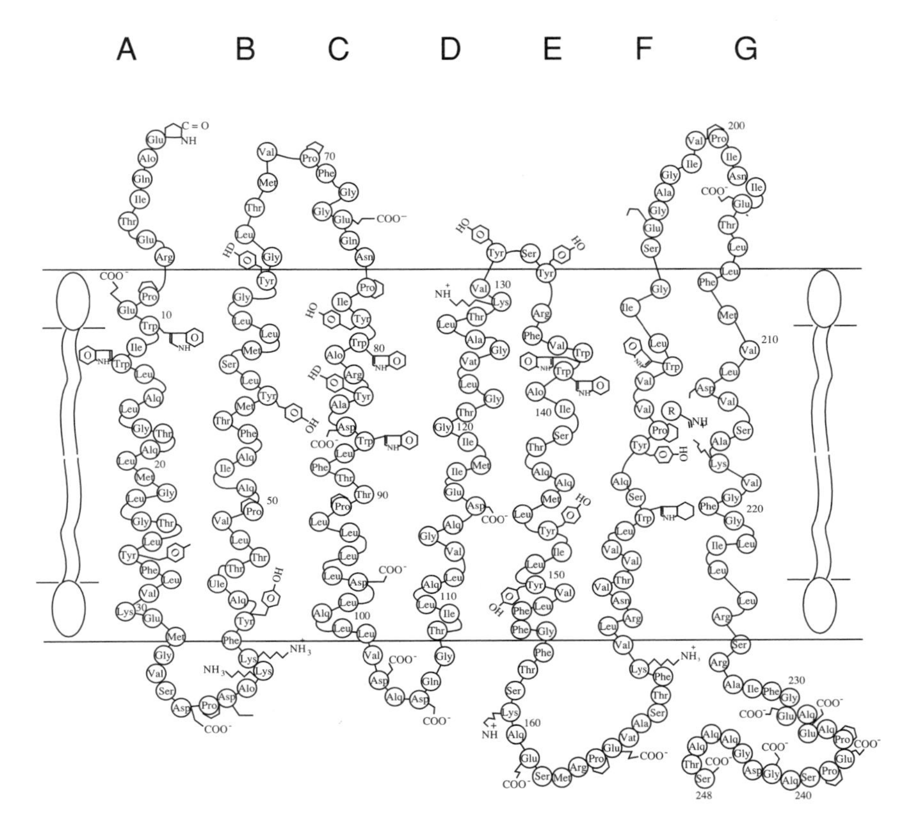

Figure 3.1. Model of the secondary structure of bacteriorhodopsin [from Ovchinnikov 1987]. The attachment site of retinal (R) is Lis216.

A to G. The length of each is 4 nm. Retinal binds as a protonated Schiff base to the ϵ-amino group of Lis216 in the G helix, near the C-terminus of the peptide [Khorana 1988]. The location of retinal in the protein depends on the state of BR. In the dark-adapted state of a BR polyene chain, retinal is tilted towards the plane of the membrane at an angle of 20–30° [Heyn et al. 1988; Henderson et al. 1990].

Knowledge about the full amino-acid sequence, and 3-D structure enabled direct investigation of BR photochemistry. Genetic engineering has proved most efficient for this purpose (see Section 3B.2).

3A.2. The Photochemical Process of BR is Reversible!

The exclamation mark in the title of this section is to emphasize the important feature of complete reversibility in BR photochemistry. In less than 5

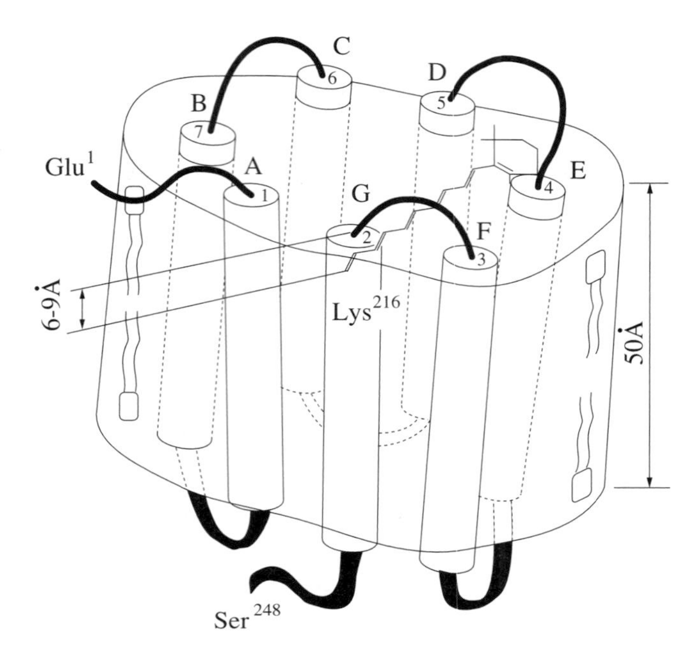

Figure 3.2. Tertiary structure of bacteriorhodopsin. Seven α n-helices, labeled A-G, extend across the segment membrane roughly perpendicular to its plane. N-terminal (Glu), C-terminal (Ser) [based on results by Henderson et al. 1990].

picoseconds after light absorption, BR retinal converts from an all-*trans* conformation to a 13-*cis* state. This concludes the photophysical process of electron migration along the retinal chain which induces its isomerization, and triggers the dark photochemical process (zero-light process). During this phase (about 10 ms under normal biological conditions), various conformational changes occur in both the retinal and protein molecules. These are reflected in a sequence of shifts in the BR molecule absorption spectra caused by the appearance and vanishing of spectral intermediates. This process is cyclic (see Figure 3.5). At the end of a photocycle, retinal returns to the all-*trans* state (conformational processes subside, and in 10 ms the BR molecule in all parameters, returns to its initial state), ready to absorb a new light quantum and launch a new photocycle. Herein lies the significant difference between photochemical processes in BR and in visual rhodopsin, the latter requiring additional energy for the completion of the photocycle (see Section 2A.2). Every radiated BR molecule pumps the average of 100 protons per second across the purple membrane, using up energy an average of 100 quanta. We say "average" because, according to some reports, 1 light quantum transports from 0.8 to 1.2 protons [Grzesiek and Dencher 1988]. We must remember that 1 PM contains 50,000–150,000 BR molecules. From an engineers viewpoint, PM is a perfectly reliable structure, which by some accident happened to be of biological origin...

In the sequence of steps in a given photocycle, protonation and deprotonation of the Schiff base occurs with subsequent proton translocation across the BR molecule. As mentioned before, the precise mechanism of proton transport is not totally clear though much of it has been revealed since 1975. Scholars still argue how many protons, one or two, participate in a single proton translocation cycle. Before 1990, it was established that the release of a proton from the Schiff base to the extracellular side of the PM, occurs when the L intermediate is relaxing via M. On the other hand, "sucking in" of a proton from the internal cytoplasmic side of the PM occurs at the stage of M decay [Stoeckenius et al. 1979] [Eisenbach and Caplan 1979] [Kouyama et al. 1988]. Alternatively, the hypothesis of a single proton participating in the cycle has been recently supported (See Figure 3.4). The most visible results on proton translocation were obtained in experiments with vesicles containing

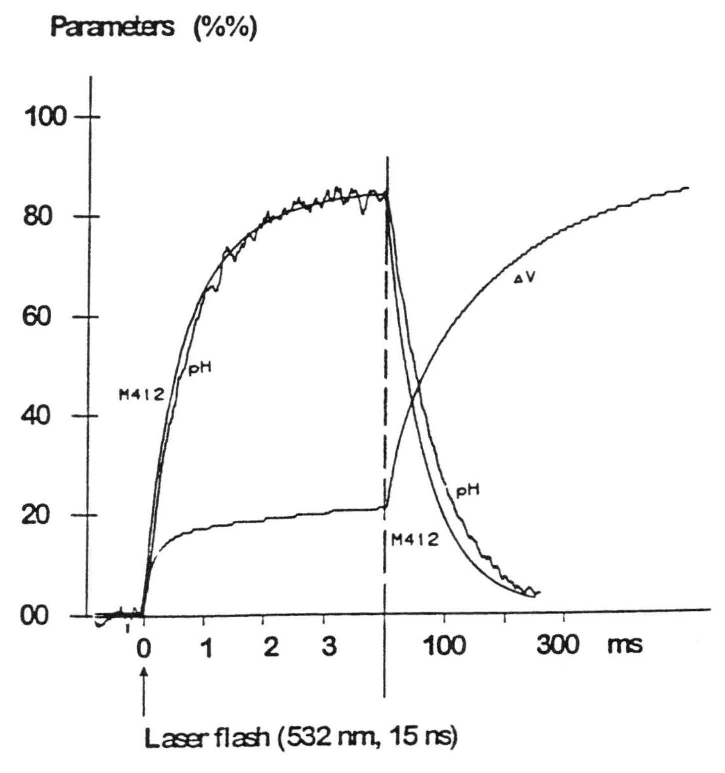

Figure 3.3. A comparison of the kinetics of formation and the decay of intermediate M412, pH changes in a PM water suspension, and electric potential generation (ΔV) in the collodion film-PM system [from Drachev et al. 1984].

liposomes and flat artificial membranes. BR molecules or PM fragments were built in, or oriented, in one direction which allowed the precise determination of the direction of proton movement through BR, and measurement of the efficiency of the proton pump [Lanyi 1984] [Khorana 1988].

In 1984, it was eventually confirmed that the formation and decay of the photocycle intermediate M, the discharge and uptake of a proton, and photoelectric responses are all related [Drachev et al. 1984]. Figure 3.3 shows the important results, confirmed by later studies, that put an end to the arguement about the origin and interdependence of pH changes and spectral and electric photoresponses.

The hypothesis about the participation of some amino acid residues forming the proton translocation chain was proposed in 1979 [Stoeckenius et al. 1979]. Experiments with bacteriorhodopsin variants helped to understand which of the amino acids are drawn into proton translocation after the light absorption though it is still difficult to name the functional role of each of them. A close-up of Asp-96 involvement in the proton translocation process is shown in Figure 3.4, as reprinted from Holz et al. [1989].

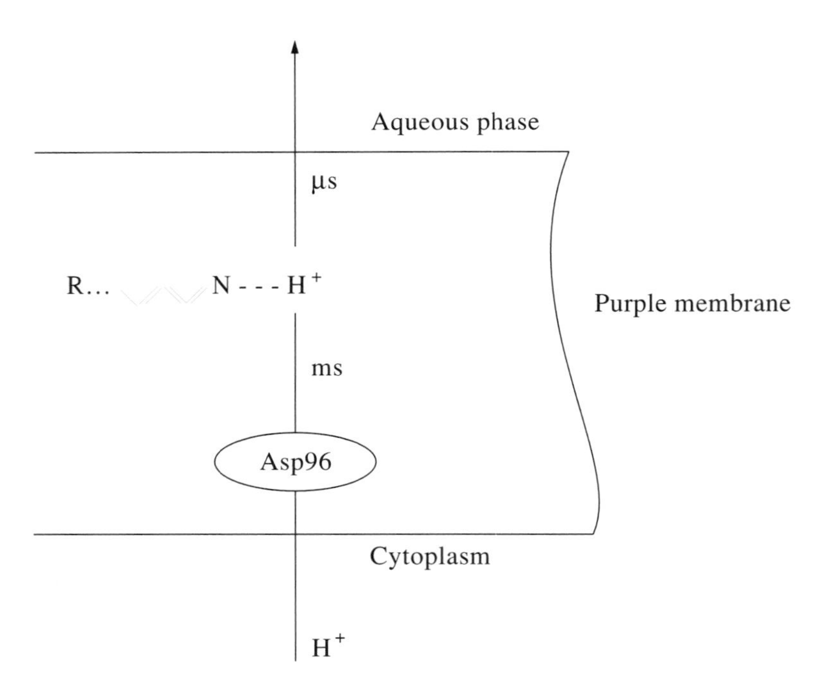

Figure 3.4. Representation of electrogenic events involved in BR proton transfer across the membrane in three steps: the first—a microsecond proton release phase; the second—millisecond reprotonation of the Schiff base; the third—fast reprotonation of Asp96 from aqueous phase [Holz et al. 1989].

Experiments with BR mutants confirm the hypothesis about proton translocation across proton-acceptor groups inside the protein. The model for this path was proposed by Henderson et al. [1990]. Upon radiation, all-*trans* retinal isomerization and protonation of a Schiff base occur; the proton is released via Asp-85 and migrates further towards the external side of the PM represented by M to N or M to BR570 conversion during a photocycle. This is one of the most widely accepted models.

As for the outbound translocated protons, in 1994 some precise experimental data was obtained concerning proton migration along the external side of the membrane when the concentration of buffer solution is less than that of the membrane layer. Earlier, proton migrations along the surface of BR-micelles had been observed [Alexiev et al. 1994]. Under certain conditions, the proton released to the extracellular side of the PM cannot transfer to the aqueous bulk phase, but continues migration along the surface of the PM. The speed of its migration is an order of magnitude higher than that of proton movement in the aqueous bulk phase [Alexiev et al. 1994]. In 200 ms, a proton in the PM suspension can traverse the path around a membrane fragment and appear on the opposite side of the membrane. Transfer to the aqueous bulk phase takes no less than 1ms [Herbert et al. 1994]. To support a photocycle, PM can employ a limited number of protons in a limited space. This must be happening in the polymeric matrix at high levels of dehydration [Vsevolodov 1988].

3A.3. Photocycle Intermediates

Since the first picture of a photochemical cycle was presented in 1975 [Lozier et al. 1975], the number of intermediates discovered has reached dozens. Figure 3.5 presents the very first photocycle scheme of 1975, and the most recent one, drawn from different reports, published before 1996. Since then, no substantial additions have been contributed to knowledge about the sequence of BR intermediates.

In Figure 3.5, the absorption maximum of each intermediate under normal conditions is given near its generally accepted Latin abbreviation. The times of decay (i.e. the transition to the next spectral form) are marked near the arrows. For some intermediates, the time of decay depends on external conditions and may fluctuate by orders of magnitude. The positions of maxima are fairly independent of external conditions and do not exceed the range of 10 to 20 nm. Upon light induced excitation of a BR molecule, and the formation of the first intermediate, other intermediates are formed at the expense of dark-dependent processes. Each of the intermediates is photoreactive, and may change its life time and relaxation pathway upon the absorption of a light quantum. Certain processes taking place during a

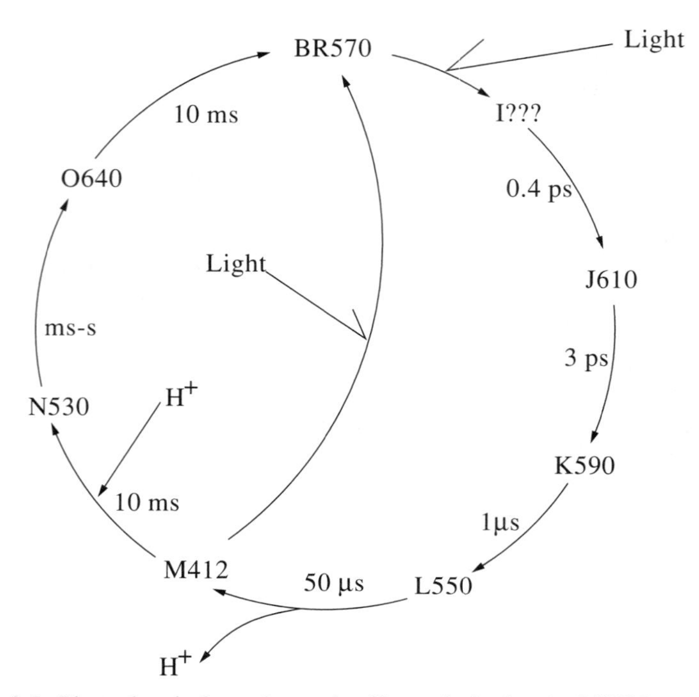

Figure 3.5. Photochemical reaction cycle of bacteriorhodopsin. M412 has two distinct forms: M1, where the Schiff base is accessible to the extracellular side, and M2, where one is accessible to the cytoplasmic side. The diagram was compiled from the works of different authors dated before the end of 1996.

photocycle are not reflected by changes in the optical range [LeGrance et al. 1982].

The structure of the intermediate M has been particularly controversial. Many authors showed that M had a minimum of two different spectral forms and life times-M1 and M2 [Hess and Kuschmitz 1977] [Groma and Dancshazy 1986] [Varo and Lanyi 1991a] [Perking et al. 1992]. Today, the generally accepted view is that the formation of M1 is related to the transfer of a proton from the Schiff base to an Asp85 acceptor. The proton is then released from another amino acid (possibly Glu204) to the external side of the membrane, and the deprotonated Schiff base reorients in the direction of the plasma channel. This stage is expressed by the transititon of M1 to M2 [Lany 1993]. At the M2 to N conversion stage, the Schiff base is re-protonated from Asp96 via the outside of the membrane.

The mechanism of reversible structural and energetic changes occurring in the BR protein during the photocycle has been well studied. Gradually more and more details are uncovered. Recently, it was confirmed that during

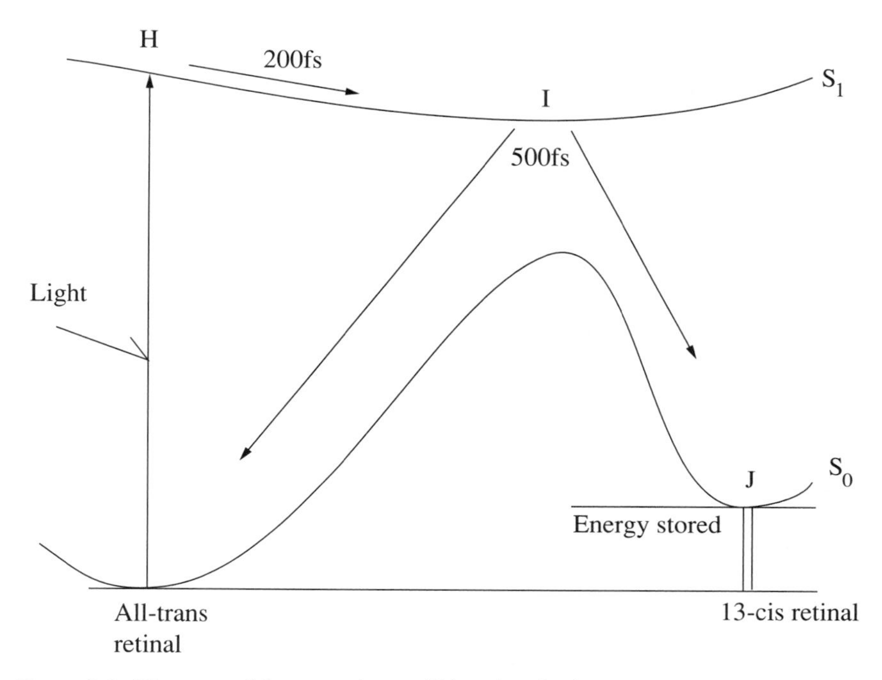

Figure 3.6. Diagram of the ground state (S_0) and excited state (S_1) energy surfaces of a bacteriorhodopsin chromophore. Absorption of a light quanta leads to vibrationally excited state H which relaxes in about 200 femtoseconds to the I state. State I can undergo internal conversion back to the BR570 all-trans state or to the 13-*cis* J intermediate (See Figure 3.5) [Ottolenghi and Sheves 1989].

proton translocation at the M decay stage, conformational changes occur in α-helices F and G near the cytoplasmic side of the PM. The ability to locate, with high precision of time and space dependent photoprocesses was made possible due to the combined application of electron diffraction, and the trapping of intermediates [Subramaniam et al. 1993].

Today, we have a more or less clear picture of the initial stage of the BR photocycle [Sharkov et al. 1985] [Mathies 1991]. Upon absorption of a light quantum at the chromophoric center of a molecule, the chromophore undergoes a transition from its ground state to a vibrationally excited state S_1, (See Figure 3.6).

After about 0.2 ps, vibrationally excited level S_1 relaxes via the lower state I. In another 0.5 ps, electron-excited state I converts to the J state of the spectrum, or returns to the ground state. A retinal molecule in the I state occurs in the 13-*cis* form, and the chromophore has a reserve of energy to launch a dark-dependent process. The quantum yield of a photochemical reaction is determined by the probability of the alternative return to the ground state in which retinal occurs in all-*trans* form [Ottolenghi and Sheves

1989]. In less than 20 years of research, the quantum yield value, as defined by different authors, oscillated from 0.3 to 0.7. According to the latest data it equals 0.64 [Govindjii et al. 1990].

A dark-dependent cyclic process starts after the transition to the first intermediate state J. The BR molecule passes through respective conformational and energetic changes while gradually losing energy. At all pre-M stages of a photocycle, retinal occurs in the 13-*cis* form, isomerizing to all-*trans* only at the stage of M to O conversion and then returns to the ground state in the same isomeric form.

In its non-excited state, BR may contain retinal both in the all-*trans* and 13-*cis* states. When saturated by light, all retinal molecules revert to the all-*trans* condition. After the suspension has stayed in the dark for 20–30 minutes, dark-dependent relaxation occurs, and 50% of the BR molecules relax from all-*trans* to 13-*cis*. The described states are called light-adapted (LA) and dark-adapted (DA), respectively. During the transition from LA to DA, the absorption maximum shifts into the blue region by 10–20 nm. At room temperature, isomerization during the conversion from 13-*cis* to all-*trans* has a quantum yield an order of magnitude less than during the backward reaction, and in weak light, virtually all 13-*cis* retinal molecules isomerize to the all-*trans* state [Kalinsky et al. 1977]. In the DA state, those BR molecules with 13-*cis* retinal also have a photocycle which does not lead to proton translocation. This cycle has been investigated by a few authors [Sperling et al. 1977] [Vsevolodov and Checkulaeva 1977] [Trissl 1990]. If BR with retinal in the *cis*-form does have a photocycle, its meaning is not yet apparent. The existence of such a photocycle has been also shown at low temperatures [Tokunaga and Iwasa 1982] [Balashov et al. 1987].

Investigations of the photochemical processes in BR at super low temperatures (which started in the 1970s) revealed an array of supplementary intermediates, and demonstrated photoreactivity in all of them [Stoeckenius and Lozier 1974] [Balashov and Litvin 1976]. A gradual lowering of temperature down to the liquid helium range inhibits and fixes all photocycle intermediates starting with M as the slowest (having the largest activation energy). Freezing of a radiated PM suspension causes the "freezing" of all photocycle intermediates. The subsequent defrosting process permits observation of the entire chain of dark-dependent spectral conversions, the establishment of their sequential order, and measurement of spectral and photoreactive abilities. The low temperature method helped to discover more than 20 intermediates appearing during the photochemical transformations of BR. They differ from the intermediates registered at normal temperatures by a slight shift in their spectral maxima and, naturally, by their longevity.

The results obtained from the low-temperature experiments make a substantial contribution to the knowledge of the photochemical processes in retinal-protein complexes, and provide a better understanding of spectral forms and structures. A brief summary of the results from the low-temperature studies of a BR photocycle is offered in the next section. When a PM suspension is cooled down to –180°C, its absorption maximum shifts from 570 to 578 nm. Green light exposure induces the formation of a special spectral product with the absorption band at 600 nm. This product is photoreactive, and red radiation shifts it to the BR570 ground state (see Section 3A.3). It corresponds to K590 of a normal photocycle, and is the first stable product in the temperature range from 4K (~270°C) –180°C. When the temperature rises to –130°C, K590 converts to L550. At temperatures above –90°C, L partially converts to M, and partially relaxes via the BR ground state in the dark [Tokunaga and Iwasa 1982]. At room temperature, similar branching of cycle pathways is observed only under green radiation.

Low-temperature conditions make it methodologically easier to observe the branching of photochemical pathways and to determine the photoreactivity of the intermediates. Balashov and colleagues conducted a detailed study on the ways of branching, and on the photoreactive abilities of different intermediates. They proposed a generalized photocycle shown in Figure 3.7 [Balashov 1995].

At temperatures below –50°C, M412 absorption shifts to 419 nm (called by authors "P419"). At temperatures above –50°C in the dark, M412 returns to the BR570 ground state. Upon light exposure, a photoinduced conversion of L and M intermediates is seen. This fact points to their photoreactivity, since both intermediates are stable in the dark. Blue light absorption speeds up the P419 several-step return to the BR570 ground state [Litvin and Balashov 1977] [Balashov and Litvin 1981]. As a first step, at temperatures below –150°C, the P419 spectrum divides and shifts from 419 to 421 and 433 nm (the intermediate forms called by authors "P421" and "P433"), which may be the result of conformational changes in a chromophore which are reverse to those occurring during the transition of P421 and P433 to the intermediary form P565. At the second stage, the P421 and P433 form converts to P565. Authors think that re-protonation of the Schiff base occurs at this stage. At the third stage, the temperature gradually rises to –100°C, P565 converts to P585, and at temperatures above –50°C, P585 reverts to the BR570 ground state. Intermediate forms P565 and P585 are photoreactive and convert back to P419 in green light. 13-*cis* BR with retinal does not form P600, P550, and P419 from the ground state.

A modern picture of the structure of the BR photocycle has been covered in the surveys [Ebrey 1993] [Lanyi 1993].

Figure 3.8, reprinted from Ebrey's review, provides a nontraditional representation of a photocycle and reflects the modern understanding of the processes taking place around the Schiff base. The statement that BR retains its photochemical properties in a wide pH range means that the photoprocess retains its reversibility. At acidic pH, the M intermediate is not observed (see Section 3A.4) and proton translocating ability disappears; at alkaline pH, the proton is quickly released and the M intermediate is retained.

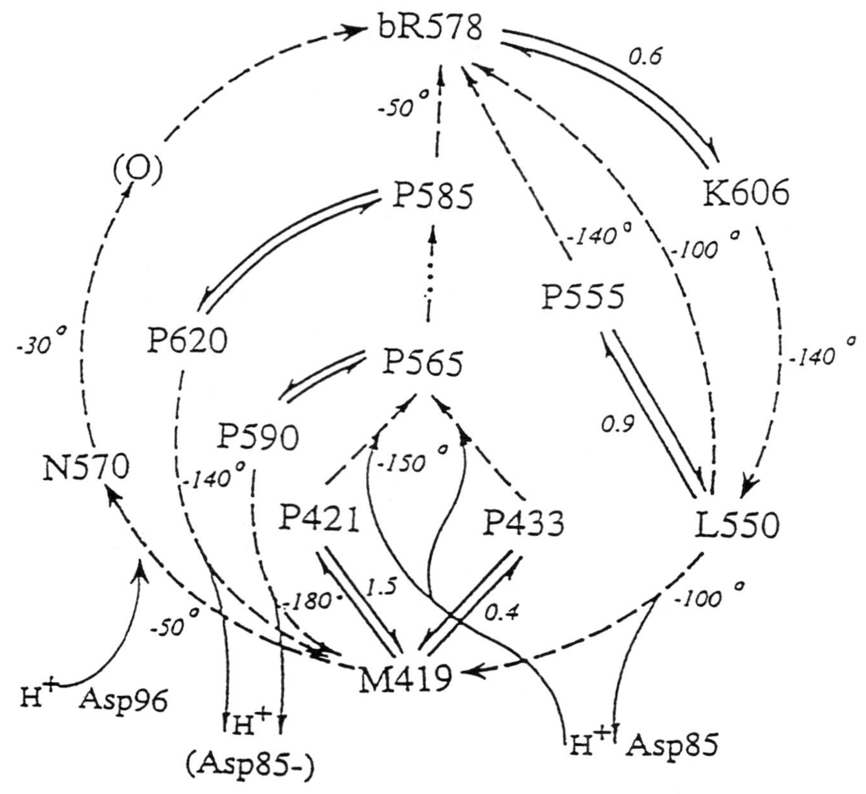

Figure 3.7. Photochemical reaction cycle of bacteriorhodopsin at low temperatures. Main intermediates of the cycle are designated by letters K, L, M, N, with wavelengths corresponding to the absorption maxima at 90K. The photoproduct M is designated by the letter P in earlier articles. Solid arrows designate a light reaction. Dashed arrows designate a dark reaction. Numbers near the center of the solid arrows are the quantum efficiency ratios of the forward and reverse reactions. Temperatures at which intermediates start to undergo thermal reactions are indicated near dashed arrows [from Balashov 1995].

It is worth noticing that the photocycle does not change in mono- and multilayers obtained using the Langmuir-Blodgett method [Ikonen et al. 1992]. This result is not unexpected since the structure of PM does not change.

3A.4. The Effects of Various Parameters on the BR Photocycle

In the course of BR photocycle investigations, purple membranes were subjected to all kinds of chemical and physical influences. Some experiments

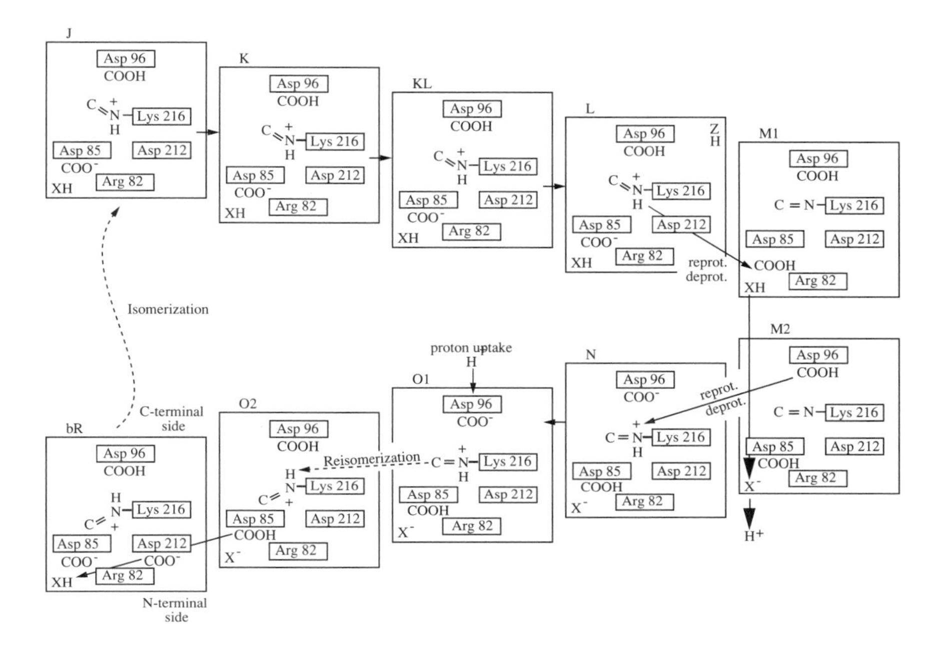

Figure 3.8. Generalized mechanism of proton transfer. Each box represents a single intermediate and position of the proton (H^+) [Ebrey 1993].

immediately gave interesting results, inspiring and broadening the scope of research, while others remained episodic. Further on, we offer a brief description of these episodes, since it is quite possible that they will re-emerge at the forefront. As for the phenomena that gave a push to big theme research (electric field effects on the photocycle, or the replacement of retinal by its analogs, or gene engineering, and so on)—they are briefly reviewed in the following section.

a. Pressure

Pressure influences the BR photocycle by altering the equilibrium between intermediates, and even by inducing the formation of them. At 2.5 kBar, the equilibrium between LA and DA purple membrane suspension preparations reversibly shifts toward 13-*cis*-retinal. Experimental conditions were staged with a PM suspension placed in 0.01M imidazole buffer at pH 7.2. For this buffer solution, the pH does not change in the range of 1–6000 Bar. A change in pressure from 1 to 6000 Bar lowers the speed of M decay by more than an order of magnitude. The M rate-of-decay constant is temperature dependent. The M412 decay rate leaps into the range of 1000–1500 Bar depending on the

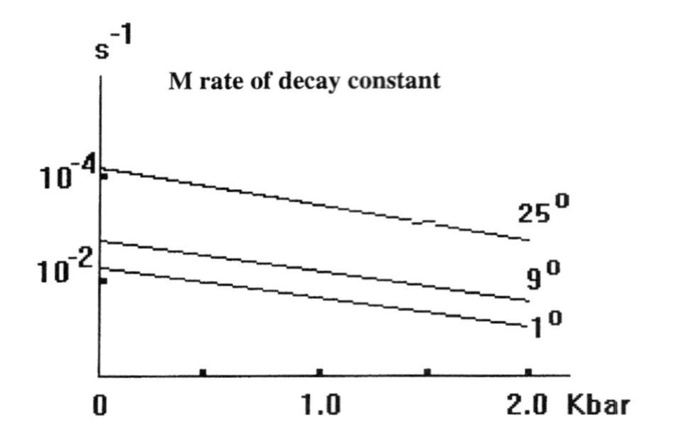

Figure 3.9. Dependence of M412 decay rate constant on pressure at three temperature values [Margus and Eisenstein 1984].

temperature (which points to a phase transition) [Tsuda et al. 1983]. The fast processes in the BR photocycle are less sensitive to pressure changes than the slow ones. The dependence of BR photoprocesses on changes in pressure resembles their dependence on the changes in viscosity. It is quite possible that the increase in the external pressure causes the increase in the internal membrane viscosity. Figure 3.9 shows the pressure-dependent relaxation rate constant of the M form according to experimental results [Margus and Eisenstein 1984].

b. Viscosity

As a rule, the lifetimes of M412 and O640 intermediates are prolonged with an increase in viscosity [Eisenstein 1982]. Viscosity is changed with additions of glycerin-like compounds to a PM suspension. The viscosity was altered by glycerin additions within a 20–80% mixture volume [Beece et al. 1981]. The dependence of photocycle kinetics on viscosity was measured for four wavelengths: 400, 500, 580, and 660 nm, corresponding to K–L, L–M, M–O, O–BR transitions in the BR photocycle. The measurements showed that only during the last two transitions was the increase in viscosity substantially inhibitive to the photoproducts relaxation process. The measurements were made at temperatures ranging from 240 to 315 K. Changes in viscosity and pressure have an analogous effect on the BR photocycle.

c. Gamma and UV radiation

The impact of pulse γ radiation at an exposure rate of several thousand Rads per impulse, induces the formation of a structured spectrum with maxima at 340, 360, 380 nm in the blue region of a purple membrane. At the same time, absorption at 570 nm weakens. The authors [Druckmann et al. 1984] think that radiation causes reduction of the Schiff base. It would take from 6 to 400 pulses at the given radiation rate to achieve a discernible effect. The author studied the effect of UV-radiation on PM in a polymeric matrix. Similar to the condition described above, the initial BR band at 570 nm vanished, and a three-peak spectral structure with maxima at 345, 363, and 385 nm was formed in the blue region. Figure 3.10 shows the spectrum obtained after 10 minutes of a PM-based film in a gelatin medium exposed to a 50 Watt UV lamp at a distance of 20 cm [Vsevolodov 1988].

d. Ionophores, protonophores, antibiotics

Ionophores and protonophores are chemical substances which change the

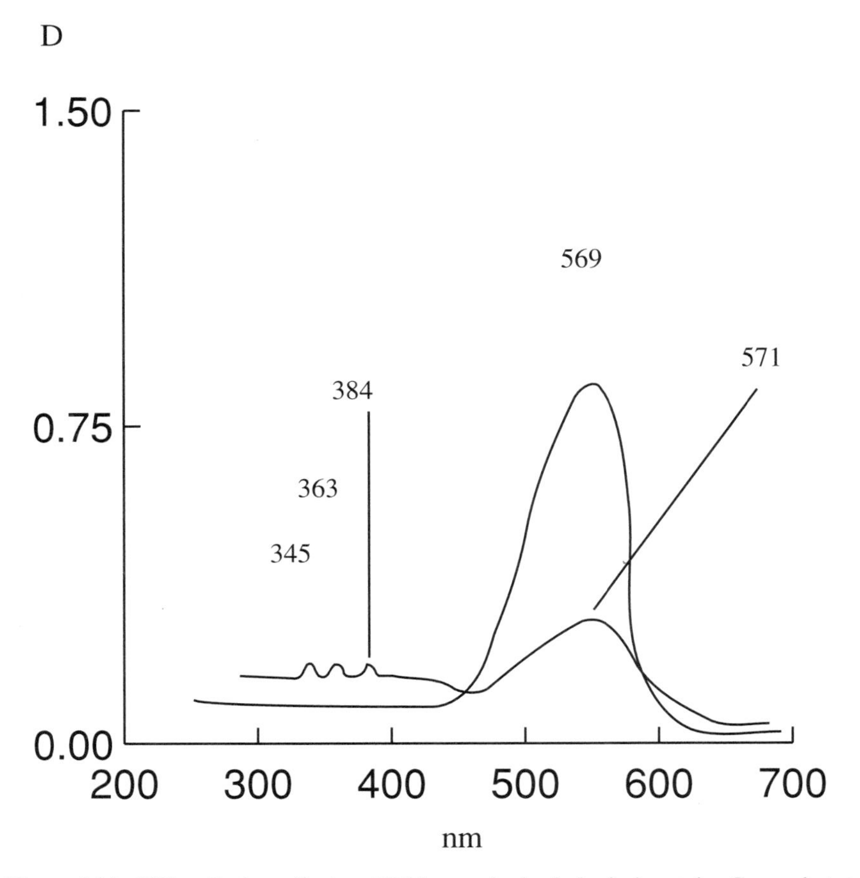

Figure 3.10. UV radiation effect on PM in a polyvinyl alcohol matrix. Ground state absorption (569 nm) vanished and shifted to 571 nm and three peaks (345, 363, and 384 nm) [Vsevolodov 1988].

activity of ions in biosystems. Gramicidin, an inhibitor of oxiditive phosphorylation, has no effect on the BR photocycle. However, valinomycin inhibits the formation of O640 and prolongs the M412 lifetime. Valinomycin and S-gramicidin substantially inhibit the decay of the M412 intermediate [Avi-Dor et al. 1979]. Bovericine strengthens the effect of valinomycin, but has no independent effect on BR photochemistry. Raman and CD-spectroscopy analysis showed that gramicidin interaction with reconstructed purple membrane may involve direct reaction with BR molecules, by-passing altogether the lipid surrounding [TuShu et al. 1981]. Experiments with ionophores supports the conclusion that only the slow phase of the M412 decay is directly connected with proton translocation, whereas the fast phase reflects stages of dissociation and association of protons with BR [Gatry et al. 1979].

e. Blue light

Blue light speeds up M decay and is proportional to the intensity of blue light [Hess and Kuschmitz 1977] [Vsevolodov and Checulaeva 1979]. The sign and amplitude of photoelectric responses strongly depends on the ratio of blue-green intensity upon simultaneous action. When the intensity of constant green light is increasing, the blue light first amplifies its photoresponse, then suppresses it. This "shunting" effect of blue light allows regulation of the proton pump power output at high radiation intensities. This is interpreted as a safeguard for the microorganism against proton exhaustion [Ormos et al. 1978]. Studies of the BR molecular response to blue light in vesicles showed that only the slow component of M-decay kinetics is affected, and at high intensities of radiation it is able to change the direction of proton translocation across the vesicle membrane [Ohno et al. 1983].

f. Ions

The inhibiting effect of lanthanum ions on phototransportation of protons across the PM was first reported by Drachev et al. [1976]. Later it was shown that with an increase of ion concentration, the M412 lifetime is prolonged. Radiation of a PM suspension mixed with silver ions by a constant yellow light at 2°C creates a spectral form with an absorption maximum at 360 nm, and a lifetime reaching dozens of seconds [Rodionov and Shkrob 1979].

Interactions between BR and the halogens, and their combined effect on *cis-trans* retinal equilibrium has been studied extensively [Fisher and Oesterhelt 1979]. In the presence of chloride ions, the equilibrium abruptly shifts to the *trans*-form. High concentrations of Na^+ ions inhibits the formation of O640 and prolongs the M412 lifetime [Sherman et al. 1976] [Tan et al. 1996].

g. Magnetic and electric fields

As shown by neuron scattering methods and polarization spectroscopy in a water suspension embedded in a magnetic field of 17,000 Gs, purple membranes orient parallel to each other [Neugebauer et al. 1977]. The orientation of the PM fragments occurs because of the diamagnetic properties of α-helical structures in BR which are oriented perpendicular to the membrane plane and cannot move or rotate within the membrane. The orientation of purple membranes can be fixed by drying-up the PM suspension. However, magnetic fields do not allow orientation of the cytoplasmic sides of all membranes unidirectionally.

A pulsed electric field on a PM water suspension was first studied by Shinar et al. [1977]. The field first affects the chromophore, altering retinal orientation by more than 20° in 200 ms. After this, for 100 ms, the orientation of the whole PM fragment is altered. The effects were observed when the applied field was not less than 10 kV/cm. In an electric field, PMs orient parallel to each other, and their cytoplasmic sides are directed in a uniform fashion. This kind of orientation stimulates a strong cooperative effect upon electrogenic activity since the protons from all membranes are ejected in one direction [Keszthelyi 1980]. The influence of electric fields on the BR photocycle will be discussed in Section 3A.7.

h. Solvents

Researchers [Tanny et al. 1979] [Eisenbach and Caplan 1979] studied the influence of various BR-solubilizing solvents, and 70 organic solvents on BR absorption spectra and photoactivity. Section 3-D describes how solubilization of BR affects its properties. Most organic solvents have virtually no effect on the photoprocesses in BR providing another proof of the extraordinary stability enjoyed by PM-embedded BR molecules.

i. Dehydration

Dehydration of a PM suspension from 100 to 1.0% relative humidity leads to essential changes in the photochemical cycle. Increasing dehydration gradually blocks the photocycle pathway through N and O intermediates. At 88% relative humidity, the photocycle ends at the M intermediate stage and then returns directly to the BR ground state bypassing O and N [Varo and Lanyi 1991]. The M lifetime, in this case, becomes extensively prolonged. Upon dehydration from 94 to 7%, the M half-life extends from 10 ms to 10 sec [Korenstein and Hess 1977]. At less than 1.0% dehydration of the PM, a new photoproduct was detected in BR, absorbing at 506 nm, whereas the L intermediate was not observed at all [Lazarev and Terpugov 1980]. Direct measurements showed that at a mass ratio of H_2O to BR less than 0.055, proton translocation is thwarted [Thiedemann and Dencher 1994]. In dehydrated polymeric layers, the spectrum maximum of the M intermediate shifts to the blue region (401 nm), while its half-life extends to dozens of seconds in agreement with the results from dehydration of a PM water suspension. However, prevention of BR conversion from DA to LA in the polymeric matrix was not observed [Dyukova et al. 1985].

j. Temperature

A temperature increase in the PM suspension causes an increase in the speed of a photocycle at the expense of shortening the M intermediate formation and decay times. Table 3.1 shows a list of M formation and decay rate constants as measured in normal and deuterated water PM suspension. In the thermal range from 20 to 50°C, the rates of M formation and decay increase by an order of magnitude [Vsevolodov and Kayushin 1976].

Studies of PM thermal denaturation in water suspensions demonstrate that gradual heating to 100°C causes distortions in the PM, and that denaturation itself is a two-step process. The first step (70–80°C) is characterized by distortions in the PM crystalline lattice; at the second stage (90–100°C), denaturation itself begins [Lazarev and Shnyrov, 1979]. In a thermal range from 40 to 80°C, BR reversibly converts to the product with a maximum at 500 nm. At higher temperatures, the maximum of a photoproduct shifts irreversibly to 380 nm. At 60°C, flat purple membranes form a hemisphere, and at 75–90°C they close up into spheres with externally-oriented internal surfaces. These bubbles can resist a temperature of 100°C for a few minutes, whereupon they disintegrate [Shnyrov et al. 1981].

Thermal stability of the PM dramatically increases in a polymeric matrix depending on the degree of dehydration. In some polymeric matrixes, the BR

Table 3.1. The M formation and decay rate constants as measured in a normal vs. deuterated water suspension of purple membranes

Solvent	Temperature (°C)	Formation rate constant (s^{-1})	Decay rate constant (s^{-1})
water	0–1	0.23×10^{-4}	0.46×10^{-2}
water + glycerol	20	0.40×10^{-4}	0.50×10^{-2}
water	20	1.20×10^{-4}	1.40×10^{-2}
deuterated water	20	0.35×10^{-4}	0.56×10^{-2}
water	50	4.14×10^{-4}	4.15×10^{-2}

molecules retain their photochemical properties at as high as 160°C [Lukashev and Bolduvin 1995].

3A.5. Photochromic Properties

In Section 2A.9, we discussed the photochromic (see Glossary) properties of visual rhodopsin. Visual rhodopsin does not by itself indicate a photochromic material because of the irreversible isomerization of retinal. As for BR, it is endowed with many of the properties specialists require from traditional photochromic materials.

While still in a live halobacterial cell, BR already behaves as a photochromic material with a brief optical information storage time. To see this, notice that the initial spectral maximum at 570 nm, in fractions of milliseconds, shifts by 150 nm toward the blue. The final event is relaxation, in dozens of milliseconds, to its initial position.

We simplified the scheme of the photocycle, leaving out only 3 intermediate photoproducts which absorb at 590, 550, and 420 nm. Each of these stores the information about an absorbed quanta in its range for a set period of time. A dish with PM suspension acts as a photosensitive cell, an element of optical memory, and a device for operative information processing. Let us further simplify the cycle, leaving out only one photoinduced product, absorbing at 412 nm. The photocycle scheme, is now simplified to the traditional scheme of photochromic materials.

Diagram 3.1

Yellow light
BR (570 nm) ⇄ M (412 nm)
Blue light or kT

As there are many ways to change the M412 formation and decay times, it is possible to produce a photochromic material with a pre-determined optical information storage time. The fragments of the PM are easily oriented on a base in the drying process, forming a satisfactory homogeneous film with photochromic properties.

When the temperature falls (see Section 3A.4), the long-lived spectral forms M, K and L are respectively blocked. The less intermediates remain, the faster the photocycle. Knowing that all the intermediates are photoreactive, we can regard every other step in a photocycle as a photochromic process, and the absorption maxima of the intermediates as the maxima of A and B forms of a photochromic process. Table 3.2 shows possible photochromic transitions of BR and their dependence on temperature. The lower the temperature,

Table 3.2. Different photochromic transitions to BR at different temperatures

Spectral forms absorption maxima for reversible transition (nm)	Temperature of intermediate stabilization (°C)	Time of photoinduced (direct) transitions (s)
$570 \leftrightarrow 610$	−190	10^{-12}
$570 \leftrightarrow 550$	−140	10^{-6}
$570 \leftrightarrow 420$	−90	10^{-4}

the higher the speed of information recording [Birge 1992]. Infusion of PM into a different transparent matrix (film-like or 3-D) does not suppress photochemical processes in BR, and allows production of a technologically convenient photochromic material with quality optical and technical characteristics [Vsevolodov et al. 1985]. These materials will be discussed in more detail in Section 4D.

3A.6. Photoelectric Properties

The difficulties encountered by scientists while measuring a photoelectropotential of BR were described in Section 2C.2. Today, such tribulations belong to the past, and the modified measuring techniques of Drachev and colleagues are applied in scientific labs all over the world [Drachev et al. 1979]. Experimental measuring of photoelectric activity is based on the electrogenic activity of BR in the bacterial membrane. Figure 2.9 depicted the measuring of a BR-generated electric potential with the help of artificial membranes. The BR molecules are orientationally built into artificial membranes that divide two compartments filled with electrolytic solutions. Radiation-induced, in a single direction, proton transfer across the membrane from one compartment to the other creates the electropotential difference which is measured by a system of registration. Some results are presented in Figure 3.11.

Upon excitation with a pulse laser, the time resolution of the system may reach microseconds [Drachev et al. 1977] and picoseconds [Groma et al. 1984)]. Kinetics of the formation and decay of the BR-generated post-light pulse potentials has several components whose forms and times correlate with the formation and decay of spectral intermediates of the photocycle. The appearance of a potential, and temporal changes in its amplitude and polarity, are related to the intramembrane and intramolecular charge and charge groups shifts, and to proton transfer across the membrane.

BR molecules are built into artificial lipid membranes in many different ways, as liposomes, or perhaps flat membrane fragments. The limitation of artificial membranes is their low stability, i.e., their short life-time and low

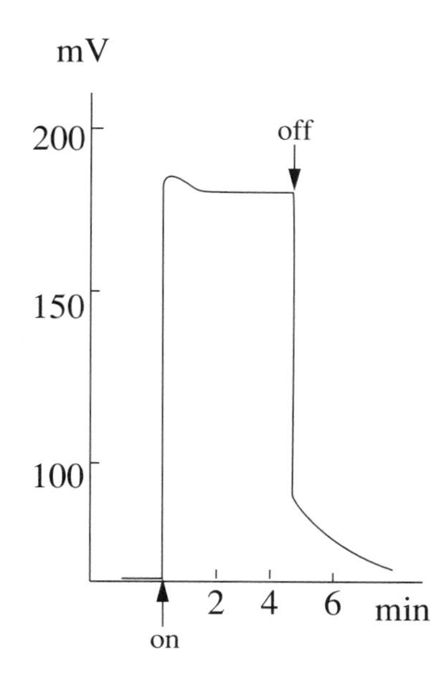

Figure 3.11. Light-induced generation of an electric potential across the azolectin-impregnated Synpore filter covered on the side with BR-proteoliposomes. "On" and "Off" refers to the switch from a halogen lamp [Drachev et al. 1979].

break-down voltage. To increase the resistance of the artificial membranes, various additions are needed [Shieh and Packer 1976]. The most stable systems were produced by BR absorbing on collodium membranes or on porous lipid-saturated ceramic filters [Drachev et al. 1979].

Maximum electroresponse in any registration system can only be obtained if the cytoplasmic sides of all PM are singularly oriented. Only in this case do cooperative effects work, adding up the proton gradients of the BR molecules. Various methods of PM orientation have been described in the literature, however, the most common ones are the methods of PM orientation in the electric field of an aerial condensor [Nagy 1978] and in an electric field applied directly to the suspension (the electrophoresis method) [Varo 1981]. PM has a large permanent dipole moment due to the difference in surface charge density on the opposite sides of the membrane. In the suspension, it reaches 140 Debeye at pH 7.00 [Tsuji and Neumann 1981], consequently the degree of PM orientation in an electric field is high.

Stable preparations with a high degree of PM orientation are produced by immobilization of PM on a polyacrylamide gel, and the subsequent polarization of the gel in an external electric field [Liu and Ebrey 1988], or by freezing the suspension [Ormos et al. 1983]. The highest degree of PM orientation is achieved through application of electrophoresis to the PM sediment on an electroconductive base. Upon radiation, some of the preparations generated a potential as high as 10V. Unfortunately, this method is not free from shortcomings. Electrophoresis causes de-ionization of the PM, it is difficult to produce superthin layers, the preparation surface area is limited, and so forth. The Langmuir-Blodgett method allows creation of mono- and multi-layers with big surface areas composed of uni-directed molecules and with a given number of layers. Unfortunately, it is not applicable to BR. The transfer of BR from a bi-layered PM to a monolayer is still a problem that has not been resolved [Ikonen et al. 1992].

3A.7. Electrochromic Properties

In 1978, Kononenko with co-workers, discovered a BR absorption band shift under the influence of an external electric field [Borisevitch et al. 1979]. When a dried layer PM suspension was placed in an electric field of 10^4–10^5 V/cm, its optical absorption increased at 630 nm, simultaneously diminishing at 540–550 nm. During simultaneous use of an electric field and red light, the amplitude of optical changes diminished. The electrochromic ability of BR was now proved. The electrochromic effect we here understand is any change of the spectrum (in response to an electric field), even if the point of invariance of the electroinduced differential spectra does not shift. Normally, (for instance, in the Stark effect) electrochromic effects suggest a shift of the invariance point since a new spectral product is being formed. At this point, we discuss electroinduced optical transmissions, and not mechanics.

Later, a more detailed study of the same phenomenon was conducted [Lukashev et al. 1980]. They showed that the electroinduced amplitude increase in the PM at 415 nm occurs only during radiation, whereas at 630 nm, the electric field affects preparations in the dark as well. The increase in amplitude is due to the accumulation of BR molecules in the M-form, which increases its lifetime. Thus, electric and light-induced spectral changes in PM do not coincide.

Lukashev and co-workers explained their results by the rearrangement of charges in polar and polarizable groups in a pigment–protein complex under the influence of an electric field. Forced orientation of such groups leads to the dielectric polarization of BR. The effect of a pulsed electric field on PM water suspensions was examined. Such studies demonstrated a PM rotational dependence by the direction and power of the field. Polarization and beam intensity penetrating the suspension depended on the value and direction of the PM rotation angle [Keszthelyi 1982]. When the pulse voltage of the field reached 3×10^4V/cm, the length of impulse prolonged from 1 to 100 ms and the concentration of the suspension grew from 1 to 20 mM (in 1–3 mM NaCl solution at pH 6 at room temperature). The reversible shifts in the optical density were measured in the range of 400–650 nm.

The effect of PM orientation in an electric field is not observed when the measuring light beam incidence angle is 54°. This angle is called "the magic angle." Under these conditions, the PM does not respond to a change in field

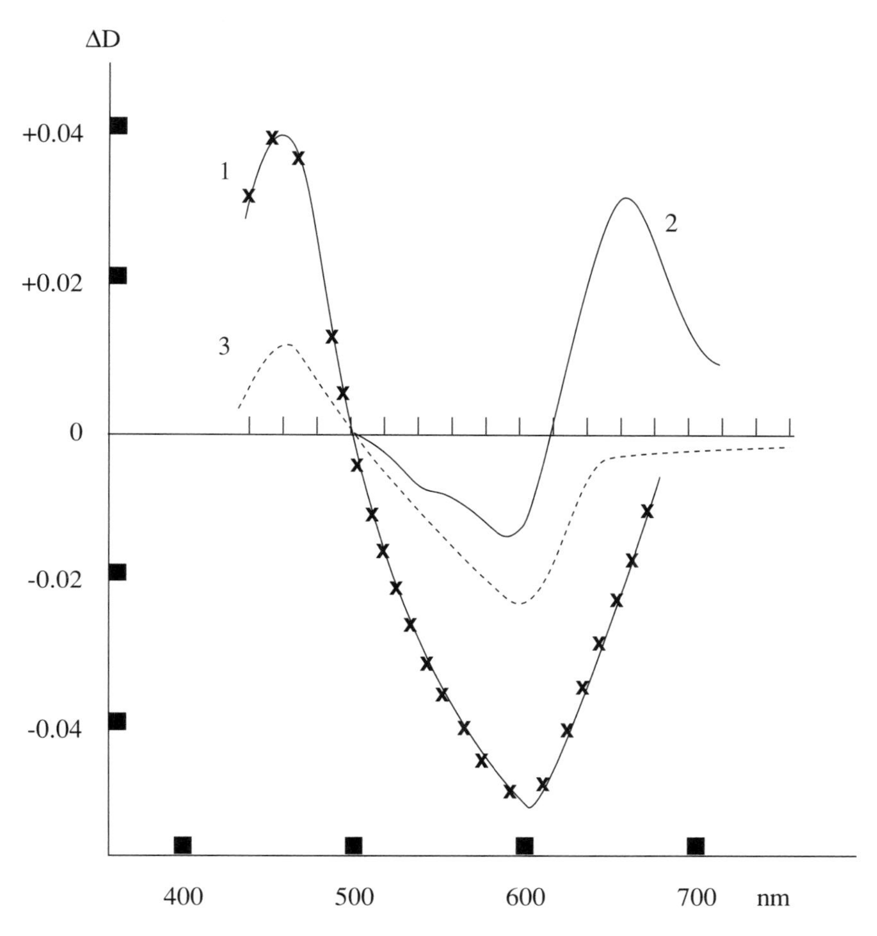

Figure 3.12. Differential spectral shifts in partially dehydrated purple membrane films, photoinduced (1), electroinduced (2) and photoinduced upon the simultaneous use of a permanent electric field (3). ΔD is the variation of optical density around the initial optical density for all wavelength measurements. The X's are points of wavelength measurement [modified from Lukashev et al. 1980].

power which permits scrutiny of its direct effect on the processes in BR molecules; proton activity, pH, etc. [Tsuji and Neumann 1981]. Those researchers conducted a detailed study of a field-induced dichroism whose magnitude did not depend on the suspension concentration, the ion strength, or the intensity of the measuring light beam. Rather, it did depend on the direction of the light beam polarization vector in relation to the direction of the electric field vector.

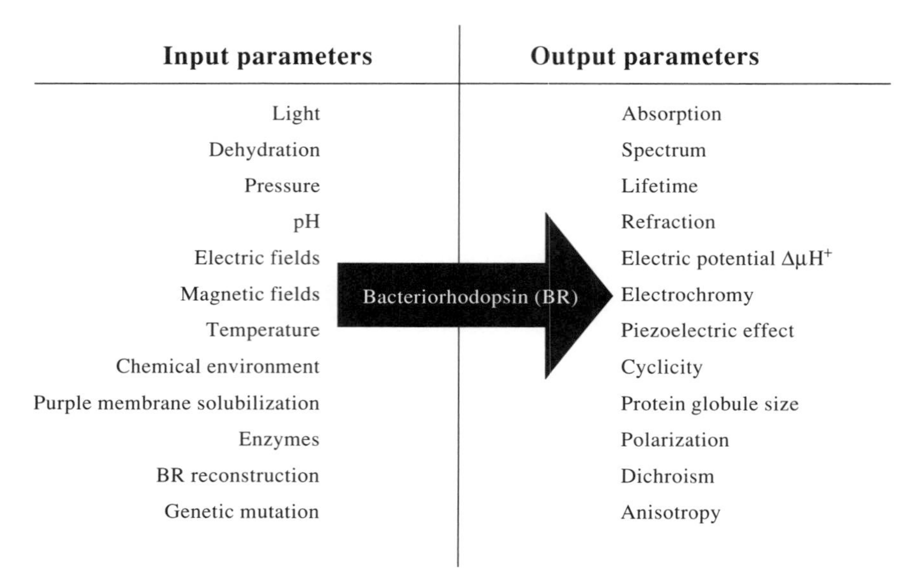

Figure 3.13. Molecule of bacteriorhodopsin as a "black box" energy converter. Changes in the environment (input parameters) are sensed by bacteriorhodopsin which implements measurable changes in its characteristics (output parameters).

3A.8. Bacteriorhodopsin Viewed as a "Black Box" Energy Converter

As Section 3A.4 revealed that native BR responds to virtually every fluctuation in external and/or internal parameters. The BR molecules are not only capable of pumping protons across membranes, but due to their ability to respond to various environmental parameters, they may be used as a kind of "universal energy converter." BR can be regarded as a "black box," in which the output signal can be modified in various ways by changing one of the parameters.

3B. Bacteriorhodopsin Variants

There are several ways to modify the performance characteristics of BR:

1. Photoinduced hydroxylaminolysis permits BR to be reconstructed by means of natural retinal replacement with an artificially synthesized analog.
2. Chemical modification of individual amino-acid groups of the BR protein, or by proteolytic cleavage of the BR molecule (see Section 2D).

3. Random mutagenesis of the halobacterium genome followed by the selection of functionally altered BR mutants.
4. Site-specific mutagenesis and expression of the mutant BR gene in *E.Coli.*

The BR analogs synthesized by means of replacement of natural individual amino acids in opsin with the others (3 and 4) are called variants to distinguish them from the analogs created using methods 1 and 2.

3B.1. Random Mutagenesis

Random, mutagenesis was suggested by Oesterhelt and colleagues in 1983 [Oesterhelt and Krippahl 1983]. In the process of random mutagenesis of the clones, new strains of halobacterias were selected in which the replacement of certain protein amino-acids modified the BR molecules endowing them with new properties [Soppa and Oesterhelt 1989] [Soppa et al. 1989]. Table 3.3 summarizes some of the results for randomly mutating BR strains.

For example, GRB-326 reduces bacteriorhodopsin in which aspartate-96 is replaced by asparagine. The duration of one photocycle revolution in such molecules extends from 50 msec to 100 sec [Miller and Oesterhelt 1990]. The replacement of Asp-85 with Glu creates two spectral forms with the absorption maxima at 550 and 610 nm, respectively, only the former being photoelectrogenically active [Butt et al. 1989]. The replacement of Asp-85 with neutral amino acids Ala and Asn shifts the absorption maximum of BR into

Table 3.3. Random mutagenesis of BR

Substitution	Maximum absorption spectra (nm)		Half-time M (ms)		Name of mutant*	Ref.
Wild–Mutant	DA	LA	Arise	Decay		
Wild type	558	568	6×10^{-2}	15	WT	1
Tyr57–Asn	555	560	M not detected		Y57N	1
Asp85–Glu	550\610	610	10^{-2}		D85E	1,2
Asp96–Asn	560	569		10^{5}	D96N	1,3
Asp96–Gly	557	566		150	D96G	1
Trp138–Arg	490	490			W138R	1
Trp10–Cys	548	562		150	W10C	1

References:
1. Soppa et al. 1989.
2. Butt et al. 1989.
3. Miller and Oesterhelt 1990.

* See Glossary and Abbreviations.

the red region, causing a 100-fold retardation of the photocycle and suppresses proton activity [Mogi et al. 1988]. By 1990, many similar data had been collected. Most authors conclude: Asp-96 and Asp-85 play a key role in the mechanism of proton translocation.

The method of random mutagenesis is described in patents [Oesterhelt et al. 1987] [patent 3730424 Germany]. The essence of it is the following: "UV mutagenesis and photodynamic bromodioxiridin killing in the presence of retinal of *Halobacterium halobium* is used to select cells that have modified BR. Cells were chosen from mutants that could not grow phototrophically under anaerobic conditions. These cells were isolated and analyzed for the presence of antibacterio-opsin cross reacting material".

Halobacterium salinarium strains (see Section 3B.3) obtained in this way are widely used in scientific research. For instance, in holographic recording, BR-326—based films show twice the energetic sensitivity compared to the wild-type BR films.

3B.2. Site-specific Mutagenesis

Site-specific, or site-directed mutagenesis (SSM) along with random mutagenesis (RM) (see 3B.1) is one of the most important methods for research between structural and functional properties of proteins (see also section 2D). This method was first successfully applied by Khorana and colleagues for the production of BR-mutants [Dunn 1987] [Nassal et al. 1987]. Selected amino acids are replaced in the BR protein by means of a chemically synthesized BR gene expressed in *E. Coli.* Although not simple, this method is being streamlined and is becoming available to many research groups [Baofu et al. 1990]. SSM allows determination of the role of each amino acid in the photochemical process in general, and in proton translocation [Hackett et al. 1987]. Of course, as Birge noticed [Birge 1990], it is difficult to understand what affects the characteristics of the processes in BR dominate: the replacement of the primary counter-ion amino acid, or the change of the protein structure around it (induced by the replacement of a nonprimary amino acid). There is still no agreement about the working mechanism of a photo-dependent proton pump in BR. There is a real hope that experimentation will sooner or later help to understand its mechanics.

In the last 10 years, different investigators assigned the primary counter-ion in the BR proton-transport mechanism to different amino acids. Despite the many interesting results using SSM, the mechanism of such replacements is still not known.

One of the first assignments for the primary counterion was Asp-212 [Birge and Pierce 1983] [Zhang and Birge 1990], then Thr 185 was suggested [Rothshild et al. 1986]. Alternatively, that three amino-acids—Asp-85, Thr-

185, and Asp-212 react with Arg-82, influencing the chromophoric process [Braiman et al. 1988]. These amino-acids have been investigated using the SSM method [Mogi et al. 1987, 1988].

The end of the 1980s marked the beginning of systematic investigations on the impact of BR amino acid replacement upon performance. For example, replacement of Asp-96 significantly inhibits the transformation of M into N in the photocycle suggesting participation of this amino acid in proton transfer to the Schiff base [Rothschild et al. 1992]. The replacement of Thr-46 increases the speed of N formation and slows its decay. Since Thr-46 is located in the B helix, and Asp-96 in helix C (Figure 3.2), the distance between them being as little as 4 Angstroms, it is possible that the α-helical structure of the BR protein is involved in the process of protonation–deprotonation [Rothschild et al. 1993].

The simultaneous replacement of Asp-85 and Asp-96 with Asn (double mutant) shifts the absorption spectrum of the ground state to 600–610 nm, and alters the spectrum and lifetime of the M intermediate [Oesterhelt et al. 1990]. The replacement of the same amino acids with Glu causes no significant changes but is of interest for research on the mechanism of the proton pump.

3B.3. How Many Strains of Halobacteria Have Been Found?

The answer is enough to provide for many interesting scientific experiments and technical investigations. Due to natural mutations, the menu of strains may become longer during the quest across the saltwater reservoirs of the earth. Table 3.5 presents the most interesting representatives of wild and mutant strains of *Halobacterium salinarium* and some of their characteristics.

3B.4. Photochromic Properties of BR Variants

Nearly all BR variants synthesized by the SSM method (see Section 3B.2), preserve the reversible photocycle, and consequently their photochromic properties (the exception is the light unstable variant in which Asp-212 is replaced). However, the absorption maxima of the ground state, the ability for light–dark adaptation, the lifetimes of the photoinduced spectral forms, and the positions of their absorption spectra, may significantly alter. The absorption maxima of the variants obtained by Oesterhelt by means of random mutagenesis (depending on the type of variant), shift from their normal position at 570 nm to the red and blue regions, and are positioned between 490 and 610 nm. M intermediate lifetimes in some variants prolong 10–100 times, and reach 10 seconds [Miller and Oesterhelt 1990].

The main goal of BR photochromism researchers is to synthesize the analog, or variant, with a maximum M intermediate lifetime. Other variables

Table 3.4. Site-specific mutagenesis BR

Substitution	Maximum absorption spectra (nm)		Half-time M (date for long-life compontent)		Proton activity of one molecule	Ref.
Wild–Mutant	DA	LA	Arise (μs)	Decay (ms)	+ one proton per s	
Ala53–Gly	544	548	1.6	15	+	5
Asp36–Asn	550	560			+ + +	1,2
Asp85–Asn	589	588			not detected	7
Asp85–Glu	556	568	2.6	350	+	7,8
Asp85–Ala	600	600	2800	230	+	8
Asp96–Ala	558	558	39	7000	+	8
Asp96–Asn	553	560	200	10000	weak	1,8
Asp96–Glu	553	561	30	900	+ + +	1,8
Asp104–Asn	550	560			+ + +	1,2
Asp115–Ala	543	543	5	130	+	8
Asp115–Asn	533	529	70	300	+ +	7,8
Asp115–Glu	536	538	130	250	+	7,8
Asp212–Asn	555	548	220	200	+	7,8
Asp212–Glu	584	581	2	850	Weak	7,8
Asp212–Ala	540	–			Light unstable	1,2
Arg82–Ala	564	566	3	60	+	7,8
Arg82–Gln	575	580	7–10	< 100	+ + + +	4,8
Arg134–Gln	551	557	10	5	+ + +	4
Arg227–Gln	551	563	10	70	+ +	4
Gly122–Cys	524	524	7	20	+	5
Glu194–Gln	541	545			+	1,2
Leu93–Thr	531	531	800		+ + +	6.9
Leu93–Val	547	553	10		+ + +	6,9
Leu93–Ala	532	532	800		+ +	6,9
Met20–Ala	551	560	1	17	+	5
Met20–Glu	550	556	4	16	?	5
Met118–Ala	474	478	11	293	Weak	5
Met118–Glu	552	556	97	51	+	5
Met145–Ala	475	477	52	159	Weak	5
Met145–Glu	550	560	8	8	+	5
Tyr185–Phe	556	573			+	1,2
Val49–Leu	551	556	11	14	+	5
Val49–Ala	549	549	167	25	+	5
Ordinary eBR	549	557	10	2.5	+ + +	7,9

References:
1. Mogi et al. 1988.
2. Hackett et al. 1987.
3. Oesterhelt et al. 1990.
4. Stern and Khorana 1989.
5. Greenhalgh et al. 1993.
6. Delaney et al. 1995 (and reference therein).
7. Subramaniam et al. 1990.
8. Otto H et al. 1990.
9. Subramaniam et al. 1991.

Table 3.5. Strains of photosensitive halobacterias

Strains	Bacterioopsin SR-I SR-II	Haloopsin	Sensoopsins	Carotenoids	Ref.
Wild	+ +	+	+	+	1
JW-5*	+ +	+		−	2
JW2N*	−	−	+	−	2
L-07*	−	+		−	3
L-33	+ + +	+	+	−	3
Flx5R*	−	+	−	−	5
FlxWH*	−	−	+	+	5
R1 mR*	+ +	−	−	−	7
ON1; ON-P	−	−	−	−	4
OD-2R*	−	−	+	−	
GRB	+ +	−	−	−	6
S9	+ +	+	+	+ +	

* Strain is retinal-deficient.
References:
1. Oesterhelt and Stoeckenius 1973.
2. Weber and Bogomolni 1981.
3. Weber et al. 1982.
4. Imamoto et al. 1991.
5. Spudich et al. 1986.
6. Ebert et al. 1984.
7. Mukohata and Sugiyama 1982.

being held constant, this provides a window for researchers to create a material for semipermanent optical memory (a very important component in the engineering of optical computers). Hampp and colleagues confirm that they have succeeded in the creation of a BR variant-based film with a memory lifetime over 24 hours. The contrast range during this time doesn't reduce more than 50% [Hampp et al. 1992, 1992a].

3C. Bacteriorhodopsin Analogs

During the history of BR research, more than 350 BR strains have been synthesized from native apo-membranes, artificial or natural analogs of retinal and their derivatives. Two methods are presently applied to the synthesis of BR analogs. One is retinal replacement in BR by its analog during photoinduced hydroxylaminolysis [Oesterhelt and Schulmann 1974]. The other method is retinal analog addition to a culture undergoing cultivation of retinal-deficient bacteria [Mukohata and Sugiyama 1982]. Both methods permit the creation of stable apo-membranes, and reconstruction of any BR analog from their base. The process of reconstruction is remarkably simple, and even elegant. A drop of colorless retinal analog is added into a tube containing the apo-membrane suspension. Both mixtures are colorless since apo-proteins and retinal absorb in the invisible UV region. In a few seconds, the colorless mixture becomes violet or purple, or yellow, or orange, or bordeau. This coloring depends only on the type of retinal analog. A large group of BR analogs allows broad-scale studies of the connection between the functional properties of native retinal–protein complexes and their structure. Nakanishi wrote, "Whenever an analog is used, it will give new results; however, if the purpose of application is not thought out carefully, these "new" results may be publishable, but on the whole lead to confusion rather than contribute to an already immensely complex field of science" [Nakanishi and Crouch 1995].

3C.1. Apo-membranes

The discovery of BR's ability to reconstitute by Oesterhelt and Schuhmann solved the problem of the BR analog synthesis 1974 [Oesterhelt and Schulmann 1974]. One of the apo-membrane (AM) synthesis methods is relatively simple, and is based on reaction with hydroxylaminolase—the interaction of retinal-protein chromophores with hydroxylamine (NH_2OH) or its derivatives [Rental and Perez 1982]. During the reaction, cleavage of the Schiff base occurs and BR retinal binds with hydroxylamine forming retinaloxime (retinal–HC = N–OH). After removal (extraction in organic solvents), nothing but AM remains. It is worth mentioning that retinaloxime removal is fairly complex and can lead significant to apo-protein losses.

The hydroxylamine reaction for iodopsin and retinochrome proceeds in the dark. However, for rhodopsin, BR and its analogs, it is light-dependent. This process is called photoinduced hydroxylaminolysis. Under strong yellow light, a photoinduced hydroxylaminolysis occurs in hydroxylamine-containing PM suspension, forming retinaloxime and AM. In AMs of this kind, protein does not have retinal in the Lis-216 area while retaining its crystalline structure. It is called bacterio-opsin (BO). The process goes according to the sequence:

Diagram 3.2.

$$\begin{array}{ll} & H^+ \\ & | \\ \text{Step 1.} & \text{retinal–HC} = \text{N–opsin} + NH_2OH \\ \text{Step 2.} & \text{Yellow light} > 500\ \text{nm} \\ \text{Step 3.} & NH_2\text{–opsin} + \text{retinal–HC} = \text{N–OH} \end{array}$$

Where:

$$\overset{\displaystyle H^+}{\underset{|}{}}$$

retinal–HC = N–opsin–bacteriorhodopsin

NH_2OH–hydroxylamine

NH_2–opsin–apo-membranes with absorption maxima at 280 nm

retinal–HC = N–OH–retinaloxime with absorption maxima at 370 nm

Infusion of natural retinal into AM suspensions leads to a nearly complete reconstruction of PM, whereas artificially synthesized retinals form BR analogs as parts of the crystalline lattice of the previous PM. If retinaloxime is not fully removed, its absorption band appears in the 360–400 nm spectral region. Apo-membranes with all-*trans* and 13-*cis* retinal isomers reconstruct only in the dark; 11-*cis* isomer-containing apo-membranes reconstitute only in the light, and, finally, in 9-*cis*-isomer-containing apo-membranes reconstitution has not been observed at all [Oesterhelt and Schulmann 1974].

Photoinduced hydroxylaminolysis can also proceed in the cell suspension of halobacterias [Oesterhelt et al. 1974]. At the infusion of hydroxylamine into a suspension, the bacterial cells do not die, but do not demonstrate light-dependent pH shifts in the medium because the absence of a chromophore blocks the BR photochemistry. The infusion of retinal (or its analogs) into a retinal-deficient halophilic suspension is effectively what reconstruction, or re-synthesis is.

Figure 3.14a shows three spectra of the described processes. Unfortunately, as mentioned, complete removal of retinaloxime presents certain difficulties, and its absorption band appears in the spectral region of 360–400 nm.

If halophilic cells are grown in the presence of nicotine, the synthesis of retinal is suppressed, forming the so-called "brown membrane (BM)" in situ on PM. BM contains BR with a modest (5%) content of cytochrome b which gives the membranes their brown tint. Protein occurs in BM in an amorphous state. After the addition of retinal, protein crystallizes and a normal PM is formed [Sumper 1982]. In the blue region of the spectrum one insignificant absorption is observed (See Figure 3.14b).

Artificial cell cultures of congenitally retinal-deficient halophils were synthesized [Sarma 1984]. The production of the apo-membrane suspension from these cell cultures is simple and similar to traditional methods of PM production. The problem of purification does rear itself since the spectrum does not show the retinaloxime absorption band. The most popular cell strains are presented in Table 3.1. A spontaneous mutant R1 mR was isolated from a standard *Halobacterium halobium* cell culture RM. It has no retinal, and the so-called "white membranes," when isolated, show no absorption of

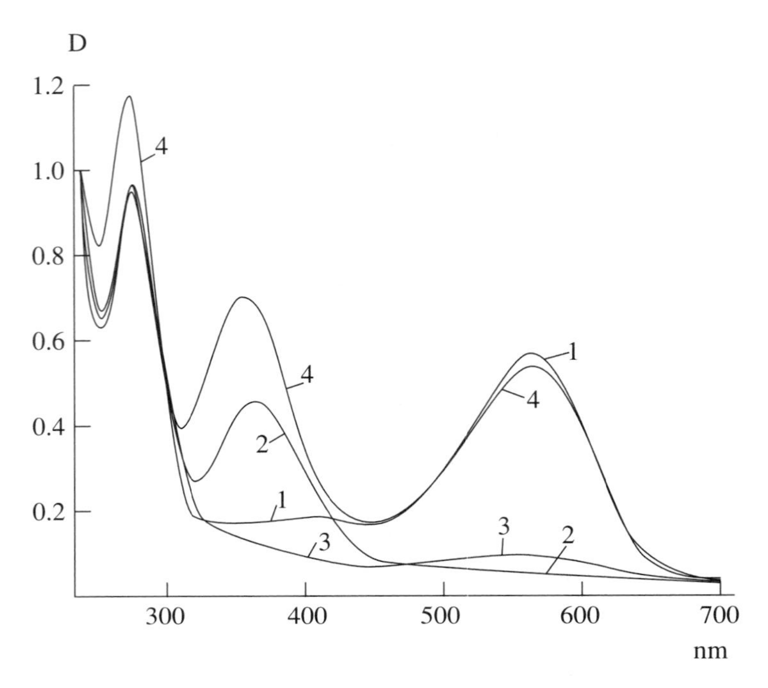

Figure 3.14. (a) Apo-membranes (2) obtained by means of photoinduced hydroxylaminolysis from the native PM of wild strain *Halobacterium halobium* (1). Upon reconstruction with all-*trans* retinal, PM with normal absorption at 560 nm is formed, leaving a retinaloxime protein-related absorption band around 360 nm (4). Apoprotein after destruction of apomembranes by strong light (3) [Oesterhelt et al. 1974].

any kind in the region of 570 nm [Mukohata and Sugiyama 1982]. Their crystalline lattice is untouched, and, after the infusion of an all-*trans* retinal, the spectrum shows the BR absorption band with no additional bands in the 300–400 nm region (See Figure 3.14c).

3C.2. Artificial Retinal–Proteins

Retinal substitution by analogs (or their derivatives) in a BR molecule, leads to shifts of varying amplitude in the spectral and photochemical properties of the new retinal-protein complex. Apo-membrane production does not distort protein structure and all the behavioral changes depend on the type and properties of the retinal analog. Reconstitution may alter certain parameters (positions of the absorption spectra for the BR ground state, and for the photocycle intermediates; the degree of light-adaptation, the lifetimes and the

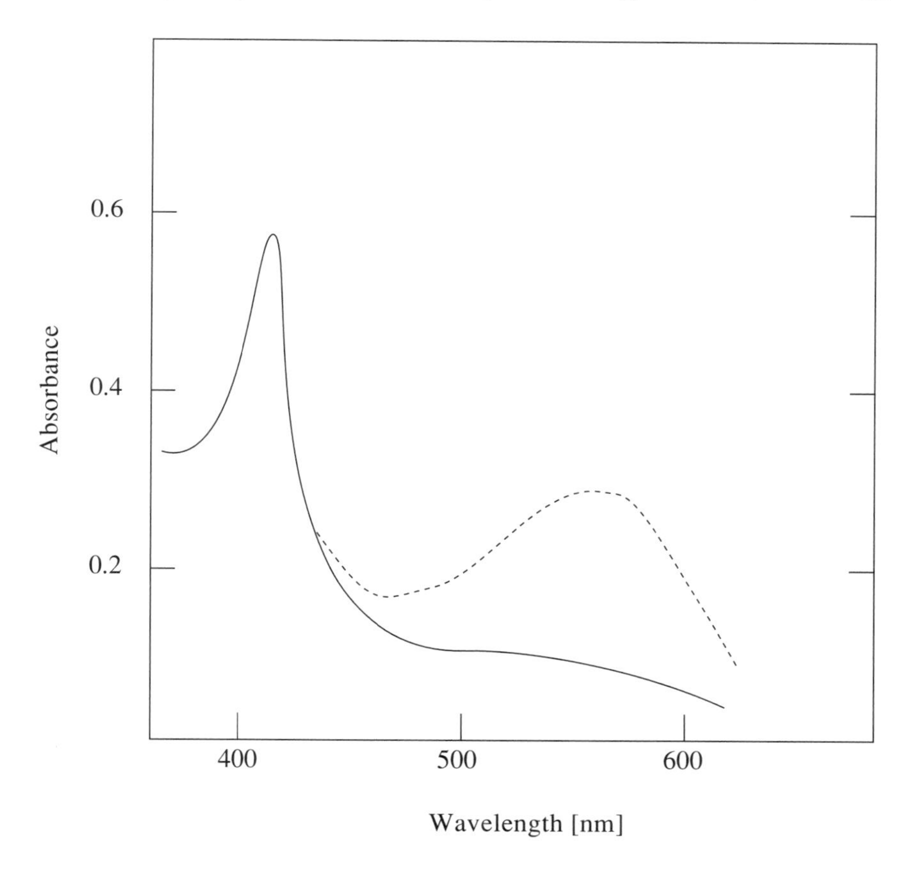

Figure 3.14. (b) Apo-brown membrane isolated from "nicotine cells" of strain R1 M1, absorption around 419 nm corresponds to cytochrome b (solid line). The PM absorption spectrum appears in 1-2 hours after the addition of all-trans retinal (broken line), however, the band at 419 nm remains [Sumper 1982].

number of the intermediates; the quantum yields, the proton activity in the BR ground state and for the photocycle intermediates; the degree of light-adaptation, the quantum yields, the proton activity etc.).

There are several ways of changing retinal properties:

1. Side-chain alteration, for example, by depletion of methyl groups [Feng et al. 1993].
2. Side-chain substitutions. Different functional groups on the 9, 13, and 14 polyene chain provide an array of BR analogs with different photo-activities.
3. Alterations in ionine moiety (acyclic and cyclic alterations). The absence of the ring on the polyene chain does not lead the main photochemical

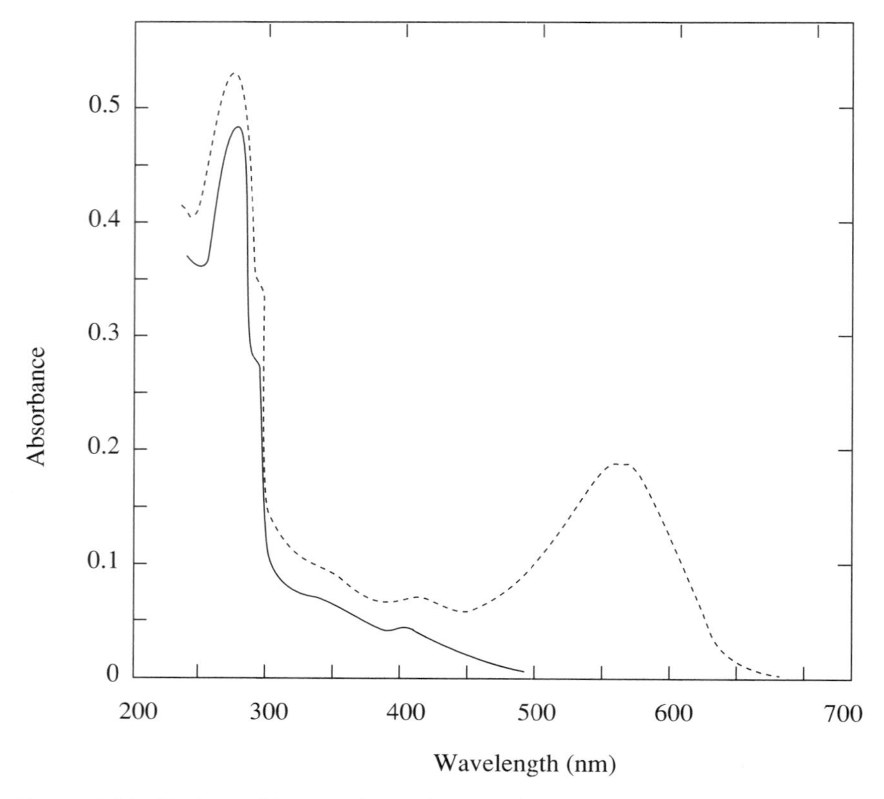

Figure 3.14. (c) Apo-white membrane isolated from spontaneous mutant R1 mR (solid line). In 30 minutes, after apomembrane reconstruction with all-*trans*-retinal at 2°C, the absorption band corresponding to normal PM appears alone (broken line) [Mukohata and Sugiyama 1982].

properties of the analog to disappear but does decrease the overall stability [Steinberg et al. 1991]. Ring modifications allow synthesis of a large family of stable, highly photoactive analogs. Adaptation to light with a maximum shift into the blue region was first demonstrated on a-retinal [Towner et al. 1980].

4. Locked configurations (means just what it says). The first analogs of this type were studied in 1983 [Feng et al. 1983]. According to expectations, both analogs were non-functional, since their *cis-trans* isomerization was blocked. Further experiments with different types of retinal chain blocking at position 13, demonstrated most of the resulting BR analogs, with rare exceptions, are non-functional. For the details of BR analogs properties dependent on the type of retinal, retinal analogs and their derivatives, the reader may refer to the brilliant review by Nakanishi and Crouch [Nakanisi and Crouch 1995].

Not all retinal analogs can form a functional complex with BO (bacterioopsin). Retinal analogs with an aldehyde group form a covalent bond with BO and form stable, functioning BR analogs. Non-aldehyde analogs make a non-covalent bond and form unstable BR analogs [Iwaga et al. 1984].

Retinal analogs, as mentioned, are easily built into halobacterial apomembranes during the growth of certain cultures. As a result, the analogs of dihydro-retinals and 4-keto-retinals produce stable and photochemically active BR analogs [Sheves et al. 1985]. Occasionally, insignificant differences occur in the absorption spectra positions of the same type of BR analogs when produced by different methods (the hydroxylaminolysis method or the method of analog cultivation with the retinal-deficient cell cultures). Table 3.6 presents the main spectral characteristics of the selected BR analogs, including their photochemical and proton activity.

3C.3. Photochemistry of BR Analogs

Photoactive BR analogs might be of interest for bioelectronics since their photocyclic characteristics vary over a wide range causing unexpected changes in the photochromic and technological properties of BR analog-based photomaterials (see Chapter 4). Let us focus on the specific properties of a few:

(a) 4-keto-retinal forms a stable retinal-protein complex of bright-yellow color [Sokolov et al. 1979]. It is one of the most fully characterized of all BR analogs. Its photocycle is unusual for its ability to proceed along three pathways simultaneously. It is possible that the 4-keto-BR analog has three independently coexisting photocycles. Each of them has a standard (native BR) set of intermediates: K, L, M, and, possibly, O. The longevity of these intermediates is nearly the same as the longevity of the correspondent intermediates in native BR, excluding the slow photocycle, in which the M lifetime is longer by two orders of magnitude [Druzhko and Chamorovskii 1995]. The scheme of the photocycles proposed by Druzhko and Chamorovskii contains two chromophores with 13-*cis* and all-*trans* retinal forms, each capable of an independent cycle (See Figure 3.15a).

It was established that the degree of light adaptation of the cycles (the level of 13-*cis* or all-*trans* concentration) depends on many factors, particularly, on the lifetimes of the M intermediates. The M-390 lifetime does not depend on pH changes ranging from 5 to 9, whereas the M420 lifetime grows proportionally to an increase of pH from 5 onward. 4-keto-BR photocycles are strongly interconnected and interdependent [Brown et al. 1992].

(b) 3–4-dehydroretinal forms a BR analog with a ground state absorption at 600 nm [Iwasa et al. 1981]. It retains a proton transporting ability, and has a minimum of three intermediates in its photocycle (See Figure 3.15b).

One of the intermediates is analogous to M since its absorption maximum approaches 430 nm at room temperature, and blue light hastens its decay (this last characteristic is the property of the M intermediate in the native BR photocycle). This analog easily fuses into apo-membranes during growth of a retinal-deficient halophilic culture,

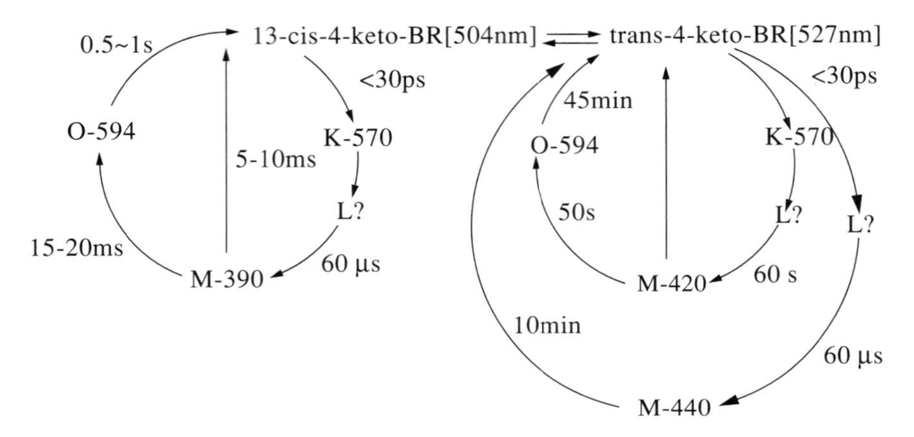

Figure 3.15a

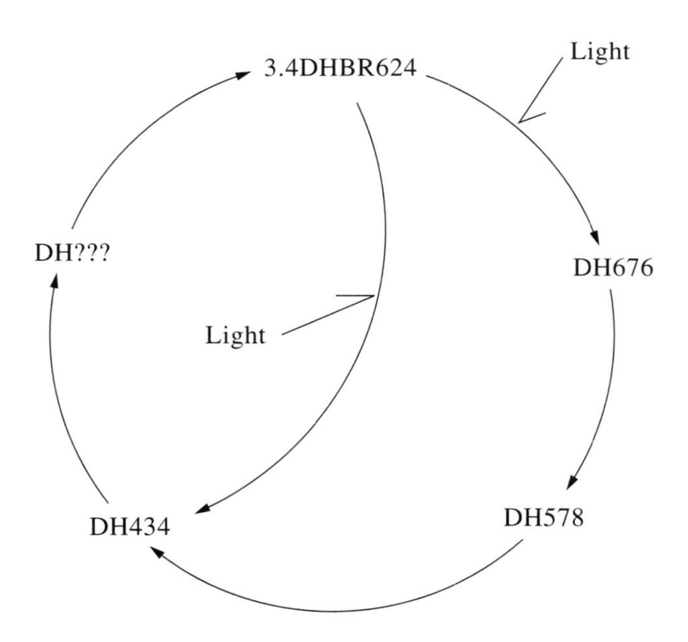

Figure 3.15b

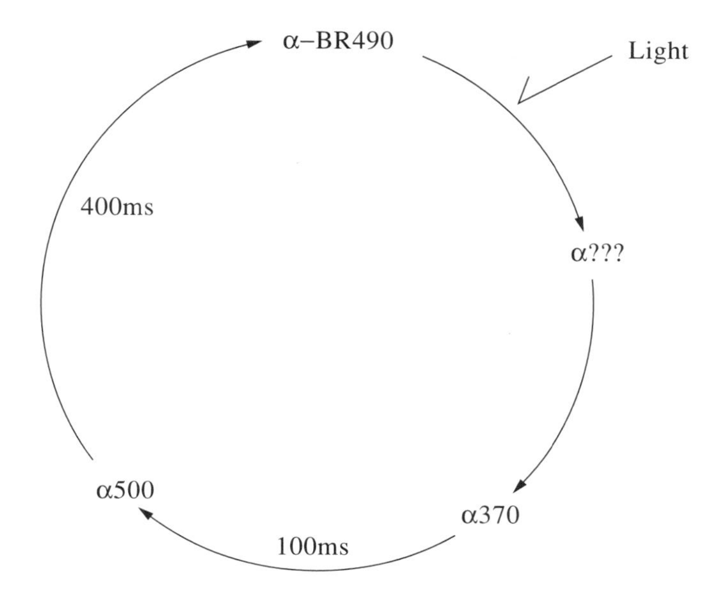

Figure 3.15c

Figure 3.15. Photochemical reaction cycles of selected BR analogs: (a) photocycles of 4-keto-BR, which coexist at the same time. The life-times of intermediates are given for pH 7.5, and room temperature [Druzhko and Chamorovsky 1995] (b) 3,4 dihydro-BR [Iwasa et al. 1980] (c) α-BR [Towner et al. 1980].

and is photochemically active. Its absorption maximum shifts to 580 nm which can only result from insignificant distortions in the protein structure, or structural differences in the crystal lattice in the suspension, and in the halophilic cells [Sheves et al. 1985].

(c) Retinal replacement for α-retinal produces a photochemically active analog. (See Figure 3.15c).

The α-form, with absorption at 370 nm, is probably M. The α-analog has a light-adaptation capacity but its maximum shifts from 492 nm (DA) to 484 nm (LA), i.e., in the opposite direction v.s native BR [Towner et al. 1980].

3C.4. BR Analogs with Infrared Absorption

Retinal-protein complexes of natural origin with absorption maxima over 600 nm have not been discovered. It is not difficult to synthesize BR-based analogs with ground-state maxima in the visible red, and near infrared regions. Azulenic retinoids are used as chromophores for such analogs. The absorption spectra of BR analogs reconstructed on an azulenic base range from 600 to 830 nm [Asato et al. 1993]. These analogs, had there been a native BR-type photocycle, could supply new photochromic materials for optical communication systems and optoelectronic technologies with cheap, IR-semiconductive lasers. Although the first "red" analogs were synthesized as early as 1988, their photochemistry is still unknown. The reason for this is a bit unusual, and has something to do with the authors of the synthesis fearing to loose their prior authority in this direction of research.

Table 3.6. The same characteristics of BR analogs in a water suspension

Name and structure* of retinal analogs	Absorption maxima DA and LA of BR analogs and M-intermediate (nm)			Half-life M intermediate (ms)		Action of protons + = 33%	Ref.
	DA	LA	M	Increase	Decrease		
All-*trans*-retinal (1)	558	568	412	0.05	20	+ + +	1,14
13-cis-retinal (54)	548	560				+ +	14
11-cis-retinal (55)	430/460	430/460				–	14
Retinal A-2 (45)	593	603	0.05		200	+ +	2
5,6-dihydrodesmethyl (49)	480	470	340			+ +	4
5,6-dihydroretinal (37)	460\475	480	360		2 min	+ +	3, 4\13
7,8-dihydroretinal (38)	440\400	460	340			+ +	3,5\13
9,10-dihydroretinal (39)	430\325		450			+ +	3\13,17
11,12-dihydroretinal (40)	280?	–	–	–	–	–	6,17
4-keto-retinal (42)	508\520	508\520		0.06	10 (30s)*	+ \ –	7,8*\14
3,4-dihydro-retinal (21)	600					+ +	9
α-retinal (52)	492\476	486\472	370			+	5,10\15
N-oxide-retinal (41)	454	459		370			5,11
3,7,11-trimethyl-2,4,6, 8,10-dodecatetraenal	528	538	370		10	+ +	12
9-desmethyl-ret. (30)	530	540				+ +	14
C-22-retinal (48)	560	560				–	14
4-hydroxy-retinal (50)	540/525	550	400				16/15
4-dimethylamino-ret. (51)	455	455	390	16			

* See structures in Appendix with bracket number.

References:

1. Oesterhelt and Stoeckenius 1973.
2. Nakanishi 1991.
3. Arnaboldi et al. 1979.
4. Mao et al. 1981.
5. Gat and Sheves 1994.
6. Gawinowicz et al. 1977.
7. Brown et al. 1992.
8. Druzhko and Chamorovsky 1995.
9. Iwasa et al. 1981.
10. Towner et al. 1980.
11. Crouch et al. 1981.
12. Muradin-Szweykowska 1981.
13. Balogh-Nair et al. 1981.
14. Stoeckenius et al. 1979.
15. Steinberg et al. 1991.
16. Sheves et al. 1984.
17. Nakanishi 1980.

Cyanide dye, used as a chromophore, allows synthesis of a BR analog with an absorption maximum at 662 nm. It is unstable, and irreversibly decomposes in light [Derguini et al. 1983]. 13-tri-fluoromethyl-retinal forms a BR analog with a maximum at 624 nm, which is highly proton active, but provides no evidence of a photocycle [Gartner et al. 1981]. This totally inexplicable fact contradicts the presently accepted theoretical view that proton activity is always reflected in spectral shifts.

A photochemically active orange analog of BR was synthesized on the base of 3,4-dihydro-retinal. It has DA and LA forms with absorption maxima at 593 and 603 nm, respectively [Tokunaga et al. 1977]. Its photochemistry is unknown, but many infrared analogs are not photochromically active. Table 3.7 presents general characteristics of some IR-analogs of BR.

3C.5. BR Analogs with Ultraviolet-Absorption

Only a few natural rhodopsins with the absorption maxima in the ultraviolet region have been discovered (see Section 2B.4). Not many BR analogs with a maximum less than 400 nm have been synthesized. Only two analogs (7,8-dihydro-BR and 9,10-dihydro-BR) have absorption maxima at 400 and 325 nm which we know exactly [Nakanishi et al. 1980] (see Table 3.6).

3D. BR Monolayers and BR Films

Monolayers of any protein plays an important role in the life of any organism and the construction of bioelectronic devices [Kuhn 1994]. It is known that BR molecules in halobacteria exist only in the purple membrane as hexagonal dimeric crystals (see Section 2C.4). Instability of the BR 3-D-structure outside the PM makes it impossible to synthesize an isolated, lipid-free BR. Thus, active monomeric BR molecules can exist only in the structure-stabilizing surroundings. For example, a monomeric BR in vesicles does not denature due to its lipid surroundings.

One must remember that PM is by nature a bilayer membrane. By saying "monolayer of the BR molecules," we are often refering to a monolayer of a protein, and not to membrane structure. The methodology of PM delipidation has been described in detail [Bayley et al. 1983]. It has been suggested that upon removal of 90% lipids, the BR molecule still perserves its ability to conduct proton translocation. Biophysicists have secret hopes that one day they will synthesize stable BR monomolecules and will assemble a 3-D crystal. The properties of this compound are going to shake the scientific world.

Table 3.7. BR Analogs with absorption in the red and infra-red region

Chromophore analogs Name and Structure (Brackets refer to Appendix number)	BR analogs absorption maxima (nm)	Characteristics	Ref.
11-methyl-merocyanine (47)	662	unstable to light	Derguini et al. 1983
13-trifluoromethyl-retinal (46)	624	proton transfer ability	Gartner et al. 1981
14-halogenated (32)	680–690		Tierno et al. 1990
Azulenics			Liu et al. 1993
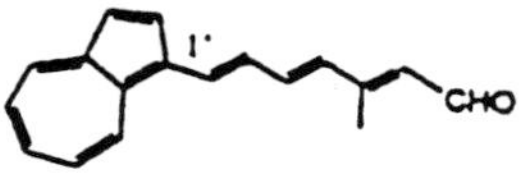	644		
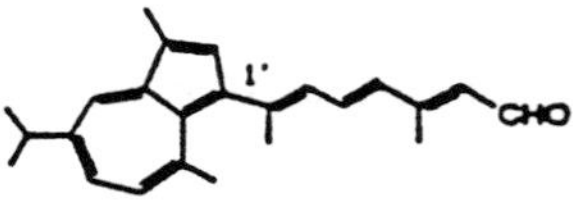	694		
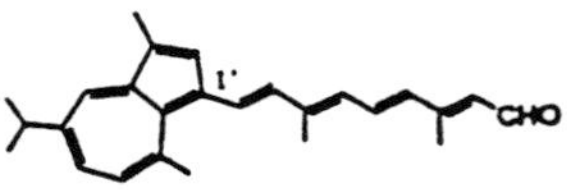	750		
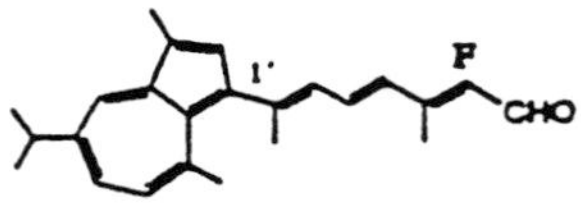	795		
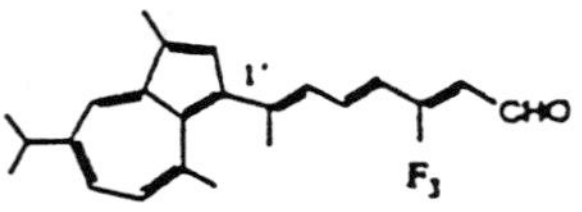	830		

3D.1. BR Solubilization

Solubilization of the purple membrane by highly concentrated detergent mixtures leads to decomposition of the crystal structure, and to the formation of unstable monomeric BR molecules which finally denature [Hwang and Stoeckenius 1977]. At indefinite detergent concentrations, the BR monomers are fairly long [London and Korana 1982]. It was shown that in detergents, BR monomers retain the main properties of a PM-inbound BR i.e. the ability for a photocycle and for light dependent proton transfer [Dencher and Heyn

1978)]. In a Triton-X-100 treated purple membrane, the maximum shifts from 570 to 550 nm for a DA form, while a red-shift amplitude of light-adapted form goes down from 10 to 5 nm [Casadio et al. 1980]. Padros et al. [1984] conducted a comparative study of the effect of five different detergents on PM suspensions. In particular, they discovered that sodium dodecylsulfate treatment leads to the appearance of an absorption spectrum with a maximum at 600 nm at neutral pH. This spectral product is photoactive and irreversibly converts to the spectral product with maximum at 440 nm when exposed to red light.

During the substitution of natural phospholipids in the PM by other types, full or partial monomerization of the BR molecules may occur. For example, PM incubation with egg phosphotidylcholine leads to the formation of average-size molecular clusters of BR. The photocycle in such clusters is slower [Bakker and Caplan 1982]. Substitution of natural phospholipids for artificial ones allows determination of how the lipids of the PM affect BR photochemical processes [Sherman and Caplan 1979].

As mentioned before, BR is a monomer in a vesicle, and in this form, is still photochemically and proton active [Grzesiek and Dencher 1988]. Preservation of photoactivity by BR monomers confirms the suggestion that the crystalline structure has a protective, rather than functional role in PM. For example, BR immersed in a non-ionic detergent (Triton X-100), interaction was shown to encumber transition from M to BR of the photocycle in BR-gelatin films. In these same films, the lifetime of the holographic grating was 2–3 times greater, when compared to the unmodified BR films (see Section 4D.3). The measured holographic sensitivity appeared to maximize in the range of Triton X-100/BR molar ratios between 15/1 and 25/1 [Cullin et al. 1995].

3D.2. BR Films

The synthesis of two-dimensional crystalline structured films of a given size, assembled from uni-oriented BR monomolecules, is a dream of many scientists. By coating such layers one over the other, it might be possible to produce a large 3-D BR crystal. Unfortunately, after isolation from PM, BR molecules denature, disallowing isolation of a homogeneous stable solution of molecular BR, and preventing attainment of this vision. Some monolayer properties may be imitated by certain PM fragments used as dimeric crystals.

Noncontinuous, mono- and multi-layers of oriented PM, are created with facility on the phase boundry in a similar fashion as "water-octane," "water-air," etc. The PM surfaces have different quantities of polarized and non-polarized amino-acid residues: the external side of the PM is more hydrophilic, while the internal (cytoplasmic) side is more hydrophobic. This property is employed to orient the PM on the border between the hydrophilic and the hydrophobic mediums (water and air). The mixture of PM and phospholipids in an organic solvent (for instance hexane) spreads on the water surface as a monolayer of phospholipids and purple membranes spontaneously oriented with their cytoplasmatic sides towards the water phase [Hwang et al. 1977]. Studies on spectroscopic and protein-transporting properties of purple membranes in such films showed no changes in the form or position of the BR absorption spectrum. The lowering of humidity in a multilayer inhibits the photocycle speed. The optical absorption of one monolayer at 570 nm wavelength equals $0.32–0.4 \times 10^{-3}$ [Korenbrot and Hwang 1980].

One of the methods of PM-oriented mono and multi-layer assembly was described in detail by Korenbrot [Korenbrot 1982]. PM multilayers assembled by this method are not ideal but do provide a sufficiently ordered 3-D array of the molecular BR convenient for electron microscopy and X-ray structural analysis [Clark et al. 1980].

The Langmuir-Blodgett method (LB) allows synthesis of high quality mono-and multilayer films from many active proteins and enzymes. These are characterized by a high degree of homogeneity in molecular distribution and orientation. The method of LB (films produced from a PM suspension and soy bean phospholipids treated in hexane) was described in detail [Ikonen et al. 1992]. According to Ikonen and colleagues, the optical absorption of a BR molecular monolayer is greater than as reported by Korenbrot [Korenbrot and Hwang 1980], and equals 0.445×10^{-3} at 570 nm wavelength. The authors explain this result by a larger number of BR molecules per unit of monolayer surface. The structure of a monolayer differs from the structure of a native purple membrane and resembles the model for a bilayer lipid membrane with an intrinsic monolayer of molecular BR. An angle between the long axis of retinal and the membrane plane must be larger than in the PM since the absorption cross-section is only 33% of that of PM. The authors particularly emphasize the high stability of BR-based multilayer films, as well as the long period of preservation of BR photochemical properties in the films on quartz. One scientist, after re-staging these experiments, remarked in a private conversation that it had been one hell of a job, and he did not wish to go back to it again. This may be the reason why, despite the inspiring results from Ikonen and colleagues, photoelectric and electrogenetic experiments are still conducted on lipid-saturated collodium films and ceramic filters.

3D.3. BR Aggregates

According to the literature, BR-aggregation is a self-assemblage of BR and BO molecules into aggregates similar to moderately sized PM fragments. During the synthesis of PM in a halobacterial membrane, naturally programmed self-assembly of PM occurs (see Section 2C.3). More often, the term "BR aggregation" refers to small groups of BR molecules assembled in a different fashion. In liposomes, BR may occur as a monomer, and it may stay in the form of triads, or small aggregations of several dozen molecules. It is not yet clear what kind of BR aggregation occurs in flat lipid membranes. Drachev (see Section 2C.2), based on his many years of artificial membrane studies, thinks that in such membranes BR may occur both as a monomer and as an aggregate. He made this privately shared conclusion after comparing the spectral response times of BR molecules in a PM suspension to the electroresponses of BR to a pulse of light in artificial membranes. Upon infusion of unsonicated PM into flat artificial membranes, PM structure may remain unchanged. In this case, the electroresponse and spectral shift times coincide. If the PM suspension is sonicated, a flat membrane may contain separate molecules of BR whose tertiary structure is stabilized by the membrane lipids. In this case, the times of electroresponses from molecular BR are much shorter. The search for publications that would give an intelligible description of the aggregated state of molecular BR in different surroundings proved futile. Still, the question is interesting: how are the experimental BR molecules assembled? The results and the interpretations of the experiments often depend on the answer to this question.

Experimenting with films or block preparations, one has to deal with large or small PM aggregates. Naturally, this must influence the results of the experiments. After the suspension has been dried, the sediment of parallel-packed layers of aggregates remains on the base. The sides of these layers may be oriented in the plane of the base in tandem or opposite directions. Upon electrophoretic precipitation, all purple membranes orient with one side to the base, which allows cooperativity in photoelectric or electrochromic experiments. The nonhomogenous pack-structure must be preserved in any case, and, to date, there is no multilayer synthesis method to provide a monolayer or monolayers of molecular BR that would be structurally homogeneous and have a large surface area.

3D.4. The Possibility of a Three-Dimensional BR Crystal

In 1981, Michel attempted the synthesis of 3-D BR crystals [Michel 1982] [Michel and Oesterhelt 1982]. Needle-shape and cubic crystal forms were obtained. The needle-shape crystals approached 50 μm across and a minimum of 3 μm in diameter. The sizes of the cubic crystals, assembled from orthorhombically modified PM (see Section 2C.4) did not exceed a few microns—which is too small for Roentgen (X-ray) structural analysis. The synthesis of crystals raised many of the previously described problems. For instance, without full lipid removal, it was difficult to obtain crystals of the experimentally required size and purity. Nobody since then has attempted to synthesize bit size BR crystals—but the dream lives...

3E. Halorhodopsin Acting as a Chloride Ion Pump

In 1979, Lanyi and Green published a paper in which they reported the discovery of a new rhodopsin in *Halobacterium halobium*. They attributed the light-driven sodium ion pump to this compound [Green and Lanyi 1979]. In yet another work, in collaboration with Oesterhelt, this new pigment was called halorhodopsin (HR) [Wagner et al. 1981]. It soon became obvious that the new rhodopsin had no relation whatsoever to the gradient of sodium ions. However, the name halorhodopsin, as it turned out later, quite accurately described its real function.

In 1982, it was established that HR molecules accomplish light-dependent chloride ion transfer across the halobacterial membrane from the outside to the inside of the bacteria, thus regulating the intra-bacterial concentration [Schobert and Lanyi 1982]. In addition to Cl^-, HR is capable of transporting other ions, and three anion binding sites I, II and III are known. Site II is more specific for chloride ions than for polyatomic ions. Sites I and III are relatively non-specific [Schobert and Lanyi 1986] [Lanyi et al. 1990].

The difference between the BR and HR photocycles has been obvious since the start of research [Green and Lanyi 1979] [Weber and Bogomolni 1981]. However, until 1982, spectral results appeared distorted due to the inability to sufficiently purify the preparation, and that led to a misunderstanding of the photocycle. The strong dependence of a molecular photochemical process in HR on the concentration of chloride ions around the molecule also impeded the study of the photocycle. It appeared that the photochemical process followed two different cycles depending on the two extremes of chloride ion concentrations [Schobert et al. 1983].

Further studies of a photochemical process in HR confirmed the hypothesis that the HR photocycle may consist of two different photocycles, depending on chloride ion concentrations in the surrounding medium [Lanyi 1984]. The lifetimes and maxima of the intermediates depend substantially on the purity of the preparation, on pH, and on temperature. Therefore, the photocycle scheme reprinted from the work of Lanyi and Oesterhelt [1982] and the one reprinted from Bogomolni et al [1984] look different. Oesterhelt and colleagues [Oesterhelt and Tittor 1989] [Hegelman et al. 1985] conducted a more detailed study of photochemical and transportation activity in HR and merged the two schemes into one.

At the instant HR behaves as a chloride ion pump, site II is activated, and the lifetimes and absorption spectra of the photocycle intermediates resemble the intermediates K, L, and O in the native BR photocycle, while the ground state absorbs at 578 nm.

The function and location of anion-binding sites I and III are not quite clear. Site III has a resemblance to its analog in pharaonis HR. When sites I and II are not bound by anions, the absorption maximum of the HR ground state shifts to 565 nm, and the photocycle loses a few intermediates (a so-

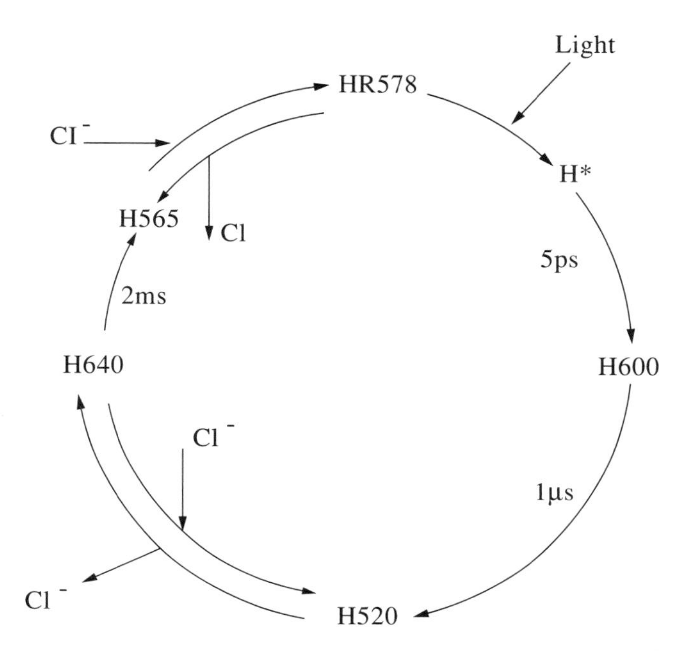

Figure 3.16. Photochemical reaction cycle of halorhodopsin from *Halobacterium halobium*. (Oesterhelt and Tittor 1989).

called "truncated photocycle"). No regeneration of a photocycle or of HR transport properties occurs when site I is anion-bound. The HR photocycle contains no M-type intermediate, which would alone participate in the proton tarnslocation in BR. Possibly, the HR photocycle has two "transporting" intermediates [Zimanyi et al. 1989].

The building of HR molecules into azometine lipid vesicles creates a system capable of photoelectric activity. Its active spectrum coincides with its absorption spectrum. Photoelectric responses are registered only in the presence of chloride ions or its analogs in the surrounding medium [Bamberg et al. 1984].

There are some interesting similarities and differences between the BR and the HR molecules. The HR molecule contains 251 amino-acid residues and has a molecular weight of 26,961 g/mol; retinal is linked by the Schiff base to a protonated Lys-242 protein. The 3-D structure is 36% homologous to the structure of BR, and consists of seven α-helical segments [Havelka et al. 1995]. During a photocycle, a chloride ion is transported from the external side of the membrane into the cell (the direction of ion translocation is opposite to BR). The cycle completes in 1–2 ms at 20°C (which is 10 times faster than BR); the

Schiff base gets protonated in the duration of the entire cycle (in BR it reprotonates during the cycle); in the HR primary structure, Asp-85 and Asp-96 are replaced by neutral amino acids.

The replacement of the key proton-transporting amino-acids at the 85th and 96th link of a polypeptide chain may be the reason for inability of HR to conduct proton transfer [Oesterhelt and Tittor 1989]. Nearly nothing is known about the mechanism of photodependent chloride ion translocation. There is only the suggestion that chloride acts as a counter-ion to the protonated Schiff base that is displaced during the isomerization of retinal [Oesterhelt et al. 1986] [Bamberg et al. 1994].

3F. The Discovery of Sensory Rhodopsins

In 1975, Hildebrand and Dencher attempted to unravel the mechanism of a phototaxis in *Halobacterium halobium*. How precisely does the organism respond to light? They hypothesized that there are two phototaxis-controlling systems in the bacteria; one absorbing at 570 nm, was presumed to be the proton pump BR, and the other, absorbing in UV (370 nm), was of unknown structure [Hildebrand and Dencher 1975]. The systems were called photo-sensory and abbreviated as PS565, and PS370. In a few years it was demonstrated that the second phototaxis-controlling system also contains a retinal-protein complex [Dencher and Hildebrand 1979] [Spudich and Stoeckenius 1979].

This hypothesis had survived untill 1982, when mutant studies revealed a new slow-cycling rhodopsin-like protein in the phototaxis of *Halobium* [Spudich and Spudich 1982] [Bogomolni and Spudich 1992]. This protein was called "slow rhodopsin" (SR) [Spudich and Bogomolni 1983], or "third rhodopsin-like pigment [Tsuda et al. 1982]. The half-life of one of the intermediates in the SR photocycle approached seconds, and was by an order of magnitude greater than the half-life of the slowest M intermediate in the native BR photocycle. "Slow rhodopsin" was renamed "sensory rhodopsin" when compelling evidence for its function as a phototaxis receptor was presented [Spudich and Bogomolni 1984]. This did not end the misadventures with nomenclature. When it became obvious that halobacteria contains not one, but two of the new rhodopsins [Takanashi et al. 1985], Spudich and colleagues offered to call them sensory rhodopsin I and II (SR-I and SR-II) [Spudich et al. 1986], which is in general use today.

Since 1982, and largely due to the efforts of Spudich and colleagues, research on sensory rhodopsins (sensory rhodopsins I and II) has greatly advanced. Spudich and Bogomolni published a few surveys where they not only summarized the experimental results on sensory rhodopsins and discussed possible mechanisms, but also drew analogies to other rhodopsins.

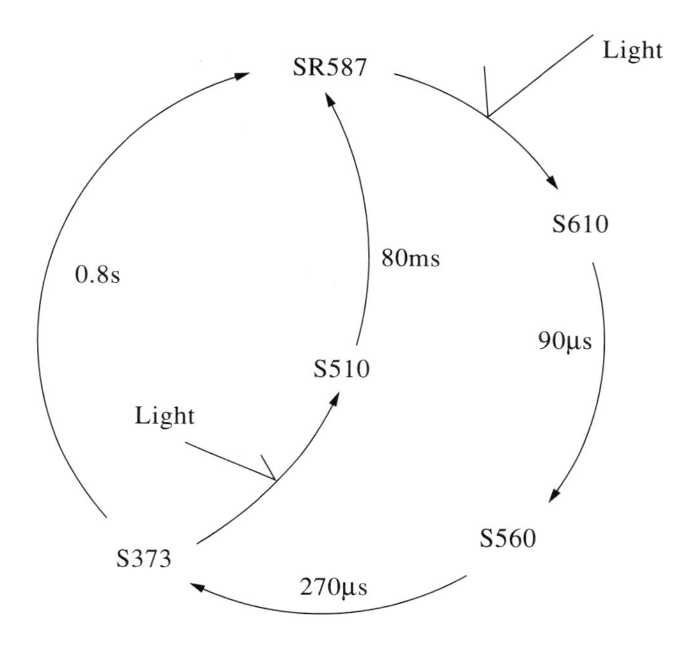

Figure 3.17. Photochemical reaction cycle of sensory rhodopsin I from *Halobacterium halobium*. (Bogolmoni and Spudich 1992).

This affords a better understanding of the role of sensory rhodopsins in the rhodopsin family [Spudich and Bogomolni 1983] [Takahashi et al. 1990] [Bogomolni and Spudich 1992]. Apo-proteins in SR easily reconstruct with retinal analogs, forming photoactive analogs of SR [Yan et al. 1991].

3F.1. SR-1—An Attractant Receptor for Orange-red Light

Sensory rhodopsin-I (SR-I) controls and directs the mechanism of reorientation in cellular motion in response to a light gradient. Compared to the photocycles of other photosensors, its photocycle is rather unusual. It has two ground states with absorption maxima at 587 and 373 nm.

The SR-I ground state has a maximum at 587 nm. The absorption of orange light quanta activates a molecular mechanism for cellular movement toward orange light, i.e., SRI-587 controls the positive taxis. Simultaneously, orange light activates a photocycle in SR-I, forming a long-lived intermediate S-373 with the absorption maximum at 373 nm. Due to its long lifetime (nearly 3,000 times longer than its formation time), S-373 photoproducts occur, in what amounts to a photostationary state and may be regarded as the second ground state of SR-I. The S-373 intermediate is photoreactive, and upon UV illumination quickly converts to the SR-587 form. During this transition, the mechanism of the negative phototaxis is activated, and the cell moves in the direction, opposite to the UV radiation source.

SR-I controls positive and negative phototaxis in halobacteria allowing it to differentiate attractive and unattractive colors of the optical spectrum. The

Table 3.8. Comparison of BR, SR-I, SR-II and VR analogs (LA-forms)

Native retinal and analogs		Maximum absorption for ground states (nm)					
No.	Structure*	SR-I	SR-II	BR	VR	Activity	Ref.
1.	Native (1)	584		570	500	Photocycles	1,2
2.	(21)	612/620		600		Photocycles	1,2/4
3.	(22)	585	470	575		Photocycles	1,3,7
4.	(23)	575		562		Photocycles	1
5.	(24)	565		565	495	Photocycles	1,7
6.	(25)	556		537		Photocycles	1, 4
7.	(26)	550/530			550	Photocycles	1/ 4
8.	(27)	525		538		Photocycles	1
9.	(28)	520		520	485	Photocycles	1,7
10.	(29)	505		511		Photocycles	1
11.	(30)	470		490		Photocycles	1, 2
12.	(14)	460		440		Photocycles	1
13.	(31)	450		450		Photocycles	1, 2
14.	(39)	360		350		Photocycles	1, 2
15.	(3)	585	513	576**		Blocked of photocycles	5, 6
16.	(33)	590	470	570		Photocycles	1,4,5
17.	(34)	550	475	550		Photocycles	4, 5
18.	(35)	480	460			Photocycles	5
19.	(36)	485				Photocycles	5
20.	(37)	486		478		Photocycles	4
21.	(38)	460		440	420	Photocycles	4,7
22.	(39)	None		343	345	Photocycles	4,7
23.	(42)			510	470	Photocycles	7
24.	(43)			542	467	Photocycles	7

* See Appendix **DA-form.

References:

1. Yan et al. 1991.
2. Spudich et al.1986a.
3. Baselt et al. 1989.
4. Yan et al. 1990.
5. Nakanishi and Crouch 1996.
6. Yan et al. 1990a.
7. Balogn-Nair and Nakanishi 1982.

results of the experiments with SR-I analogs provided indirect evidence of SR-I as a cellular signal mediator [McCain et al. 1987)] [Yan and Spudich 1991]. Some retinal analogs prolong the S-373 lifetime and increase the photosensitivity of the cell phototaxis. Sensory opsin of a halophilic mutant F1X3R was reconstructed with a series of retinal analogs for the discriminative study of their spectral properties [Spudich et al. 1986]. The absorption maxima of the same retinals, reconstructed with BR opsins or with SR-I, differ by not more than 10–20 nm.

There is a similarity in appearances between the BR and SR-I photocycles. For example, S-373 occupies the same place in the SR-I cycle as M does in the BR cycle, and, also similar to M, speeds-up its transition to the ground state if activated by blue light. Neither proton transfer across the cell or vesicle membrane, nor membrane hyperpolarization, occur in the SR-I photocycle [Spudich and Bogomolni 1988]. No definitive theory exists on how a photosignal controls the motion of flagellae [Spudich and Bogomolni 1992] [Spudich and Lanyi 1996].

One of the few existing hypotheses is based on the analogy with the mechanism of chemotaxis in eukaryotes. A protein chemically analogous to the *E. Coli* membrane flagellae ("motor-controlling" protein), was discovered in *Halobacterium halobium*. This protein, called methyl-accepting phototaxis protein-I, is extracted simulataneously with the sensory rhodopsin I chromophore from three different SR-I mutants [Spudich EN et al. 1988].

In 1992, the htrI gene encoding this methyl-accepting protein was cloned [Yao and Spudich 1992]. HtrI, as a protein is now called, was found to interact with SR-I, and is proposed to govern transduction signals from the receptor to the flagellar motor [Spudich 1994] [Krah et al. 1994] [Jung and Spudich 1996]. Another hypothesis suggests that the cyclic nucleotides, G-protein and Ca, are involved in the photosensory transduction mechanism [Schimz et al. 1989]. This hypothesis suggests an evolutionary link between visual and bacterial rhodopsins (see Section 2C.5). The results of SR-1 reconstruction with 24 different retinal analogs have demonstrated the existence of a general photo-recepting mechanism for archaebacteria and vertebrates [Yan et al. 1990].

3F.2. SR-II—A Repellent Receptor for Blue-green Light

The SR-II absorption maximum coincides with the region of maximum solar energy (~ 500 nm). Upon an increase in light intensity, and when the cell has a sufficient oxygen supply from the surrounding medium, the SR-II-connected flagellae mechanism moves the cell to the lower light intensity area.

Fast intermediates of the SR-II photocycle were studied at low temperatures [Iamamoto et al. 1988]. Other intermediates were also studied at room temperature [Tomioka et al. 1986] [Scherrer et al. 1986]. The photochemical cycle of SR-II is shown in Figure 3.18.

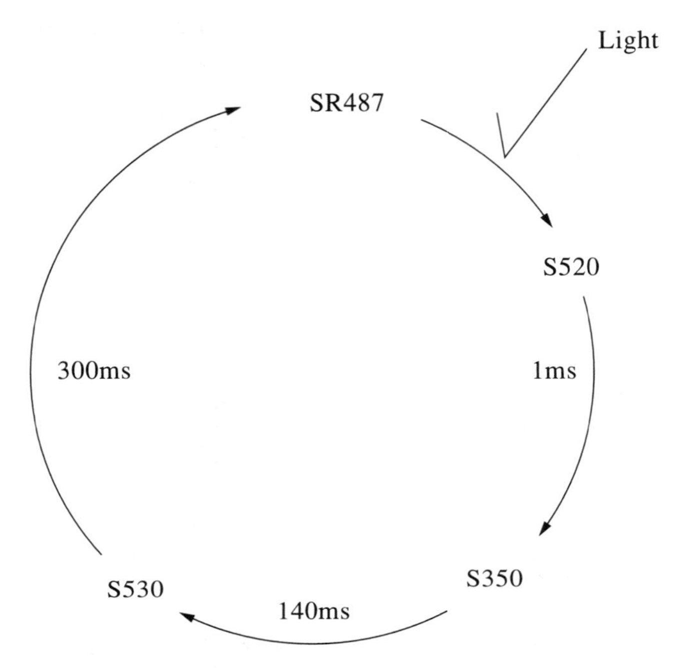

Figure 3.18. Photochemical reaction cycle of sensory rhodopsin II from *Halobacterium halobium* (Bogolmoni and Spudich 1992).

It is similar to the BR and SR-I photocycles. The SII-530 intermediate can be identified with the product O in the BR photocycle, while SII-350 corresponds to the intermediate M. Similar to SR-I, SR-II has no electrogenic properties, and no ion pump of any kind.

3G. Phototaxis of Phoborhodopsins

Phototaxis are defined as the orientation and/or directed movement in response to a light gradient. This phenomena is common among bacteria and microorganisms. The photoreceptor system of the organism can utilize at least three properties of light for orientation: intensity, polarization, and spectral composition. The photoreceptor itself must work in a wide range of intensity changes, be spectrum discriminative, and be able to distinguish the instructive signal from noise. In the preceeding chapters, we described the natural solution for the first two problems. The isolation of the signal from the noise depends more on the general behavior and construction of the microorganism, than on the type and construction of the photoreceptor. For example, noise sources for bacteria are rotational and lateral diffusion, convectional environmental flows, light fluctuations (for instance, induced by ripples on the surface of a reservoir) and the noise in the photoreceptor itself created by various internal fluctuations (potential on the membrane, biochemical processes etc.). The cell opposes the noises in different ways. For example, by probing with rotational and vibrational movements. If the frequency of a single cell vibration does not coincide with the frequency of light, the averaged signal of the correct direction of the cell motion is worked out between the difference of mechanical and light oscillations. There are some other methods of noise fighting with the development of light-focusing antennas or special light focusing devices [Foster and Smith 1980].

Phototaxis as a natural phenomenon has been investigated since 1817, and has been described in monographs and reviews [Macnab 1984]. From the onset of the research it was evident that a phototaxis controlling photoreceptor must have photosensitive molecules for the primary event of photon energy conversion into flagellar motions of the cell. *Halobacterium halobium* were the first to have these molecules identified as retinal–protein complexes.

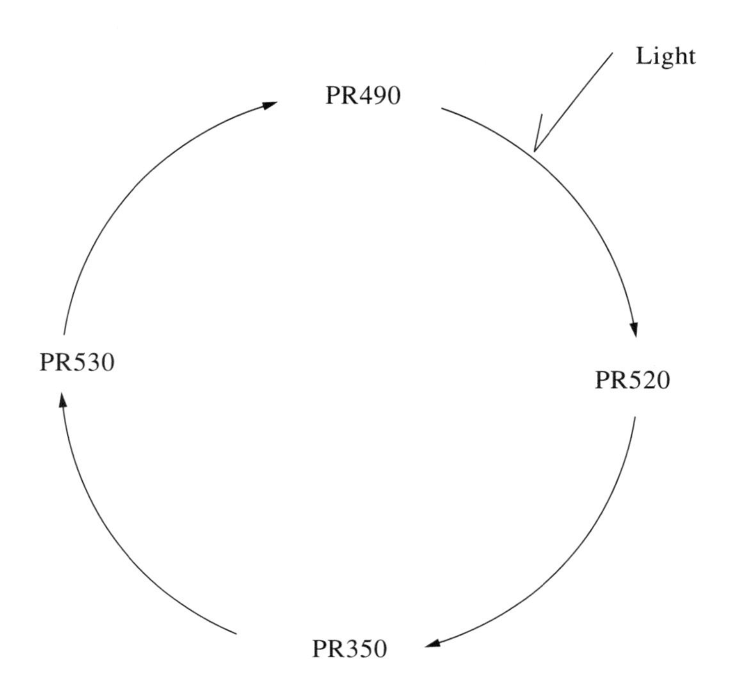

Figure 3.19. Photochemical reaction cycle of phoborhodopsin (PR) from *Halobacterium halobium* [modified from Imamoto et al. 1991].

3G.1. Phoborhodopsins Isolated from Halobacteria

At least four retinal–protein pigments have been discovered in *Halobacterium halobium* (see Section 2C.1). The fourth rhodopsin-like pigment controls solely negative phototaxis of the bacteria and is, therefore, called phoborhodopsin (PR) [Tomioka et al. 1986]. After a sufficient quantity of PR had been isolated from ON1, ON-P mutants (mutants depleted of all other rhodopsins), it was thoroughly characterized. The absorption maximum of PR is positioned at 490 nm. It has a photochemical cycle with at least two intermediates with spaced absorption spectra maxima (at 350 and at 530 nm), and possibly one more intermediate absorbing at 520 nm, expressed at low temperatures [Imamoto et al. 1991]. The possible scheme of the PR photocycle is shown in Figure 3.19.

The authors draw an analogy between BR and PR intermediates, but isolation and purification of this retinal–protein is extremely complex, making it too early to discuss the precision of the results. Moreover, the similarity between BR and PR photocycles, such as the position of the PR ground state absorption spectrum, is closer to the rhodopsin spectrum of vertebrates than to BR. Besides, PR concentration in the cell membrane of the wild-type *halobacterium* slowly decreases in the process of cultivation and bacterial growth, while concentrations of three other rhodopsins (BR, HR, and SR) in the same membrane increases.

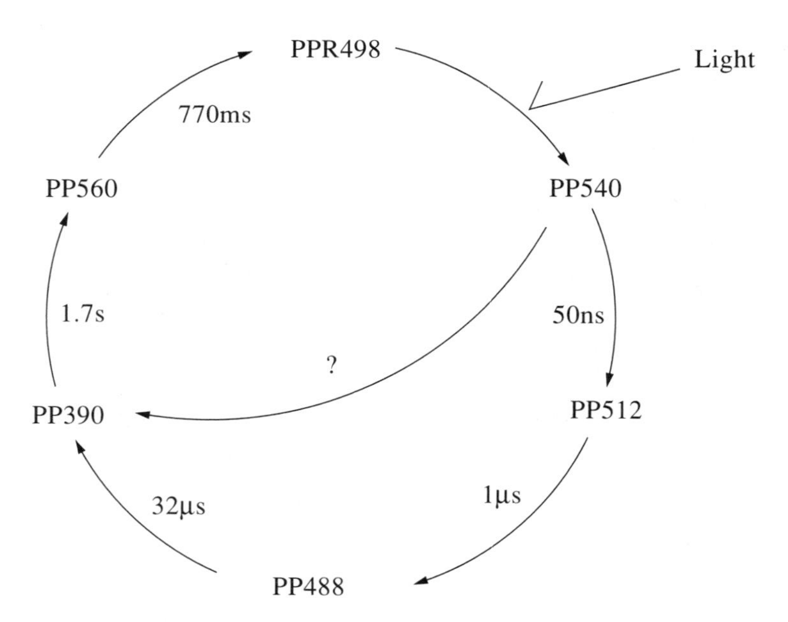

Figure 3.20. Photochemical reaction cycle of *Pharaonis phoborhodopsin* (PPR) from *Natrobacterium pharaonis* [modified from Hirayama J et al. 1992].

Knowing that one of the SRs also controls negative phototaxis, we may ask: why does one bacterium need two different rhodopsins to control the same function? It doesn't. There is only one rhodopsin named differently by different authors.

3G.2. *Pharaonis* Phoborhodopsin

In 1986, yet another retinal-protein complex was isolated from the long-known *Natronobacterium pharaonis* [Bivin and Stoeckenius 1986]. Preliminary investigations showed that its photocycle was similar to the photocycle of PR from *Halobacterium halobium*, differing only by the positions of the absorption maxima of the ground state and intermediates. To distinguish it from PR, it was called PPR (*Pharaonis phoborhodopsin*) [Tomioka et al. 1990]. Unlike PR, the new phoborhodopsin was stable against detergents, and could be easily purified by means of column chromotography. This permitted the study of the PPR photocycle by low-temperature spectrophotometry [Hirayama et al. 1992]. PPR absorbs in the ground state at 498 nm, i.e., at 10 nm greater than in PR. Unlike the PR photocycle, the first (photophysical) intermediate K consists of at least two components with different spectral forms and maxima, each being photoreversible. Added to K, there are at least two more intermediates, called M and O, absorbing at 390 and 560 nm. The photocycle scheme is shown in Figure 3.20, and is generally similar to the PR photocycle.

In the dark-adapted PPR molecules, retinal occurs in the all-*trans* form. In the light, at the stage of M intermediate formation, up to 90% of retinal converts into 13-*cis*. During thermal relaxation of M to O, retinal returns to its initial all-*trans* form and the isomerization cycle is fully completed without spending external energy [Imamoto et al. 1992]. Thus, PPR, similar to PR, has a fully reversible, BR-like photocycle [Tomioka and Sasabe 1995].

3H. Yellow Protein Isolated from a Halophile

In 1985, Meyer isolated a protein of brillant yellow color from the halophilic bacteria *Ectothiorhodospira halophila* [Meyer 1985]. The protein, weighing approximately 15 kDa, with an extinction coefficient of 48,000 M^{-1} cm^{-1}, was capable of reversible discoloration in the 446 nm range. The time of decay of the primary form after light exposure was a few msec while regeneration lasted about a second. The authors suggested that the chromophore of this protein is similar to rhodopsin chromophores, but by relaxation time, it is closer to sensory rhodopsins (see Section 3F). The protein crystallizes upon evaporation of the suspension. The hexagonal crystals are in the space group

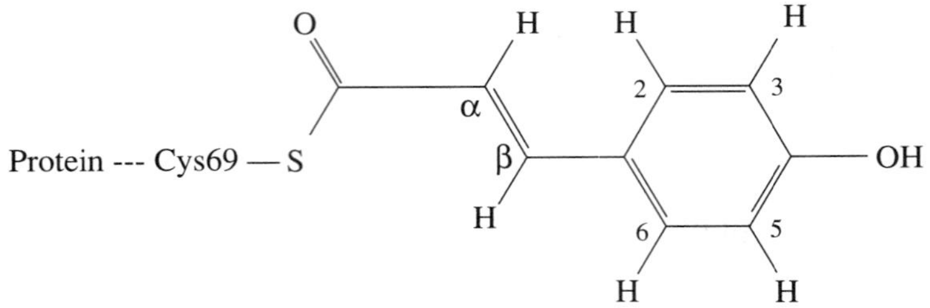

O H H H
2 3
α
Protein --- Cys69 — S β OH
H 6 5
H H

Figure 3.21. The chemical structure of the yellow protein chromophore linked to Cys69 of apoprotein via a thiol ester bond.

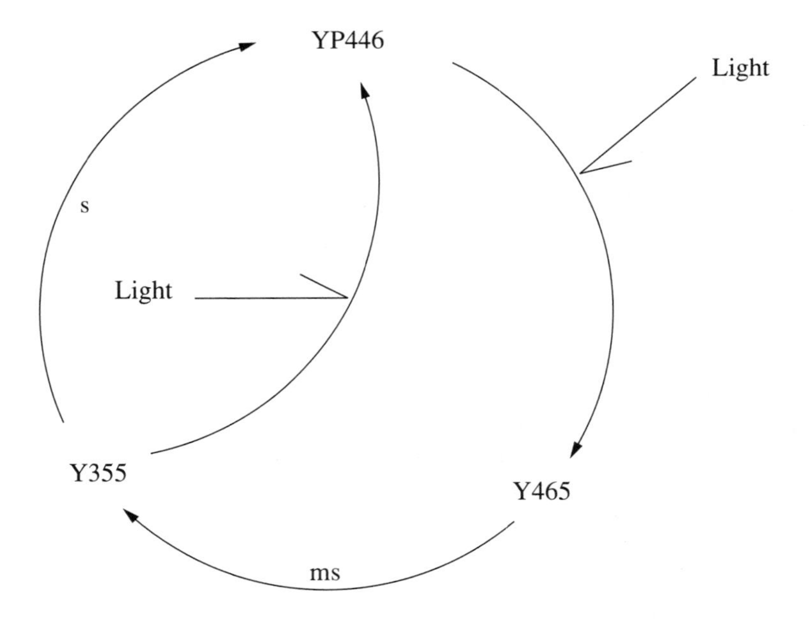

Figure 3.22. Photochemical reaction cycle of photoactive yellow protein from *Ectothiorhodospira halophila* [modified from Hoff et al. 1995].

P6.3 with unit cell dimensions approximately $66 \times 66 \times 41$ Angstroms [McRee et al. 1986].

This photosensitive protein was identified as a photochrome in 1994 [Hoff et al. 1994, 1994a]. It must be responsible for the new type of negative phototaxis of bacteria, since the PYP absorption band is close to the action spectrum of phototaxis. Its chromophore, a p-cumaric acid molecule, is bound to a Cys residue through a thiol ester (See Figure 3.21).

The authors thoroughly studied the PYP photocycle [Hoff et al. 1995]. Figure 3.22 shows an outline of PYP phototaxis. Upon absorption of blue light, *trans-cis*-isomerization of a chromophore occurs.

This results in the formation of a single short-lived form with an absorption maximum at 465 nm and a lifetime less than a ms. Another longer lifetime form with an absorption maximum at 355 nm also appears. This last form is photoreactive and relaxes into the ground state under UV radiation faster than in the dark. During a photocycle, reversible conformational changes occur in the protein. These changes involve less than 30% of protein chains while the process must be unfolding-refolding. Such significant changes must be necessary for generation of the signals which control the organisms response.

3I. A Summary of Properties of BR and BR Variants

This section contains a summary of the primary properties of BR and BR-like proteins distinguishing them from other natural proteins.

BR:

1. Molecular weight 26,000 Da
2. Extinction coefficient BR of 66,000 1/mol cm
   Extinction coefficient M of 45,000 1/mol cm
3. Protonated Schiff base of *all*-trans retinal, attached via Lis-216, embedded inside a highly polar binding site
4. Fully reversible photochemical processes: practically unlimited cyclicity
5. Quantum efficiency of the primary photoreaction around 0.7
6. Extinction coefficient of 63,000 1/mol cm
7. Formation of a 9-*cis*-retinal at acidic pH
8. Stable under light illumination for many years
9. Photosensitive region from 350 to 650 nm.
10. Light-driven proton pump
11. Generates proton-motive force to 300 mV across the membrane
12. Size of one BR molecule: $5 \times 5 \times 5$ nm
13. Preservation of photochemical activity:

— in artificial lipid membranes and gels
— at low temperatures up to 70 K
— in the process of dehydration in liposomes and PM
— pH range from 3.0 to 9.5

PM:

— 2-D-hexagonal crystal from trimers of BR-molecules
— Uniaxially oriented BR-molecules in PM
— PM patches have diameters from 1 to 5 μm and thickness of 5 nm

— Permanent dipole moment
— Stable under the following conditions:
temperature up to 80°C (in a PM suspension), up to 180°C (inserted in a polymeric matrix)
from pH 0.00 to pH 12.00
high ionic strength
presence of more proteases and some organic solvents
mixing with transparent water-soluble and same organic-soluble polymers
dehydration to 0%
mixing with detergents
most hydrophylic polymers

HR:
— Light-driven chloride pump
— Light-regulation of intracellular NaCl concentration
— It has a UV absorption intermediate
— Double photocycle

SR: Two different types SR-I and SR-II:
— SR-I red absorption of ground state
— SR-II blue absorption of ground state

PR:
— Phoborhodopsin from *Natrobacterium pharaonis*, but one is generaly similar to the SR-II from *Halobacterium salinarium*.

Analogs BR:
— Absorption maxima between UV and IR
— Lifetimes of intermediates like M from ms to days
— Self-assembly
— 4-keto-BR: Triple cycle

Variants:
— Very long M lifetime in a water suspension
— Prognostication properties
— Similarity of structure:
(see Figure 3.23)

3J. From Electronics to Bioelectronics

The reader must have noticed how often technical terms have been used in the descripton of the properties of BR, BR analogs, and their derivatives. In a sense, retinal-protein complexes largely resemble modern photoelectronic devices invented over the last 100 years.

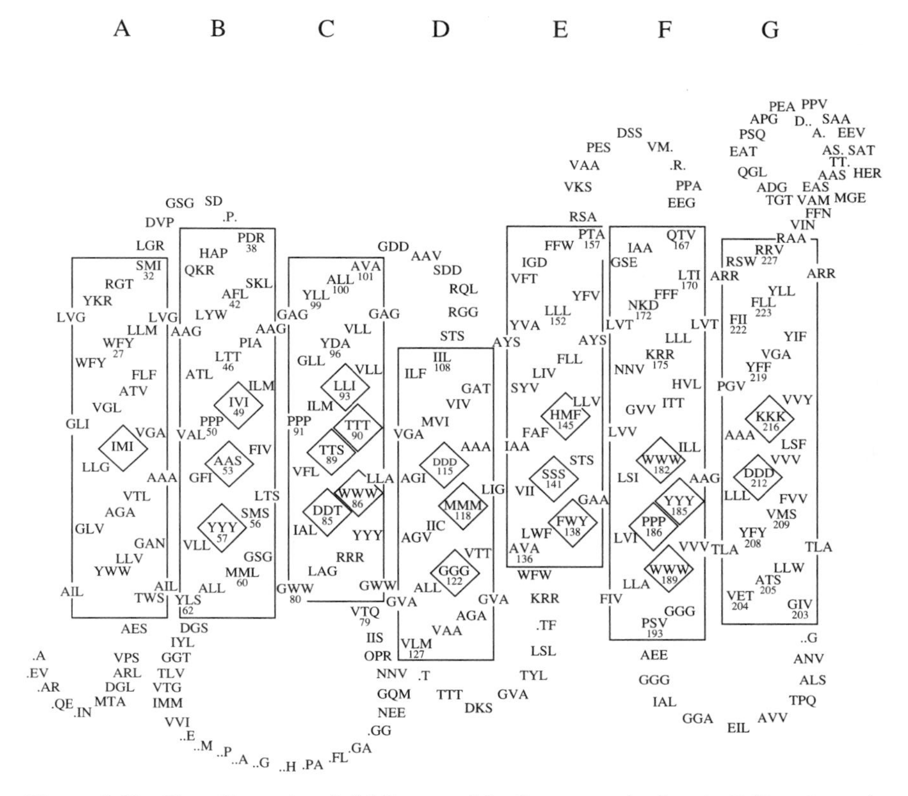

Figure 3.23. Two-dimensional folding model of sensory rhodopsin I (first letters), bacteriorhodopsin (second letters), halorhodopsin (third letters). Numbers of the aminoacid residues and A–G refer to seven transmembrane α-helices corresponding to the BR model [from Bogolmolni and Spudich 1992].

What is demanded of an electronic device component? A 100% efficiency, a high signal to noise ratio, a high speed of signal processing and/or transmission, small size, stability under normal conditions (of humidity and temperature), longevity, and compatibility with the previous generation of components.

Today, nobody argues that BR in the PM, and PM in a polymeric matrix satisfy the requirements for longevity and stability, and are compatible with many modern electronic systems. The time for a hybrid between electronic systems has arrived. Those components that can serve as links in the transfer from electronic to bioelectronic and monoelectronic systems need to be isolated and fully characterized.

Figure 3.24 shows a photodiode with a PM-based film as a photosensitive

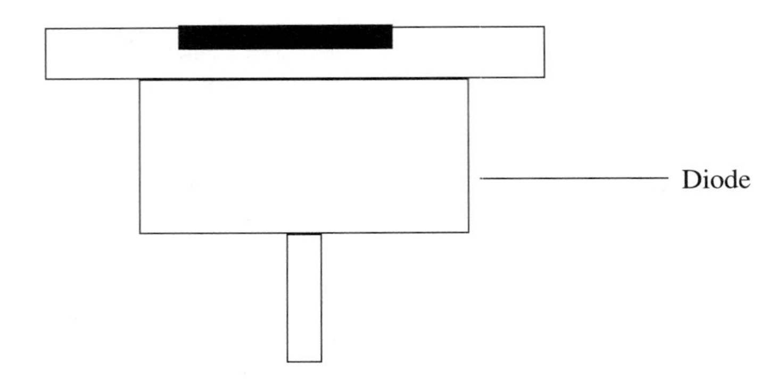

Figure 3.24. Hybrid of a semiconductor diode with bacteriorhodopsin [Trissl H-W 1987].

element. Its response speed to a laser impulse reaches 100 psec [Trissl 1987]. The reader will find other examples in Chapter 5.

Recalling the history of energy converters, from steam engines to vacuum tubes, we see that efficiency increase has always been a function of time. Table 3.9 shows the values of photocurrents generated by BR and PM in different

Table 3.9. Efficiency of BR in artificial systems*

System	Specific resistance (Ohms cm^2)	Photocurrent density (mA/cm^2)	Efficiency
1. PM + membrane from lipids	10^8–10^{10}	1×10^{-6}	10-8–10^{-6}
2. BR + proteoliposomes	2×10^{10}	10^{-6}	2×10^{-6}
3. BR + proteoliposomes + membrane from lipids	10^6	4×10^{-5}	5×10^{-8}
4. PM + gel	2×10^2	5×10^{-3}	5×10^{-8}
5. PM + artificial membranes	10^2	10^{-2}	2×10^{-8}
6. PM dry films by shunting resistance	10^7	1.5×10^{-5}	2×10^{-7}

* From Chamorovsky et al. 1990.

systems and their efficiency [Chamorovskii et al. 1990]. Obviously, such components are not profitable, however, they are not meant for application in their present state. What we have now is a transition period to wholly biological systems. For this reason, parameters like efficiency and signal/noise ratio are not of primary concern now; this is the time for unraveling the general principles governing the behavior light-sensitive proteins and their application to optoelectronics. The fine-tuning will come later.

Despite the cautionary tone of the last paragraph, it should not be forgotten that a photoreceptor that amplifies photon energy by a thousand times, resembles the working mechanism of a modern transistor complete with high efficiency and signal/noise ratio. It still has to be decided whether there are significant differences between the functioning mechanisms of the natural biomolecules, and those of the artificially synthesized organic molecules. It is still not clear what the priority is in biological processes: conformational or charge division processes.

There are many examples of technological areas where BR might play a leading role:

- Biosensors
- Photoelectrical devices
- Photochromic materials
- Laser technology
- Light switching
- Optical memory
- Pattern recognition
- Piezoeffects
- Interferometry
- Dynamic holography
- Reprography
- Neural networks
- Artificial photosynthesis
- ATP generation on reactors
- Desalination of seawater
- Medical diagnosis

In Chapters 5 and 6, we will attempt to show how BR may be employed. The next chapter briefly summarizes the history of photosensitive materials and explains why scientists and technicians search for new photomaterials. The next chapters also offer an explanation why companies producing photomaterials fund this research despite the fact that excellent Kodak films are sold around the globe.

4

Photosensitive Materials for Use as Optical Memory

In the Introduction, we touched upon the main advantages of optical memory against other types. Table 4.1 offers a comparison of different methods of information recording and storage. As a whole, the table shows that optical recording is superior to the others. Only the human brain (and possibly not only human) stays ahead. Optical memory is divided into permanent and temporary categories. Artificial photomaterials for permanent recording of optical information have a rich history, and it can only be regretted that the relation of even one-hundredth of it would fill the size of this chapter. Their younger brother, the reversible photomaterial for temporary recording, has a shorter and somewhat sadder history—since from their very conception they have remained in the shadow of their older brother.

Photomaterials may be produced as films or layers on bases, or even as spatial elements of cubic, and other shapes. Photosensitive composition of photomaterials may be different and can consist of chemical, organic, biochemical, and biological compounds and their mixtures. The most common are, like a century ago, the photomaterials based on the particles of silver halogens. Photopolymers, photothermoplastics and other silver-depleted photomaterials, are more often used in reprography and polygraphy. Non-silver photomaterials owe their existence to technological progress, but their application depends largely on economic factors.

Photomaterials are used in technical systems related to optical information processing and storage. The simplest and best known are cameras. An analogy between photographic cameras and the eye was discussed briefly in Chapter 2. Other devices are less known to the general public, not only because this particular field of research belongs to scientific technology rather than to the popular science domain, but also because many application packages still have not been realized. The following chapter is descriptive and has no references to the literature. Those willing to know more about the history of photography may refer to [James 1967]—in the 30 years that have passed since its publication, nobody has written a better one. For information on photomaterials specifically designed for holographic recording, refer to Gunter and Huignard [1988, 1989].

4A. Photographic Films

The popularity of photography, technnical and artistic, is well-supported. One of the reason for this is the expansion of visual possibilities of the human eye. The main areas of increase are:

1. The ability to register a wider spectral range- from X-rays to the distant infrared (from 10 A to 1 mm), which permits color vision inaccessible to the ordinary eye;
2. Greater information density—the number of elementary details on the photograph may run into millions. Due to its psychomotoric mobility, the eye cannot fixate on such a number of details simultaneously;
3. The ability to register dim light sources. At large exposures, photographic film registers the stars, nondistinguishable by the naked eye;
4. The registration of fast-going processes. Special films of superhigh sensitivity detect microsecond-scale processes. As a reminder, the eye cannot separate two processes if they last less than 0.06 s.

Photographic imaging has many other advantages. Convenience and multiple usage, the simplicity of scale changing and unrestricted copying, geometrical precision and longevity of storage, just to name a few. In cinematography, photofilm may register movements in one time scale and reproduce it in a different one. As optical memory devices, photofilms are able to record and preserve images with more precision and more details than visual or graphical memory. Besides, visual memory is limited by the time of human life, whereas a photograph has no storage restrictions (if in 100 years a photograph bleaches, it can be recopied).

World production of photofilms and photopaper is constantly growing, and the production of affordable videocameras does not slow it down. Its tendency to increase leaves no doubts, that the industry of non-silver and uncommon photomaterials is also expanding (see Section 4A.2). This is due to two factors—the expensiveness of silver, and the new demands for the properties of the new photomaterials applied to reprography, polygraphy, holography, and other more recently invented fields. A third reason is that modern processes of image development and fixation include a wet stage, which is not applicable to most technological devices, and is expensive. Besides, halogen-silver materials create one unsolvable problem: their resolution decreases with an increase of sensitivity. This ratio depends on the size of the silver monocrystals. In other words, sensitivity is proportional to the size of the crystal, but the bigger the crystal the smaller the resolution. It is worth remembering that this dilemma is also valid for many types of non-silver photolayers.

The mentioned disadvantages have not yet affected the popularity of halogen-silver photomaterials. In artistic photography and cinematography, it still holds priority. There are at least four reasons for this:

1. the highest sensitivity among the photomediums,
2. the ability to change sensitivity, resolution, contrast range, and the interval of registered brightness, in a wide range, depending on the needs of the customer,
3. high-quality color and color tint representation,
4. the ability to register radiation of the entire optical range.

On the other hand, new technologies require new photomediums with sometimes unusual properties. For instance, holography requires photofilms with a resolution of 5000–10000 lines per 1 mm. Reprography requires dry copying methods. Big copying demands require a large quantity of cheap photofilms for microfilming, and for photosensitive paper to reprint information in facsimile form. Naturally, economics is also important. The faintest opportunity to replace expensive halogen-silver materials with cheaper ones will be accomplished in a moment. The main types of silver and non-silver photomediums and their applications are shown in Table 4.1. Photochromic materials that do not require developing are capable of multiple regisration of optical information in the exposing process. This class of photoregistrators is of particular interest to bioelectronics and will be discussed in Section 4B.

4A.1. 150 Years of Photography

In 1835, Louis Daguer invented image fixation on a photosensitive layer of iodine silver. In 1839, the description of a photoprocess called a daguerreotype was published. This marked the birth of photography. Strange as it may seem, silver is still the main component of the photolayer, although image recording by means of non-silver processes was attempted first in 1824, and these attempts have continued without success ever since. Let us put the main stages of silver photography development in chronological order:

1835 — invention of the daguerreotype (L. Daguer, France)
1841 — introduction of the positive–negative process and use of the glass base (F. Talbot, England)
1871 — production of halogen–silver emulsion (R. Medoks, England)
1873 — discovery of spectral sensitization (H. Fogel, Germany)
1881 — appearance of chloride–silver photopaper
1882 — appearance of flexible bases (I. Boldirev, Russia)
1897 — first color image is obtained (G. Lipman, U.S.)

Table 4.1. Basic differences between optical and magnetic records

Characteristics	Methods of record	
	Magnetic	Optical
Information density	0.7–0.8 Gbyte	4–5 Gbyte
Record types:		
binary	Yes	Yes
text	No	Yes
graphic	No	Yes
picture	No	Yes
holographic	No	Yes
color	No	Yes
Speed readout	10–20 ms	100–200 ms
Parallel copy	No	Yes
Storage information	5–10 years	40–50 years
Resistance to influence of electromagnetic fields	No	Yes
Cost (relative units)	1.0	0.01

1907 — serial production of color photoplates Autochrom (brothers Lumier, France)
1912 — appearance of color photopaper
1935 — multilayer color photomaterial Kodachrome (U.S.) and Agfacolor (Germany)

In fact, 1935 concludes the century-long era of discoveries and basic inventions in the field of halogen-silver photography, and begins 50 years of improvements of photocameras, photoprocesses, and sensometric characteristics of photomaterials. The history of silver photography is full of paradoxes. For example, the interferential method of Lipman, or Lumier's raster process for color image production, first invented and put into use at the beginning of the century, still surpasses modern methods of color printing in the quality of image reproduction. Prohibitive high costs, and the complexity of the process hindered further development and introduction into general practice. Another oddity is that heliography was invented 10 years before photography by the Frenchman Josef Nieps. In reality, that was the first process of image registration on a photosensitive material. However, it is not regarded as a photographic process since image exposure on a layer of Syrian asphalt gives a relief of image, good for rubbing, not for watching, i.e., Nieps invented lithography.

Non-silver photomaterials and photoprocesses which appeared simultaneously with the silver ones, continue to actively improve and develop. The chronology of these processes is as follows:
1824 — heliography (J. Nieps, France)

1842 — cyanotopy (D. F. Gershel, England)
1923 — diazotopy
1939 — xerography (Ch. Karlson, U.S.)
1950 — wide production of xerographic apparatus
1952 — thermography (process "Thermofax")
1954 — electrography
1958 — vesicular process (R. Notley, U.S.)
1962 — photothermoplastics
1971 — migration process (U. Goff, U.S.)
1974 — electrophotographic organic films
1983 — AMEN process

The chronology of non-silver processes is more diverse and involves a wide range of discoveries in different fields of science, from organic chemistry to modern physics. Non-silver processes and photomaterials lack the universality of silver ones. This may be why in art, and other branches of technical photography, silver processes are without competition. On the other hand, in poligraphy, holography and systems of optical memory, non-silver materials make up to 80% of all applied materials, and in reprograhy—up to 90%. Table 4.2 highlights some of the differences between silver halide and non-silver photomaterials.

Table 4.2. Silver halide and non-silver photomaterials and photoproceses

Process type or material	Sensitivity (J/cm^2)	Sensitivity region (nm)	Resolution (mm^{-1})	# of cycles	Development process
Silver halide					
traditional	10^{-8}–10^{-5}	X-ray–IR	50–500	1	Chemical
holographic	10^{3}–10^{-1}	UV–IR	< 2000	1	Chemical
Dichromate gelatin	10^{-1}	350–500	3000	1	Chemical
Electrographic	10^{-7}–10^{-4}	700	500	10^5	Electric
Thermoplastic	10^{-5}– 10^{-4}	400–1100	< 1000	200	Heat
Electro-optic	10^{-4}–10^{-2}	300–450	500	100	Electric
Magneto-optic	10^{-2}	700	300	100	Electric
Photopolymer	10^{3}–10^{-1}	UV–650	< 1500	1	Chemical
Photoresistor	10^{-2}	UV–500	< 3000	1	Electric
Photorefractive crystals	10^{3}	350–550	10,000	10^3	None
Liquid crystals	10^{-1}	700–IR	10,000	10^3	None
Photochromic					
inorganic	10^{-1}– 10^{2}	450	5000	10^3	None
organic	10^{-2}– 10^{-1}	< 520	5000	100	None
Biochrome–BR	10^{-3} –10^{-2}	350–700	> 10,000	10^6	None

4A.2. The End of Halogen-Silver Films?

Fifty years of non-stop work on halogen-silver photomaterials have not exhausted the possibilities for their improvement but have shifted priorities. Nowadays, the main goal is to lower the initial cost of the material by decreasing the quantity of silver, by reducing the amount needed after image development, and by improving and simplifying the image-obtaining technology. Apart from economic problems, the prospectives for further technological progress have been outlined: optimization of layer sensitivity; high-density packaging; maximum, continous distribution of microcrystals, and utilization of flat or microlens-shaped microcrystals. According to scientific forecasts, these improvements will permit silver materials to remain at the top for another 25 years. After this, the production of silver-based films will be limited to special orders. Silver has no future, and the time is approaching when the smallest amount of silver will make the photofilm unprofitable.

Twenty five years leaves enough time to improve the existent non-silver photomaterials and photoprocesses, and to develop new ones. Some of them have almost fully replaced silver in reprography and systems for copying microfilm. It is worth mentioning that microfilming (micrography) is one of the most effective modern devices for technical processing and storage of visual information. In the fields of polygraphy and holography, the competition for photomaterials is still going on. Let us briefly describe the prospects for non-silver materials application in the mentioned fields:

1. The world annual growth of original microfilms and copies in micrography has reached 30%. In the U.S. alone, up to 40 billion exposures are created, copied, and magnified to the desired size including those on paper. The tendency to increase use of non-silver photomaterials has purely economical grounds. Against the forecasts, the appearance of optical disks barely affected this tendency. Recording density and cost of optical disks and microfilms are almost the same. Hence, advantages and disadvantages are defined by specific customers demands and rarely intersect.
2. In polygraphy, the competition between the two types of photomaterials has been going on for decades. Halogen-silver films are used where the registration of the primary or tone image is required; in all other cases, cheap non-silver films are preferred.
3. For holographic recording, halogen-silver materials are not optimum. Their low (even theoretically) diffraction effectiveness, high granularity, and limited resolution led to the appearance of non-silver photofilms designed specifically for holographic demands. These are: photothermoplastics, photochromes, films on a bichrome gel etc. Unfortunately, the holographic method of information storage still remains within the boundaries of lab research, although theoretically it contains more information than traditional photography.

4. Optical memory systems for CM require photomediums with the highest possible resolution and, frequently, with the highest photochromic properties. As experience shows, non-silver photomaterials have more potential. Nowadays, optical disks with a thin metal layer are inseparable from computer memory. Of course, such a layer can hardly be called a photosensitive medium and is nothing but an additional component of memory. All that accounts for operative or temporal optical memory still remains at the initial stage of development and introduction into industry. Obviously, such an introduction demands reversible (photochromic) photomediums with high optical and technological characteristics. Similar optical elements for temporal and operative memory for CM will be discussed in detail in section 4B.2.

What is the ideal future photomaterial for optical information recording? First, it must register all light and registration medium-related characteristics of a light beam. This task can be performed using at least five methods:

* amplitudal modulation methods based on photoinduced changes in the absorption of the medium. The mechanism of registration begins from a photochemical reaction to the evaporation of thick layers of metal.
* phase modulation methods based on photoinduced changes in the medium's refractive index. The mechanism of registration may also depend on the change in the transparent or nontransparent layer thicknesses, or on magneto-optical and electro-optical effects.
* polarization methods, based on photoinduced dichroism (the conversion of isotopic medium into an anisotropic one), depending on the type of incident light polarization.
* the color method, based on the possibility to obtain numerous color shades from three basic colors. The more color shades the photomedium can register, the more information it will hold.
* registration of light beam direction—possible in any transparent photoregistering mediums shaped as thick layers or spatial elements.

Apparently, the photomaterial able to register optical information by means of all the described methods would be an ideal photoregistering material. If this photomaterial, along with the above mentioned properties, has photochromic as well, the possibilities of its application will be virtually unlimited. The next section offers a survey of photochromism and photochromic medium problems.

4B. Photochromic Materials

Unlike photography, photochromism didn't have to be invented. Many centuries ago, people made the observation that the colors of some organic and inorganic compounds irreversibly change in light. Photochromism as a phenomenon is defined as the reversible, light-induced transformation of a substance, leading to a change in the position or shape of its absorption spectrum. The photochromic cycle follows the reaction sequence:

Diagram 4.1

$$A\ (\lambda_1) \underset{h\nu_2 \text{ or } kT}{\overset{h\nu_1}{\rightleftarrows}} B\ (\lambda_2)$$

A photochromic compound in initial state A, with absorption maximum λ_1 and energy $h\nu_1$, converts into state B with absorption maximum λ_2. After a certain time, which depends on the characteristics of the compound and on the external conditions, the compound returns to initial state A utilizing thermal energy kT. If the B state is photoactive, the return is via energy $h\nu_2$. Timing, the number of direct and return transitions, depends on the type of compound, and the mechanism of photoreaction. Such properties attracted the interest of researchers of photochemical processes, optical information processing, and optoelectronics specialists. Different from the photographic process, the photochromic process involves no amplification; one light quantum excites no more than one photoactive molecule, therefore, the quantum yield of photochromes never exceeds 1 [Hoshino and Koizumi 1972].

By now a numerous number of inorganic, organic, and biological photochromic materials have been discovered and studied. Photochromes can be liquid, gas-like, aerosol-like, amorphous and solid (films, spatial elements, and crystals). Photochromic films and glasses are the most popular photochromes due to their convenience and ease of synthesis.

Application possibilities for photochromic materials are wider than for their photographic counterparts. They can be used in optical computers and other systems of operative information processing. Specifically, for protection of visual organs (including an atomic bomb explosion), improvement of contrast in reprographic works, as optical filters in laser interferometry, etc. (see Section 4B.2). However, all these prospects largely belong to the future. The only pride photochromists have had for years are photochromic glasses called "chameleons." The reason for this discrepancy lies in the history of photochromism studies and photochromic materials development.

4B.1. 35 Years of Photochromic Materials Research

According to historians, Alexander the Great's warriors had to wear special hand bands saturated with a chromophoric vegetable solution. The change of the bands color in the sun signaled the onset of collective and military maneuvers [Brown 1971]. At the start of this century, the reversible color change of a substance was called phototropism, and in 1950, the term photochromism was coined. The year 1960 marked the beginning of serious scientific research on the origin and mechanisms of photochromism. During the past 35 years, an enormous number of compounds and materials have been discovered and synthesized more or less corresponding to the scheme of "photochromism" as shown in Diagram 4.1. The three principal characteristics defining the effectiveness of a chromophoric system are: the absorption spectra of the main (ground state A), the photoinduced (B) forms, their quantum yields, and cyclicity. The closer the quantum yield is to 1, the more sensitive the photochromic material. The greater the spacing of the positions of A and B maxima, the greater is contrast range. The more "write-erase-write" cycles the better. As for resolution, all photochromes, with no exceptions, possess resolution many times higher than the most capable halogen-silver compounds. This is due to the small size of the photochromic molecules which does not exceed a few dozen Angstroms [Jackson 1969].

In the process of development, new demands are constantly made on the properties and optical characteristics of photochromes. For instance, their application in holographic systems requires the additional important parameter of holographic effectiveness. In operative systems for information processing and optical computational machines, the important factor is to minimize the formation and decay times of the A and B forms [Dorion andWiebe 1970]. By the beginning of 1980s, a large number of organic and inorganic photochromes had been studied though researchers failed to find or synthesize a photochrome that could be widely applied to technology. This discouraged many photochrome fans, funds were cut, and many research laboratories closed. It is strange that the researchers failed to find a universal photochrome that would combine all the required characteristics since the short history of photochromism research had witnessed many discoveries about different photochemical processes. The ensuing section will give a general idea about the central photochromic processes discovered and described during 35 years of research.

4B.2. The Physics and Chemistry of Photochromic Processes

The most typical photophysical and photochemical processes in photochromic materials are the following:

* *Cis-trans*-isomerization around various double bonds like $>C=C<$, $-N=N-$ – etc. Stilbenes, indigoid pigments, and azobenzoles are the best studied examples of this type.
* Valence isomerization, resulting from redistribution of p and s electrons

causing inter-atomic angles and bonds to change without altering the binding sites of atomic groups and atoms. An example of this type are the well-studied spiroperans.

* Heterocyclic photodissociation as a reversible break-up of covalent bonds which create discolored forms. Examples of such photochromes are leicoderivatives of triarylmethane pigments and other spirocompounds.
* Heterocyclic dissociation—after the break-up of covalent bonds, couples of radicals are formed. Due to high reactivity, photochromism follows this scheme solely at low temperatures.
* Tautomerization—transport of one atom from one part of the molecule to the other. Sometimes hydrogen transport is followed by the splitting off of a proton resulting in a color change. The number of photochromes following this scheme is large, however they are highly reactive under oxidizing conditions (for example, anilines, pyridines, etc.).
* Photodimerization is observed in anthracene, tetracene, acridisium salts, etc. The advantage of this type of photochrome is the long lifetime of their photoinduced state. They can be quickly returned to the ground state by heating up to 150°C.
* Oxidization-reduction photoreactions are similar to intermolecular electron transport in donor–acceptor complexes, and to reversible electron transport within one molecule. The formation of complexes via charge transport summarizes reactions of this type. Photochromes of this species have not been studied well enough.
* Triplet–triplet absorption also may cause the reversible coloration of some aromatic carbohydrates in a polymeric matrix. The lifetime of triplet states may reach dozens of seconds.

None of the above described processes involves energy conversion , charge migration to neighboring molecules, enzyme catalysis, or any other amplifying reactions. Therefore, quantum yields of such reactions theoretically do not exceed 1, and in reality are much smaller.

After the discovery of photochromism in molecular complexes of biological origin, photochromic materials, and systems were divided into three types—inorganic, organic, and biological. Biological ones are conveniently divided into bio-organic (single molecular complexes, like retinal-protein) and biological (systems like rods and cones of the eye retina, light-response systems, etc.).

4B.3. Organic and Inorganic Photochromic Materials

Inorganic photochromes includes a large group of compounds in the form of thin metal films, crystals, thermoplastics, segnetoelectrics, and glasses (see Table 4.3).

An example of an inorganic photochrome is silver halogenyde incorpo-

Table 4.3. Inorganic photochromic materials

Materials	Energetic sensitivity (J/cm^{-2})	Holographic efficiency (%)	Information storage time	No. of cycles
Magneto-optic	10^{-2}–10^{-1}	$< 10^{-2}$	years	from tens to millions
Photochromic crystals liquid:				
a. Nematic	10^{-4}–10^{-3}	50	years	tens
b. Cholesteric	10^{-2}	50	years	hundreds
Photorefractive crystals:				
a. $LiNbO_3$	1	20	weeks	millions
b. $Bi_{12}SiO_{20}$	10^{-3}	25	weeks	millions
Photothermo plastics	10^{-5}	10	years	tens
Polycrystals	10	1–3	years	hundreds

rated in a glass matrix. Unlike the photographic process, upon absorption of light, and the dissociation of the halogenyde silver molecule, the Cl atom does not diffuse, but remains at the silver atom. This happens because the glass matrix is more rigid than the gelatin one. The appearance of free atoms of silver causes the increase of absorption, and the glass darkens. After the light intensity decreases, the molecule regenerates via a thermoprocess, and the glass lightens.

The main disadvantage of inorganic photochromes is their low energetic sensitivity. Furthermore, their primary absorption maximum lies, as a rule, in the UV range, which does not permit the use of cheap light sources, including helio-neon and semiconductor lasers. Low sensitivity necessitates the use of powerful irradiation which often leads to fast photodissociation of a photomaterial, and decreases its cyclicity.

The class of organic photochromes is practically unlimited because modern organic synthesis is an inexhaustible source of photochromic molecules and complexes. Most photochromic reactions are based on four elementary photochemical reactions—photooxidization-reduction, photodissociation, photoregrouping, and photodimerization. The most common and well characterized organic photochromes are presented in Table 4.4. Not all chinones and chromens have photochromic properties [Chibisov 1966] [Heyman 1969]. Photodimers have high cyclicity and low energetic light-sensitivity [Ferguson 1974]. Spiropiranes, conversely, have good light-sensitivity, but low cyclicity [Smets 1983] [Lisyutenko et al. 1990]. Comparing the characteristics from the tables, we may conclude that no single type of organic

Table 4.4. Organic photochromic materials

Photochromes	Energetic sensitivity (J/cm^2)	Holographic efficiency (%)	Information storage time	No. of cycles
Spiropyrans	10^{-2}	8–10	hours	hundreds
Quinones	1	1	seconds	hundreds
Photodimers	10^{-1}	10^{-1}	days	thousands
Thioindigoids	1	1	months	tens
a-Salicyldena-nilin	10^{-1}	10	minutes	thousands
Stilbenes	10^{-1}	10^{-1}	nanoseconds	millions

or inorganic chromophore has the properties required for technological use.

As for bio-organic chromophores, there are several reliable photochromic systems, one of which is chlorophyll. Photoreaction of chlorophyll during photosynthesis is fully reversible, and the molecule has multiple oxidization-reduction reactions. The primary stage of photosynthesis is easily reproduced under laboratory conditions.

Figure 4.16 shows the absorption spectra of a reaction called "The Krasnovsky Reaction" [Krasnovsky 1966]. Needless to say, it is not only a photochemical reaction, but a good model for a bio-organic photochrome.

The eye rods contain another photochromic bio-organic system. It was not easy to find an application for the photochromic properties of chlorophyll and visual rhodopsin (VR) since their properties are only fully manifested *in vivo*. For a long time, it seemed that direct technological application of bio-organic photochromes was out of the question. However, the situation changed with the arrival of bacterial rhodopsins and retinochrome-like molecules. Before we start the description of their technological capacities, let us survey the problem of photochromic materials at its current stage. We will also explain the question mark in the title of the next chapter.

4B.4. Is There a Future for Photochromic Materials?

In the 1970s, photochromism was proclaimed to have brilliant future. In the 1980s, the interest in photochromes ebbed since the universal photochromic material had not been found. Tables 4.3 and 4.4 clearly show that while having high light sensitivity, spiroperans have low cyclicity, and that the crystals with high cyclicity and long memory have low sensitivity, and so on. The position of the absorption band of most inorganic photochromes in the UV and blue spectral regions further encumbers their practical application. Only halogenyde-silver photochromes are successfully used (photochromic glasses are widely applied to protect eyes from drastic illumination changes). Other great ideas, born in the 35-year period, remain nothing but ideas. Progress had significantly slowed down by 1985. Fortunately, it didn't stop altogether.

What was that brilliant future predicted for photochromes?

The diversity of photochromic materials and their properties discovered before 1985 gave birth to many ideas for technological inventions. Here are the most interesting ones:

1. The means of eye protection from excessive light. Most organic photochromes are colorless in their ground state and color under the action of UV light. Unfortunately, insignificant cyclicity inhibits their application as a protectant. Thus, photochromic halogenyde-silver glasses are without competition.
2. Eye protection from a light flash generated by a nuclear explosion requires the absorption in the visible and UV spectrum ranges, with density no less than 3-D per less than 100 msec. The return to the ground (transparent) state must happen in not more than 10 sec. As far as we know, there hasn't been much progress in the study of this problem. However, the convenience of photochromic materials in this case is self-evident.
3. To protect from laser radiation, absorption must lie in the laser spectral range with optical density at less than 6–9 D in less than a nanosecond. In other words, it is more difficult to protect from the laser radiation than from the nuclear one.
4. Some photochromic types may be used as photomaterials when the image must be obtained in the process of exposure, i.e. at a real time scale. However, the majority of photochromes have the initial absorption spectra in the UV and blue spectral ranges, while upon light exposure, their absorption spectrum shifts to the red range. In this case, the image is negative. Furthermore, this operation requires additional light sources because even the light sensitivity of spiropyrans, the most sensitive photochromes, is two-three orders smaller than the sensitivity of a common photofilm. The fixation of a stable image is not impossible, but complex, and requires additional, frequently compicated and expensive, means of fixation. In photolithography, some organic photochromes are used in the controlling processes [Patent 3869292 U.S. 1977], and in electrography spiroperan-based photochromes are used as sensibilizers [Patent 4029677 U.S. 1977] [Patent 3609093 U.S. 1971].
5. Some photochromic materials are used in actinometry [Patent 3597054 U.S. 1971] and dosimetry [Patent 3609093 U.S. 1971] for measuring intensities and distribution of the UV, X-ray, and gamma-radiation in space.
6. Photochromic films are ideal for measuring a photoimage contrast range.

Their usage replaces and makes cheaper the labor-consuming process of photomasking [Dorion and Wiebe 1970]. An example of a contrast range increase in aerophotography and microfiche is described in Section 6D.

7. It was 1956 when the application of photochromes as elements of optical computers was suggested [Hirshberg 1956]. According to calculations, the use of photochromes as elements of operating memory of extralarge-capacity has many advantages against the traditional systems used in modern computers.
8. Holographic methods of information registering have greater capacity and write–read speed compared to optical recording in binary code (such as photography). They can write up to 100 different holograms on one and the same area of a photosensitive medium. The principles of holographic memory in the creation of photochromic materials has been described in the literature [Mikaelian and Bobrynev 1974]. Devices for recording and selective reading of information on photochromic crystals have been realized [Blume et al. 1974].
9. The principles of dynamic filtration of optical print is used for recognition of images, and for associative search for the assigned information. This method is to a certain extent the imitation of brain activity. Assembly schemes for optical image filtration is described in many works.
10. Photochromic materials-based devices for information visualization and mapping in real time on large screens have a number of advantages: they have molecular scale resolutions; high input/output signal contrast, and the capacity for selective erasing with full or partial replacement of information. There was a time when these devices received a lot of attention, especially from military specialists.
11. The systems described in 10, due to their high resolution, turned out very useful for cartography, project, and draft drawing.
12. Some photochromes can be used for decorative, camouflage, and even cosmetic purposes.

The above examples constitute only a small part of the possible application of photochromic materials. As the reader might have noticed, most of them remained in the form of hypotheses. Why? There are two unresolvable disadvantages of organic and inorganic photochromes that have prevented them from becoming popular. The first of these is the low energetic sensitivity which is one order lower than the theoretical one, and three orders lower than the photographic one. The second disadvantage is their low cyclicity, i.e., a low resistance to nonreversible reactions. This is why after 35 years of persistent research of hundreds of photochromic types, and in spite of numerous interesting and useful suggestions for their application, the today of photochromes has no tomorrow.

However, the question mark in the title of this chapter signifies hope, the hope inspired by bio-organic photochromes which appeared upon the waning of interest in organic and inorganic photochromes. This lack of interest undercut the development of bio-organic photochromes. Too much effort and money had been spent during these 35 years, and the rekindling of interest will take a lot. All the above goes into explaining why the appearance of biochromes has remained virtually unnoticed.

4C. Construction of Layered Purple Membranes

A single purple membrane may be regarded as a photosensitive film of micron scale and monomolecular thickness. To make a photolayer or a photofilm of any size out of such fragments takes a bit of work. The easiest way is to pour the PM suspension onto a glass plate and let it dry. Upon image exposure, the print is positive. Lifetime of the print depends on many factors (see Section 4D for details). PM may be also regarded as a microphotoelement, i.e., as a light converter into electric potential and/or ion current (ion pump). The base dried PM suspension forms a macrophotoelement. Its photoelectric, as well as photographic effectiveness depends on many factors: on the type of base, on the method of imposing the PM suspension upon it, on the temperature and humidity of the surrounding medium, are some examples. As we frequently emphasized, the properties of BR molecules in the membrane of a live bacteria are preserved while they stay in the PM in the form of a crystalline monolayer. Thus, photolayers from the PM can function indefinitely long. Herein lies the interest paid to PM films since the first years of BR research.

4C.1. Air-dried Layers on Supports

A drop of water suspension of PM, dried on a glass, can serve as a model for the simplest photochromic film. The photochemical cycle of the BR molecules in such films proceeds without interruption, and its lifetime depends solely on the humidity. Upon gradual water evaporation, and gravity, the plane of each PM orients almost in the base plane. The number of membranes oriented with their external side to the base approximately equals the number of those oriented with their cytoplasmatic side to the base. Such films are convenient models for the study of noncooperative optical effects. PM orientation can be made more parallel by centrifuging [Clark et al. 1980]. This method permits the obtainment of dense films, pills or even blocks which may lose their photochemical and mechanical properties (i.e., they may crack and fall from the base) because of low humidity and, especially, at many g.

Photoelectric response and electrochromic effects are difficult to observe in films with nonoriented PMs.

4C.2. Layers with Oriented Purple Membranes

Orientation is not important for the studies of spectral transformations in PM films, but it is for measurement of photoelectric or electrochromic properties of BR in mono or multi-layer films. If there is cross orientation of equal numbers of participating BR molecules, the cooperative effect disappears.

All methods of PM orientation can be divided in two groups, those with and those without an external electric field. Between 1974 and 1984, about 10 methods of uni-directional orientation of PMs were developed:

* on black membranes, i.e., in thin artificial lipid membranes [Drachev et al. 1974]
* at electrically charged black membranes [Dancshazy and Karvaly 1976]
* at thin teflon films [Trissl and Montal 1977]
* at a lipid-impregnated filters [Drachev et al. 1978]
* in an electric field, or transiently in a low ionic strength solution [Keszthelyi 1980]
* dried on a conductive glass [Varo, 1981] [Varo and Keszthelí 1983]
* absorbed onto a planar lipid bilayer [Fahr et al. 1981]
* between heptane/water interfaces [Trissl 1985]
* by the method of electrophoresis [Maksimichev et al. 1984]
* immobilized in polyacrylamide gel [Der et al. 1985]

The possibility of oriented BR introduction into thin artificial lipid membranes was first demonstrated by Skulachev's group [Drachev LA, et al. 1974]. The authors used the PMs incorporated into proteoliposomes. This structure (proteoliposome-lipid membrane) showed good orientational characteristics, but was unstable and had a short lifetime. The problem of stabilization was resolved in different ways: for instance, by the addition of polystyrene [Shieh and Packer 1976] or using an orientation created by using an electrically charged black membrane [Dancshazy and Karvaly 1976]. Such films were convenient for registering photopotential characteristics, but the photopotential did not exceed dozens of mV which meant that not all the PMs were oriented correctly [Ormos et al. 1978].

The photolayers with optimum oriented PM were obtained using electrophoretic deposition of a PM water suspension on a conductive base. These are the most effective films, and they are very convenient for the study of the external field influence on the structural and functional properties of the BR molecules. Electrophoretic mobility of PM was demonstrated [Kantcheva et al. 1982]. The method of production of such films was first suggested in 1984 by Maksimichev and colleagues. It is based on the ability of the PM fragments to orient in an external electric field [Keszthely 1980]. Figure 4.1 shows the cell for the production of highly oriented PM fragment films.

Glass plates with a transparent electroconductive SnO_2 coating are used as a base and serves simultaneously as the cathode. The other electroconductive

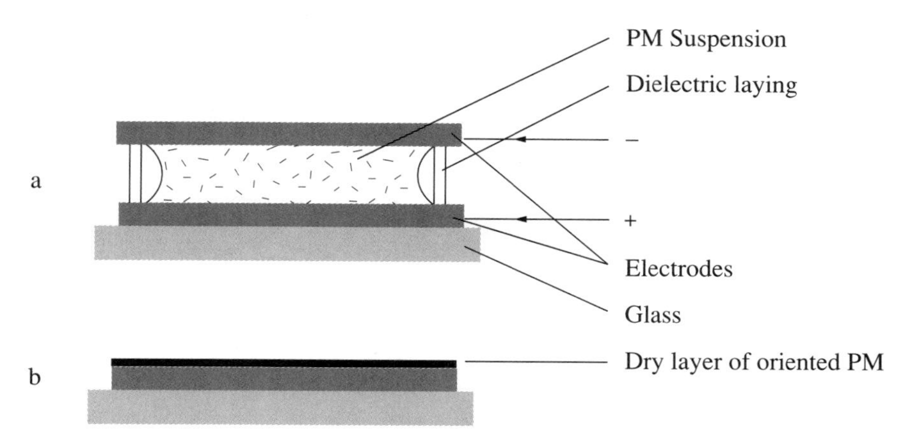

Figure 4.1. (a) A plan to obtain highly oriented PM multilayers using electrophoretic lodging from a de-ionized suspension at the conductive surfaces. (b) After the sedimentation of the PM at the semitransparent conductive surface and drying, the second electrode is applied or coated by dusting at the dry layer of highly-oriented PM.

plate, 1 mm from the cathode plate, serves as the anode. The predeionized and dark-adapted PM suspension is incorporated between the plates. A potential of not more than 3–6 V is applied to the plates. Upon electrophoresis, the unidirected layer of PM forms depending on the potential on the upper or bottom plate (electrode). The PM suspension must be deionized with an ion concentration not exceeding 10 μm otherwise constant dipolar moment compensation occurs on each PM, and their orientational abilities vanish. The resulting layer of oriented PMs is dried at room temperature and 60–70% humidity. As a rule, the square surface of the film is not greater than 1–2 cm^2 at a thickness of 10–12 μM. At this thickness, optical density reaches 2.0 D at 570 nm. The form of the absorption spectrum is similar to native BR.

For experimental purposes, the surface of the film is sprayed with aluminum or nickel to form a thin layer which acts as the second measuring electrode (the first is the conductive layer of the base). These preparations provide beautiful results for the studies of electrochromic and photoelectric properties of the BR molecules. The electric scheme of measurements is shown in Figure 4.2.

In the dark, the potential difference between the conductive base and the electrode, pressed to the surface of the oriented PM layer equals 1V, and in the light it reaches 10V in some preparations [Maksimichev et al. 1984]; which is 1–2 orders greater than in nonoriented layers. Electrophoretic orientation is higher than any other method [Nagy 1978] [Hwang et al. 1977]. An action spectrum coincides with the absorption spectrum of dry PM films. The action of a specifically directed external electric field causes bathochromic shift of the absorption maximum towards 630 nm. Figure 4.3 shows the kinetics of

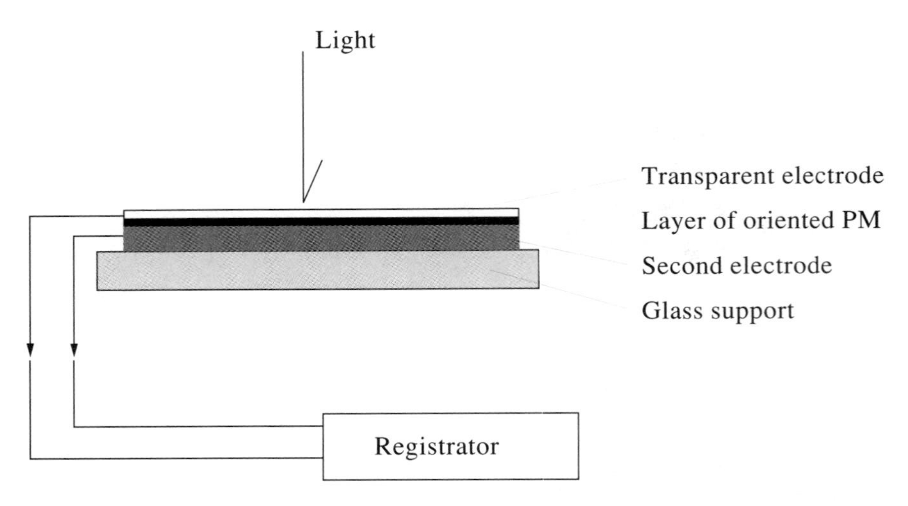

Figure 4.2. A photoelectric design based on a layer of highly-oriented PM.

photoinduced signals for oriented films. The movement of a proton in such films is, most likely, directed but localized, due to the dehydration of PM in the film. This indirectly confirms that the speed of M relaxation significantly depends on the direction of the applied electric field.

4C.3. Purple Membranes in a Polymeric Matrix

Solutions of water-soluble polymers with a PM suspension, mixed in a certain proportion, provide a mixture that permits easy production of technologically convenient films on solid or flexible bases. The force of gravity in a liquid layer of polymer, and the process of drying makes PM orient multidirectionally along the surface of the base. The result is a temporally stable photosensitive film whose mechanical properties depend only on the material of matrix, and whose photochromic properties depend on the type of matrix and many other factors like humidity, pH, chemical additions, etc., [Djukova et al. 1985]. It is also possible to produce films based on water-non-soluble polymers; however, this process is more complex, for it frequently requires high temperature conditions, while the possibility of partial denaturation of opsin is not excluded [Zubov et al. 1987].

As we said in the previous chapters, the films with unidirected PM orientation may be produced only in electric [Korenstein and Hess 1982] or magnetic [Neugebauer et al. 1977] fields. In a polyacrylamide gel placed in an electric field, the polarization of the PM is satisfactorily high. The polymerization of the gel results in the formation of a rigid film, or block, with unidirected immobilized PM [Der et al. 1985]. The kinetics of photocurrent with high temporary resolutions are fixated on these blocks [Liu and Ebrey 1988].

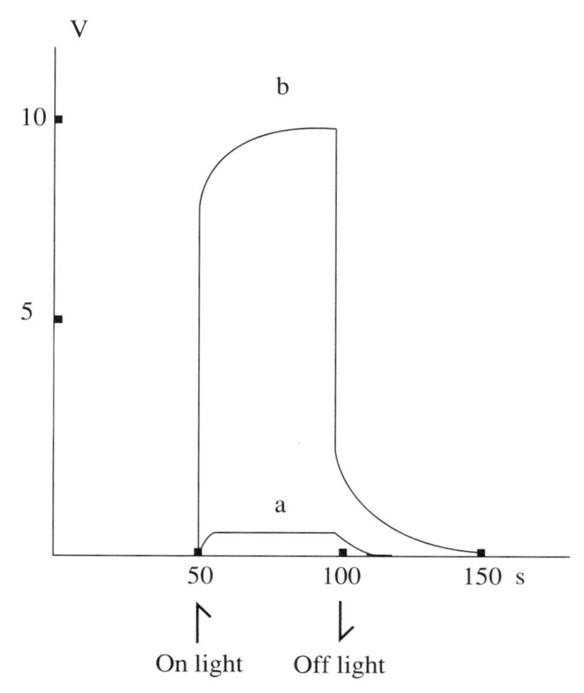

Figure 4.3. Photoinduced signals: (a) in non-oriented PM layers (b) in oriented PM layers [Maksimichev et al. 1981].

At present, we know about ten methods for the production of photoactive BR molecules incorporated in polymeric layers. They can serve as technical components in many optoelectronic systems. Characteristics and prospectives for application of a few of them will be discussed in the following sections.

4D. Biochrome: A Bacteriorhodopsin-based Photochromic Film

To distinguish the photochromic material based on photosensitive proteins and protein complexes from the known photochromes of inorganic and organic origins, we suggested the name of "biochrome." The prefix "photo" is omitted because in these proteins, both *in vivo* and *in vitro*, photochromic precesses proceed not necessarily under illumination. Biochromic material based on BR molecules will be called Biochrome-BR. In the very first years of Biochrome-BR research, an important distinction of this film was discovered: unlike other photoregistering mediums, it can simultaneously fixate all parameters of a light beam (amplitude, phase, frequency, and polarization). Besides, it turned out that Biochrome-BR had no competitors

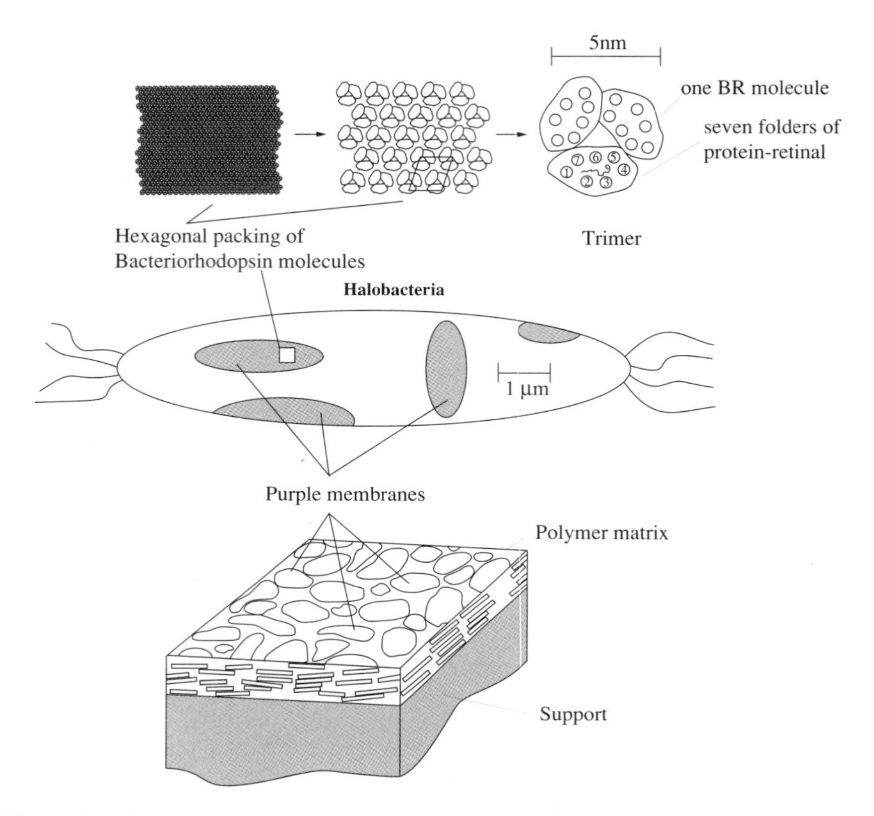

Figure 4.4. Some close-up views of Biochrome film. As the upper part of the picture shows, the photosensitive layer is based on the regions of cytoplasmic surface of *Halobacterium halobium* called purple membranes (PM).

by such parameters as energetic sensitivity and cyclicity (see Section. 4D.2).

The technology of Biochrome films production is simple and resembles that of common photofilms and photoplates with halogen-silver emulsion. Let us describe two laboratory methods of their preparation:

The first and the simplest one: a liquid polymer solution (polyvinylalcohol, gelatin, polyvinylpyrolidone) is mixed with a magnetic mixer and a water solution of PM for 5–7 minutes at 30–35°C. If necessary, chemical additions are added to the suspension or into the mixture. Mixing continues for another 5–10 minutes. The mixture is then pasted on the base with the help of an applicator, or is poured onto a preassigned surface. The base is slowly cooled to room temperature, and the layer dries at normal humidity (50–60%) for 10–30 hours. Schematic sections of the produced film is shown in Figure 4.4.

The drying time depends on many factors like polymer concentration, thickness of a layer, type of chemical additions, humidity, and temperature.

The thickness of the layer depends on the polymer concentration and can be pre-determined. For gelatin, it lies in the range from units of microns to 100–200 μm. Optical density depends on the PM concentration in the mixture and on the thickness of the layer. At maximum (200 μm) thickness, it does not exceed 2–2.5 D. PMs distribute parallel to the base in the process of dehydration which is confirmed by electron microscopic photographs (See Figure 4.5). The quality of the external surface of the film is not high.

The second, more complex technology is applied when a higher quality of

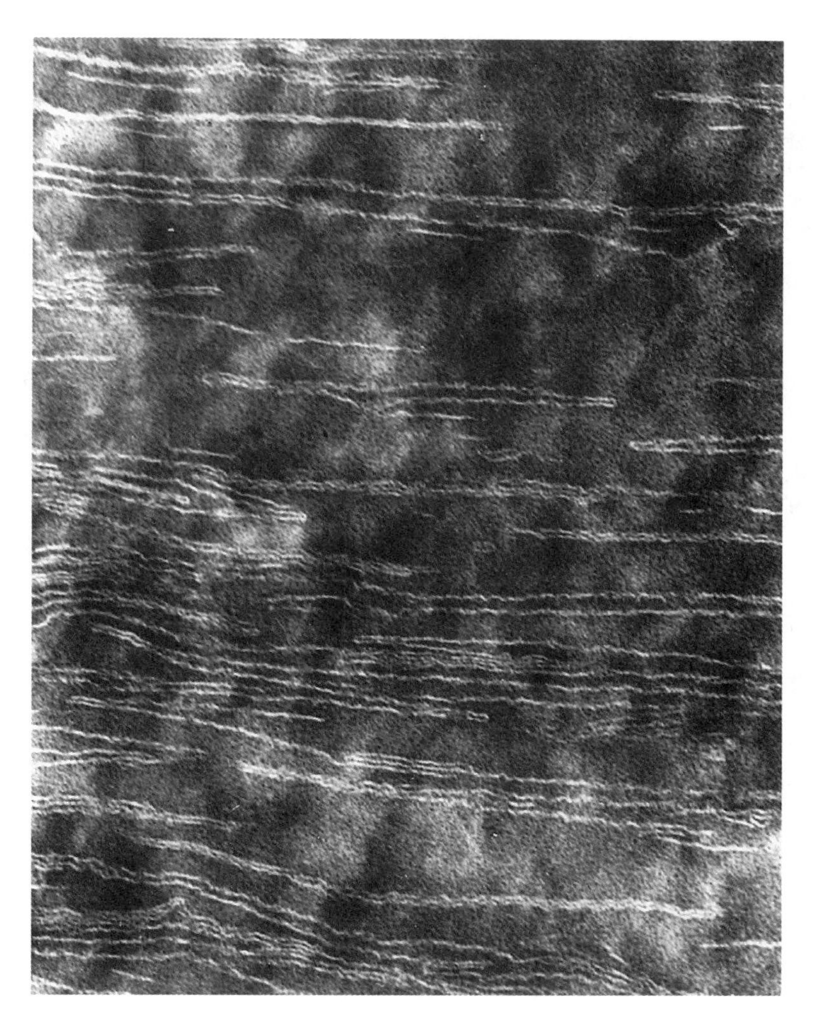

Figure 4.5. Distribution of PM fragments in a gel layer. The orientation of every PM plane, in the plane of the base, occurs automatically in the process of drying the PM suspension, or its polymeric mixture [photograph courtesy of Viktor Popov].

the external film surface is required (in holographic research). The initial mixture is prepared as described , but gelatin is used only as a polymer. The mixture is then inserted between two glass plates with high surface quality. One of the plates (most frequently, the upper one) must have a hydrophobic coating (for instance, a thin layer of silan). Instead of the glass plates, any other highly flat-parallel, hydrophobic material can be used. After the addition of the mixture, both plates are cooled to 8–10°C. In a few hours, the so-called casting phase occurs and the gelatin and upper hydrophobic plate are removed. The film is dried as described above. Figure 4.6 shows the scheme of a photochromic process in Biochrome-BR film. It differs from the traditional scheme of a photochemical process in BR by its formation times, and the M intermediate lifetime. Other optical characteristics of biochromic films depend on the mechanical and optical characteristics of the matrix.

The technologies of biochromic BR-based films preparation are described in patents [Vsevolodov et al. 1983] [Dyukova and Vsevolodov 1996]. Other types of commercially sold photochromic films were developed by Robert Birge and George Rayfield in the U.S., and Norbert Hampp in Germany. Attempts to invent a BR-based photographic film have not yet succeeded [Vsevolodov et al. 1991].

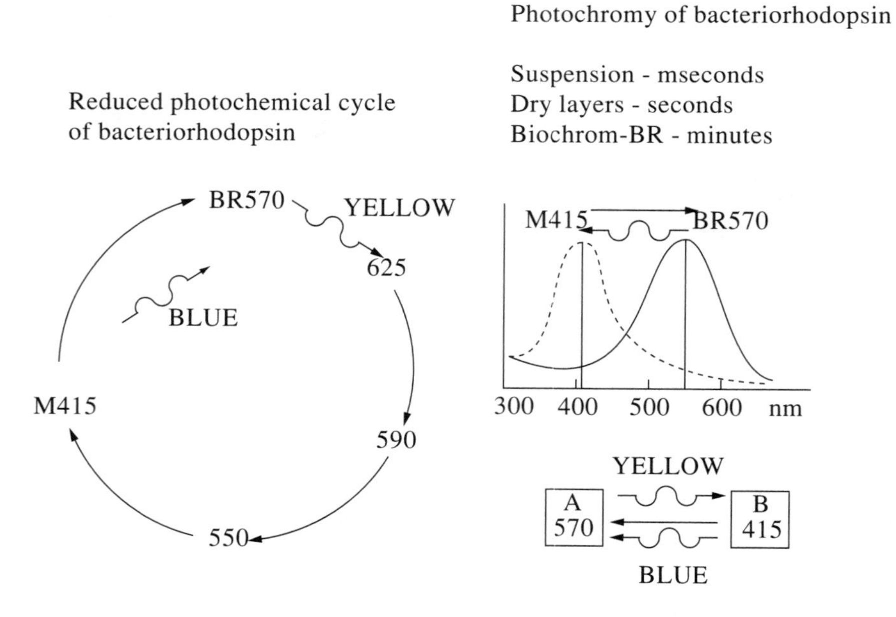

Figure 4.6. The photocycle and spectra of the primary and photoinduced form of the photochromic process in Biochrome-BR film. The time of decay of M form in the dark depends on the optical memory of the film, the degree of dehydration, polymer type, chemical additions, and other conditions.

4D.1. The Discovery of Biochromic Films

The story of the appearance of the first biochromic films, like the discovery of BR (see Section 2C.1), was largely the result of a coincidence. I worked at the Institute of Biophysics, the USSR Academy of Sciences (see Section 2C.2), and from 1975 onwards was actively studying a PM suspension photochemical cycle using a huge flash-photolysis aggregate which I had assembled. From the works of Oesterhelt and Stoeckenius, I knew that the addition of ethers into a PM suspension prolongs the lifetime of the BR photocycle. I distinctly remember a student, Feodor Isangalin, walking from one room to the other, demonstrating a tube filled with violet liquid which turned bright-yellow when he brought it up to the lamp, and resumed its violet color when he protected it with his hand from the light. The liquid was an ether-saturated PM suspension. The photochromic effect was obvious. At that time, however, nobody paid attention to that. Before 1976, I was involved in the routine research of photochemical properties of BR in the form of a PM suspension [Kayushin et al. 1974] [Vsevolodov and Kayushin 1976] [Vsevolodov et al. 1974]. In 1977, I got interested in the study of the M intermediate of the BR photocycle in dry PM films. Suspensions were coated onto a slide and placed into a dish where external conditions humidity, vapor saturation of volatile organic liquids, temperature, and other parameters could be easily altered. In ether vapors, the photocycle slowed down and shifts of amplitude and kinetic characteristics of M intermediates could be observed in the second, or even minute range. Instead of a powerful xenon flash-lamp, we used a common filament lamp with electromechanical shutter. However, after awhile, the overdried film would crack and become loose. After a few months of torture, I decided to fixate the film with some polymer.

At least two coincidences helped me at this point. In my search for a polymer, I wandered into the lab of Boris Gavriliuk who at that time was looking for a new replacement material for burnt and injured human skin and, therefore, had to have a collection of various water-soluble polymers. It was late in the day, and he was dressed in a street jacket with many pockets. I vividly remember how, upon listening to my request, he produced a flat flask out of the inside pocket, took a tube from the shelf, poured some transparent thick liquid from the flask into the tube and handed it to me. This was a concentrated solution of polyvinyl alcohol. Why at that very moment it happened to be in the flask, and the flask inside his pocket, remained a mystery. That was the first coincidence. The polymer immediately improved the quality of the experiments. Polymer based films were resistant to external influences, convenient for the introduction of different reagents, and preserved the photochemical properties of BR for unlimited time. The results were easily replicable.

The second coincidence happened in spring 1978 during the study of the microstructure of the film. I was using a common optical microscope where the lighting didn't work right. The light was so weak that to distinguish anything was totally impossible. While trying to fix the lighting, I was taking the slide with the film in and out, and noticed at some point a small yellow spot in the center of the film which disappeared in a few seconds. I repeated the "experiment" a few times: the spot appeared if the film remained in the microscope no less than 5 seconds. Evidently, the film behaved photochromically. However, this was not the main point. Photochromism of BR molecules had been known since the first publication of a reversible photocycle scheme. What was new is that the photochromic effect was manifested in the film at a very small light-excitation intensity.

I demonstrated a few films to Valery Barachevskii, the best specialist on photochromism in the former USSR, whose book I had read before our meeting. The first testing of a new photochromic film following the standard testing procedure was conducted in his laboratory. To everybody's surprise, the films showed excellent photochromic properties and good holographic characteristics. I was lucky to meet Barachevskii. During the first few years, he was the only specialist in the USSR who immediately had faith in the new photochromic material and held it in high esteem. Barachevskii's viewpoint at that time, followed the unwritten law that technological innovations could come only from the West, was very important. Some of my colleagues at that time looked at my results with disbelief and attempted to convince me to drop this study. The first-year results turned out so interesting that Genrikh Ivanitsky, our director, included my work into the leading group. I got a group of young assistants, and the work rushed forward at double speed.

Curious things started to happen when we attempted to get a patent for the new film. The USSR Patent Bureau asked for a reference to a prototype which didn't exist. My friend, a high-class chemist Boris Mitsner (at present in the U.S.) suggested synthesizing a few retinal analogs in order to reconstruct several BR analogs on their base, obtain new films from those, and receive a patent for BR-analogs-based films; take that patent as a prototype and, eventually, obtain a patent for the native BR film.

This was accomplished in 2 years [Vsevolodov et al. 1983, 1985]. Since then, the studies of the optical properties of films have not met serious obstacles, save the psychological barrier, which people tend to experience when encountering something new, and which some of my colleagues have not yet overcome.

By 1980, we had a large array of photochromic films, both BR- and BR analogs-based. However, I saw that one small group like ours couldn't carry out the whole amount of research, particularly in the aspect of optoelectronic and holographic properties. That's why, having not yet received a single patent for our films, we started offering our preparations to all the largest institutes and universities of the former USSR that had previously expressed a spark of interest in them.

The researchers of the Moscow Institute of General Physics were the first ones to produce our films, and their preliminary results were published in 1981 [Bunkin et al.1981]. However, it was at the Kiev Institute of Physics of the Ukrainian Academy of Sciences, at the holographic lab of Dr. Marat Soskin, where the films attracted real and continuous interest. The first results on the holographic properties of biochromes were obtained in this lab. The beginning of this work was preceeded by my meeting in the fall of 1982 at one of the scientific conferences with a young physicist, Helena Korchemskaya, who had just joined Marat Soskin's group upon her graduation from Kiev University in 1981. She started active research on the holographic properties of films. In the first 2 years, her colleagues in the lab did not express any interest in the new holographic medium. Some even laughed at her preoccupation with some membrane scales which did not deter them from helping their young colleague. The atmosphere of free scientific creativity, created by Marat Soskin, played a positive role, and in 1983, the first paper on the photoinduced anisotropy of BR was published [Burykin et al. 1983]. Many years later, she wrote me a letter, saying that at that time her lack of scientific experience had helped her not to pay attention to the horrible quality of the first preparations, not to be bothered by the disbelief of scientific authorities, and eventually had allowed her to demonstrate the ability of the membrane scale films to full spatial polarity reversal of the wave front. This was a significant result. Since then, the attitude toward biochromes began to shift, and gradually she succeeded in luring most of her colleagues into her work.

Once again, the above-mentioned rule was confirmed: the best results belong to those who do not posses enough knowledge to prove to themselves the impossibility of these results. The results were so amazing that our films soon became the object of studies at many research institutes of the former SU, both academic and military ones. At that time, the academician Yuri Ovchinnikov, well known to bio-organic researchers all over the world, created an academic project, called "Rhodopsin," which was funded by the state (see Section 2C.2). Our group received additional financing and welcomed a new colleague, a young biochemist Tatyana Dyukova, whose scientific career since then has been inseparable from biochromic research. It took time for the name Biochrome to take root. All my friends knew of my passion for the sound of names. First I called the films Biophoto, hoping to invent a universal photographic material. Later I came to like Biochrome better, and started using it first in scientific presentations, and then in papers. Robert Birge wrote about the Rhodopsin project in *Scientific American*: "Many aspects of this ambitious project are still considered military secrets and may never be revealed. We do know that the Soviet military made microfiche films, called Biochrome, out of bacteriorhodopsin" [Birge 1995]. Truly, at first it seemed that Biochrome could be applied in microfilming as a microfiche of the highest resolution. However, our experiments of 1982, at one of the Soviet military institutes, proved it too expensive for these purposes—but ideal if used as an intermediate carrier to improve the contrast of old microfiche films. This work, like many others, had to be halted for reasons related below.

The research on bacteriorhodopsin as a possible component of microelectronic devices was, naturally enough, curtained by secrecy. Normal scientific espionage was developing between the U.S. and the USSR. I was far from playing a patriot, but never rejected new strains of halobacteria or PM suspensions, secretly smuggled from the U.S. and Germany by the Soviet Intelligence Service. I knew from Ovchinnikov, that American intelligence workers were doing the same to the USSR. In 1992, when I started working in the U.S. on a contract with The National Institute of Standards and Technology, I was not surprised to learn that our principal investigator, Howard Weetall, had a visitor from the FBI, looking for information about the reasons for our arrival. Moreover, my American colleagues were not hiding the fact that the composition and technology of Biochrome preparation had been stolen from us in the 1980s. These games never gave me negative emotions. The more scientists are able to work in the field, the faster the research moves forward. As for the priority, if one is worried about that, it is defined by the time of publication...

Our work on BR had particular appeal to both the U.S and Soviet military. It was no secret that Moscow was circled by many institutes doing military research. One of them had an agreement with us to create some sort of overcoat which had a photochromic BR layer. Such a layer would function as camouflage, being earth colored but in the event of a nuclear detonation the BR layer would become transparent leaving only the white surface of the coat to reflect the light. This would last for only a few seconds after which the BR reverts back to its protective earthy color.

By this stage in our research we had already generated BR solutions of various colors. Upon mixing these solutions we were able to obtain a solution of dirtyish color which became almost transparent under high intensity light. We began working on this project but due to the secrecy surrounding this military institute, I was never able to be present at the simulated detonations the lab was conducting. According to the researchers at that institute, the initial results were inspiring. However, elementary calculations showed that the price of such a protective coating would have been larger than that of a coat made of gold! For this reason, it was decided that the first prototype would be presented for testing to one of the Russian James Bonds. This exotic direction of our research came to an abrupt halt after certain political and managerial shifts had occured at the Institute of Biophysics.

The majority of the rumors about BR military applications in the U.S and USSR remained unconfirmed. For example, in Russia I often heard that the famous American spy plane, undetectable by radar, the Stealth bomber is covered by a BR layer. I have not heard this since I have been in the states. Ironically, back in 1992, Howard Weetall of NIST, was asked by an FBI agent whether or not BR was produced by the kilo in Russia. According to my estimate, the coating of one Stealth would require no less 10–20 kilos of dry weight BR, but this of course must be a coincidence.

According to another rumor, this time emanating from American sources that a synthetic aperture radar optical processor using BR was constructed in the USSR. I have never heard anything about this, but it is possible that this rumor originated from some articles in popular science magazines published in Russian and mistranslated. For example, back in 1984, I wrote a paper for a Russian journal called *Microprocessor Systems and Resources*. The paper was devoted to BR applications toward creating a biocomputer and it contains a figure with a 3-D processor based on BR cube [Vsevolodov 1984]. This figure was no more than an idea at the time. Incidentally such a cube was actually created probably independant from my publication by Robert Birge 7 years later [Birge and Govender 1991].

For some reason, nobody speculates in the U.S. about our attempt in Russia to create large optical disks (4–5 mm in diameter) for archival conservation of databases and processing. This work was begun in collaboration with one of the Ukranian institutions. This work was real and generated great interest from the military and lasted from 1980–1985 at which point it was destroyed by the same evil forces I have already described.

In the period from 1979 to 1992, we experienced many rises and falls, hopes and disappointments, sometimes of a strange and inexplicable nature. For example, academician Ovchinnikov, who had been very helpful at the beginning, suddenly quarreled with our director, Genrikh Ivanitsky, and, using his uncontrollable power, attempted to destroy our Institute. In the course of this meaningless and ambitious action, our group lost the opportunity to receive equipment and chemical reagents from the enormous reserves of Ovchinnikov's Institute, and citations of our paper vanished from the journal that were under his control. It is still unclear to me, why in 1987, after the fall of director Ivanitsky, in the hopeless battle against the all-powerful Ovchinnikov, the new director Evgeny Fesenko who was far from being the last specialist in the field of photochemistry of visual rhodopsin [Fesenko 1985], started his new appointment by closing all our grant themes, thus cutting off the funding. In my opinion, this was another meaningless and ambitious action. Three years of his rule attenuated our research for a few years.

Nevertheless, all these misfortunes, which are, as I now realize, typical for Western science, couldn't stop the expansion of scientific interest in biochromes. Beginning in 1985, the independent researchers of photochromic films based on BR and its variants started appearing in the West. Felix Hong and Robert Birge in the U.S., Norbert Hampp in Germany, Aaron Lewis in Israel, just to name a few. Thus, soviet scientists Ovchinnikov and Fesenko, as it had many times happened before in Stalin's times, succeeded in harming only Soviet science and, fortunately, not global science.

4D.2. Optical Characteristics of Biochrome

The properties of biochromic films were first measured by Vladimir Kozenkov, a researcher at Barachevsky's lab, in 1979 (report of testing of BR films, nonofficial publication). The results showed unexpectedly high values for sensometric characteristics.

Energetic sensitivity was not worse than in the best spiroperan-based photochromes, and cyclicity was by orders higher. Furthermore, the absorption band of BR covered the red region, which from the first days permitted use of a low-power helio-neon laser for holographic research. In addition to its optical characteristics, biochromic film proved extraordinarily (for a protein material) long-living and resistant against various external influences. The first films, preserved since 1978, still have not lost their properties. The process of film preparation is relatively simple, ecologically clean and absolutely harmless, while halophils are not pathogenic. The biomass of halophilic cells is grown in standard bioreactors, resembling the cultivators for the production of green seaweed. For the production of films, the standard technology for halogen-silver photofilms and photoplates production is used.

Table 4.6 shows comparative characteristics of the most famous inorganic and organic photochromes, and of Biochrome-BR. Let us review the definitions of the main optical characteristics of all photochromic materials:

1. Spectral light-sensitivity is determined by the absorption spectra boundaries of the ground (A) and the photoinduced (B) forms within the limits of which photochromic processes are induced.
2. Energetic sensitivity is measured in J/cm^2 (or, *vice versa*, cm^2/J) and determines the energetic quantity of illuminance necessary to obtain the difference of optical density 1-D at the wavelength of registration.

Table 4.5 Characteristics of Biochrome-BR film

Parameters	Values	Remarks
Photosensitivity between 460–660 nm	~ 0.8 mJ/cm^{-2}	Depends on chemicals added
Resolution	>250 lines/mm	
Spatial resolution	>5000 lines/mm	Holographic date
Diffraction efficiency	1–7%	Depends on optical density
Cyclicity	>1000000	
Time of info storage	from 0.01 s to 10 min	Depends on matrix and supplements added
Contrast	$K = 0.67$ ($\lambda = 412$ nm)	
	$K = 0.95$ ($\lambda = 570$ nm)	$K = \Delta D_\lambda / D_\lambda$
Shelf-life	$>$ 10 years	Under normal conditions

Table 4.6. Comparison of photomaterials for holography (characteristics are given for their maximum values)

Material	Spectral range (nm)	Sensitivity (mJ/cm^2)	Resolution (lines/mm)	Diffraction efficiency (%)
Silver halide	400–700	10^{-5}–10^{-1}	< 500	5–6
Dichromate	350–500	10^2	> 10^3	100
Crystals	350–500	10^3–10^0	10^3–10^4	25
Liquid crystals	300–500	10^2	10^4	50
Photochromic	400–700	10^1	10^3–10^4	1–2
Biochrome-BR	400–700	10^{-1}	> 10^4	5–7

3. Resolution is determined by the size of a photosensitive microelement and is measured by the number of resolution lines per 1 mm film length. Photosensitive elements of all photochromes have molecular scale sizes, therefore, all photochromes have high resolutions of over 10,000 lines/mm (resolution of the best, special purpose silver halide photographic films does not go above 3000 lines/mm).
4. The lifetime of a photoinduced form is determined by the energetic intensity of the incident light, by its spectral composition, and by the type and composition of photochromic material. The formation time of a photoinduced B form, as a rule, depends on light intensity, on the time of decay, and on the type of a photochromic material.
5. The cyclicity value is determined by the resistance of a photochrome against irreversible photo- and thermo-processes. One cycle is a one-time write-erase-write process without the loss of sensitivity or other optical properties of the material.

As seen from Table 4.6, Biochrome-BR energetic sensitivity is nearly an order higher than in other photochromes and is close to the theoretical value (see Section 4D.3). Its cyclicity is virtually unlimited.

4D.3. Holographic Properties

This book cannot hope to describe the frontier of holographic research, and the following passages assume a basic understanding of this area. The motivated reader is referred to the overviews presented by Collier [1971] and Hariharan [1984]. Holographic achievements are regularly surveyed in "The Proceedings of the International Society for Optical Engineering." Application of holography to modern optoelectronics, optical systems of information processing, and so on, requires new types of holographic photo-mediums, particularly the so-called dynamic holographic mediums, allowing recording at the real time scale. Such mediums define what biochromes are.

The main characteristic of holographic sensitivity of any photomaterial is its diffractional sensitivity. The quality of a hologram depends on this value. Diffraction sensitivity is proportional to the diffraction efficiency (η) which is determined as a ratio of the brightness of a holographically reproduced image, to the brightness of the initial image (the exact definition is the following: η is the ratio of the intensity of a hologram-diffracted wave of the first order to the intensity of the wave illuminating the hologram during the reproduction of the image). High resolution values and signal-to-noise ratio are also important. These two crucial parameters depend on the size of the photosensitive centers of a photomaterial, and on the quality of the matrix. As we said, in all photochromes, including biochromes, such elements have molecular-scale sizes, i.e., are significantly smaller than the wavelength. Thus, the guaranteed resolution of biochromes may go over dozens of thounds of lines per mm, and the value of signal to noise ratios depends only on the optical quality of matrix.

A short paper published in the Soviet scientific journal *Letters to JETP* (*Journal of Experimental and Theoretical Physics*) gave a start to the intensive research on diffraction efficiency [Bunkin et al. 1981]. The researchers from Valery Savransky's lab experimentally showed the photoinduced modulation of the refraction coefficient in the process of the BR photocycle, i.e., proved

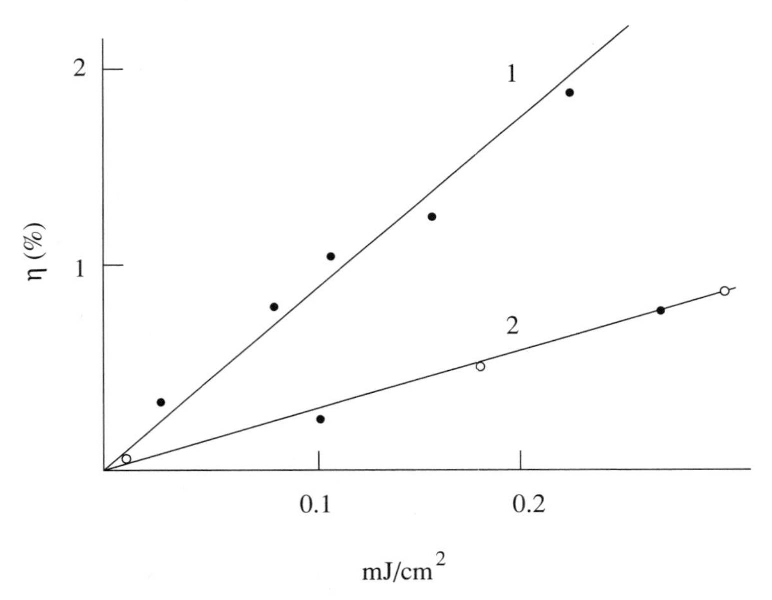

Figure 4.7. Diffraction efficiency of retinal-protein films as a function of the energy density of the recording beam. (1)-bacteriorhodopsin film: (2) 4-methoxy-bacteriorhodopsin film [Bunkin et al. 1981].

the ability of biochromes for dynamic phase holographic recording of optical information.

However, the real breakthrough in holography started a little later when photoinduced anisotropy, known in scientific literature as the Veiger effect, was discovered in Biochrome BR using ellipsometry and holography [Burykin et al. 1983]. Stationary and kinetic characteristics of photoinduced anisotropy were studied, and the possibility of biochrome application in polarizational holography was shown [Vsevolodov et al. 1984].

Further studies showed that the holographic characteristics of biochromes suffices for their application as effective holographic mediums in many directions of optical holography. The main characteristics of Biochrome-BR, and the range of applications were defined:

1. Sensitivity. Values of energetic sensitivity lie in the range 10^{-3} J/cm^2. The speed of photochromes is evaluated by the formula:

$$S_\lambda = \frac{96\ \Delta DE_\phi}{\epsilon\varphi\ K}$$

   where:
   ΔD — the optical density change at wavelength λ
   E_ϕ — the energy of a photon
   K — the amplification coefficient
   ϵ — the molar extinction
   φ — the quantum yield
   For Biochrome-BR at $\Delta D = 0.2$, $E_\phi = 2.5$ eV, $\epsilon = 6.3 \times 10^4$ I/mol.cm, $\phi = 1$, $K = 1$, the value of S_λ approximates the theoretical one and equals 7.5×10^{-4} J/cm^2.
2. Dynamic photoanisotropy (nonlinear Veiger effect). The action of plane-polarized light guides the optical axis in Biochrome film, which is parallel to the light polarization and vanishes after the light is turned off [Burykin et al. 1985]. This is the difference between the dynamic Veiger effect and the classical one. Figure 4.8 shows the experimental stationary values of dichroism birefringence. The sign of the guided dichroism depends on the wavelength of both the excitation and the testing light.
3. Reversal of the wave front. The model of the co-beams interaction was used in the studies. In the experiments with circular-polarized co-beams, the full spatial-polarizational reversal was observed [Vsevolodov et al. 1986]. Figure 4.9 shows the dependencies of the reversed wave intensity on the ellipticity of the signal wave.
4. Self-excitation and optical bi-stability. Due to its exceptionally high speed and the absence of nonreversible, destructive processes, Biochrome-BR can be used as a highly effective optical-nonlinear medium. At the wavelength of the helio-neon laser (633 nm) and a film thickness

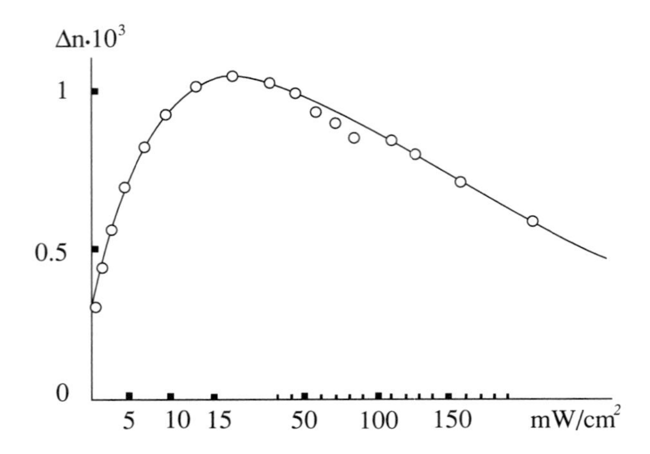

Figure 4.8. Dependence of the stationary value birefringence ($\Delta n = n_{\parallel} - n_{\perp}$) on the intensity of excitation light [Burykin et al. 1983].

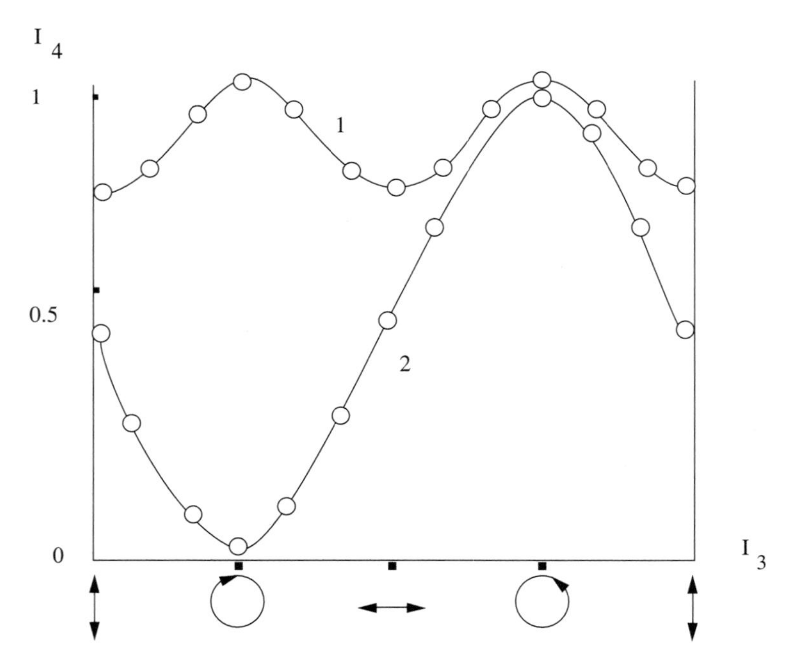

Figure 4.9. Dependence of the intensity of the reversed wave I_4 on the ellipticity of the signal wave I_3 (pumping waves I_1 and I_2 are circularly polarized, and rotate in the plane of the film towards each other (1) or in one direction (2)).

of 100 μm, the direction of polarization turns up to 20, the intensity of the incident beam being less than 10 mW/cm^2. This corresponds to an enormous value of self-rotation, reaching 2.5 cm/mV [Bazhenov et al. 1985].

5. Polarization-holograpic properties. Recording and read-out of holograms was accomplished in the process of recording by coherent beams from single-mode helio-neon laser at the wavelength of 633 nm. The diffraction efficiency is standardized as the effectiveness of the grating, recorded by mutually perpendicular directions of two plane polarized light beams. Its values are close to each other—which makes a drastic difference from the small number of classical mediums for the recording of such holograms-there the spread in values reaches a 30-fold ratio [Vsevolodov et al. 1986] [Hampp et al. 1991].
6. Holographic sensitivity. It is evaluated by the formula η/E where η—is the diffractional effectiveness, measured in %, and E is the exposure in J/cm^2.
   For Biochrome-BR, at an optical density of 0.4D at 633 nm, the holographic sensitivity value is 10–15 mW/cm^2. η is proportional to the value of the refractive index change (for scalar holograms) or to the value of nonguided birefringence $\Delta n = n_{\parallel} - n_{\perp}$ (for polarizational holograms), depending on the intensity of the excitation light. Scalar η in the first preparations of Biochrome-BR reached 4%; polarizational 2% [Burykin et al. 1983]. A comparison of BR-films containing WT of BR and BR mutant (D96N), and the influence of different modifications on the diffraction efficiency is shown in Figure 4.10 [Oesterhelt et al. 1991].
7. Polychromatic recording of holograms is possible due to the wide absorption band (from 460 to 700 nm). Experimentally, the possibility of color hologram recording at Biochrome-BR was shown by Kostilev and Vsevolodov [1985].

The above described properties of biochromes are now ready for wide application in the scientific research and technical supplements (see Chapter 6). As for object holography, the possibilities of registration of transparent and non-transparent object holograms in the dynamic regime were demonstrated in 1984 by Vladimir Poltoratsky [Vsevolodov and Poltoratsky 1985] [Vsevolodov et al. 1989]. That was the first and the last time when I had a chance to observe the slowly dissolving 3-D holographic image of the Soviet two-kopeck coin. Two years later the USSR itself dissolved.

4E. BR Analogs for New Photochromic Films

The reconstruction of BR analogs from native BR is described in Section 3C. By now a few hundred of them have been synthesized. The preparation of

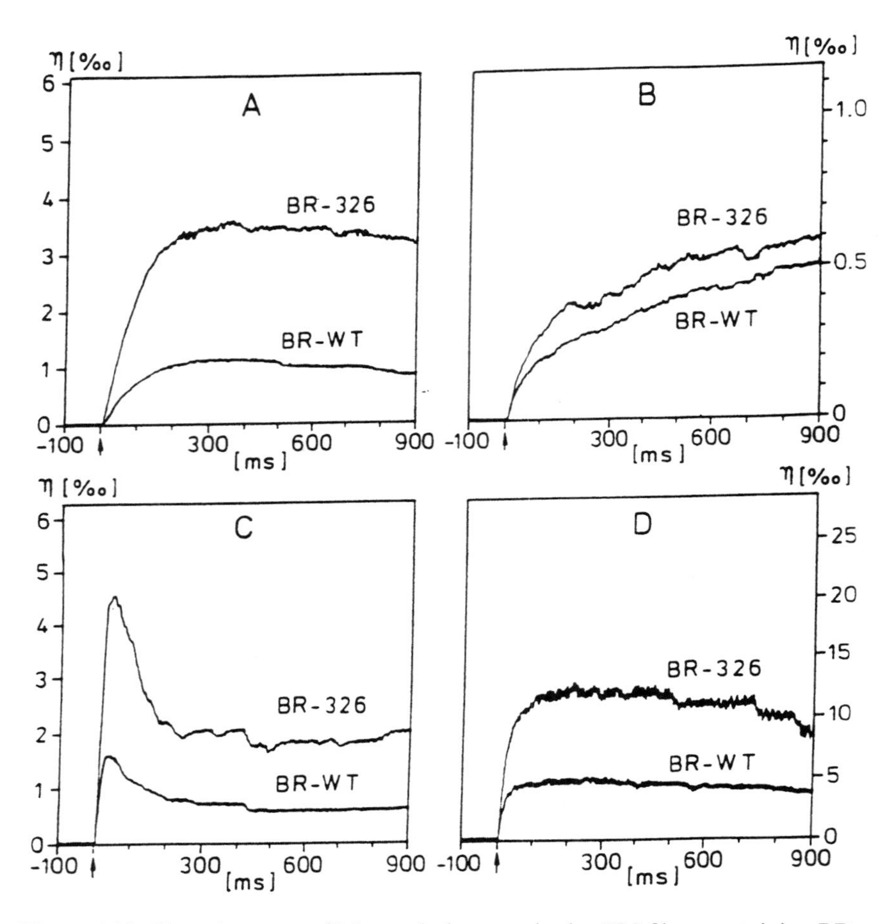

Figure 4.10. Growth curves of M-type holograms in dry PM films containing BR—wild type (BR-WT) or mutant BR-326 at different ratios of writing (I_W) and reading (I_R)/ pumping intensities of beams. (A) I_R = 20 mW/ cm^2, I_W = 15 mW/ cm^2 : (B) I_R = 20 mW/ cm^2, I_W = 145 mW/ cm^2: (C) I_R = 210 mW/ cm^2, I_W = 16 mW/ cm^2 : (D) I_R = 210 mW/ cm^2, I_W = 150 mW/ cm^2 . The recording beams (I_{W1}) and (I_{W2}) were switched on at the indicated time (↑). η ‰ ≈ 0.1 % of the diffraction efficiency [Hampp et al. 1990].

photochromic films on their base is similar to that of native BR, described in Section 4C. Naturally, films with photochromic properties can be produced only on the base of those analogs that have a reversible photochemical cycle. The absorption spectra of different analogs cover the range from 350 to 850 nm. Thus, theoretically, it is possible to produce films of any color, except green. By mixing the suspensions of carefully selected analogs, it is possible to cover the entire color range. However, kinetic forms of lifetimes of the participating analogs' photoinduced forms must be similar. Besides, the constants of retinal analogs binding with opsin must be also the same. It is not difficult to create the close to black and white film variant. The first set of color films was prepared in 1979, however, later for different reasons, the research concentrated predominantly on the films based on 4-keto-BR (see Section 4E.2)

4E.1. Properties of BR Analogs

Table 4.7 shows selected values for characteristics of biochromic films based on five different BR analogs for the visual range and two hypothetical values

Table 4.7. Characteristics of a few BR-analog films

BR-analogs: Names and structures*	Maximum absorption spectra of photochromic transitions** (nm)		Storage time	Ref. and remarks
	A-form	B-form		
Wild BR (1)	570	405	Seconds	See Sec 4D.2
4-keto-BR (42)	510	410	Tens of minutes	Vsevolodov, Dyukova 1995 in references therein
3-methoxy-BR (44)	540	405	minutes	Vsevolodov, Dyukova 1995 in references therein
3,4-diH-BR (45)	593	410	minutes	Vsevolodov, Dyukova 1995 in references therein
5,6-epoxy-5,6-diH-BR (43)	460	370	minutes	Vsevolodov, Dyukova 1995 in references therein
Azulene-BR	650–830	?	?	Hypothetical (see Sec. 3C.4)
9,10-diH-BR (39)	325–343	?	?	Hypothetical (see Sec. 3C.5)

* See Appendix for bracket number.

** See Section 4B.1.

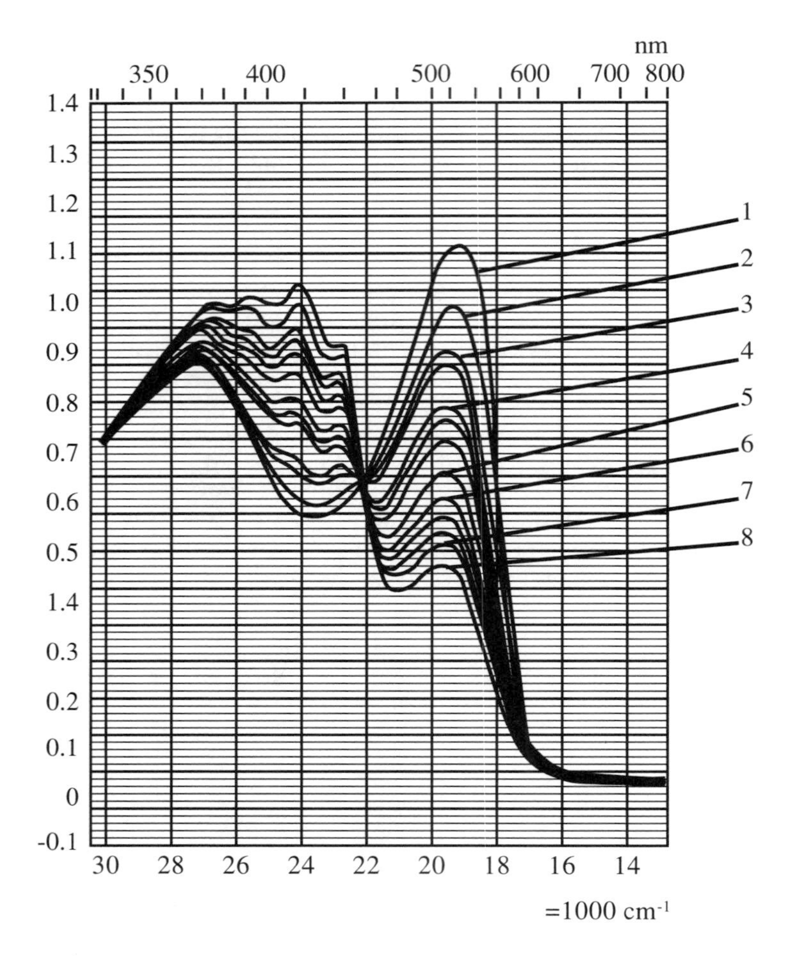

Figure 4.11. Photoinduced spectral transformation of 4-keto-BR immobilized onto gelatin film. 1. initial spectrum. 8. spectra after constant illumination (> 500 nm, 10 seconds). 7-2. spectra of the film, recorded after 30 s, 1.5 min, 1, 4, 10, 70 h, respectively.

for the UV and IR. Energetic and holographic sensitivity of most analog-based films is lower than in Biochrome-BR, but the reasons for that have not been investigated. Figure 4.11 shows the shifts of the absorption spectra of two 4-keto-BR-based films.

Some variants of BR-analog films (see Section 3B), particularly, such as BR-326, have been successfully used as holographic recording mediums. Their quantum yield is high, approximately 0.7, and they can be used in the spectral range of 400–670 nm. Unfortunately, their diffractive effectiveness does not exceed 1% [Hampp et al. 1990] (See Figure 4.10).

It is possible to change the position of the BR absorption maximum without reconstruction. It is known lowering the pH to 3 and less, the PM suspension changes from violet to blue, and the absorption maximum shifts from 560–570 nm to 600 nm [Tsuji and Rosenheck 1979]. The photocycle sustains, but the maximum of M intermediate shifts by 20 nm to the red region [Kabayashi et al. 1983]. Films based on such suspensions demonstrate the photochromic effect with limited cyclicity. Depending on the preferences of a researcher, "acid" BR may be called "analog" or a "derivative" of native BR.

4E.2. 4-keto BR Films

4-keto-BR analog-based biochromic films have been studied more carefully than the others, because 4-keto-retinal is characterized by nearly 100% reconstruction, and a very long lifetime of the 4K420 intermediate. The absorption maximum of the ground form in gelatin film is 510 nm, i.e., it is

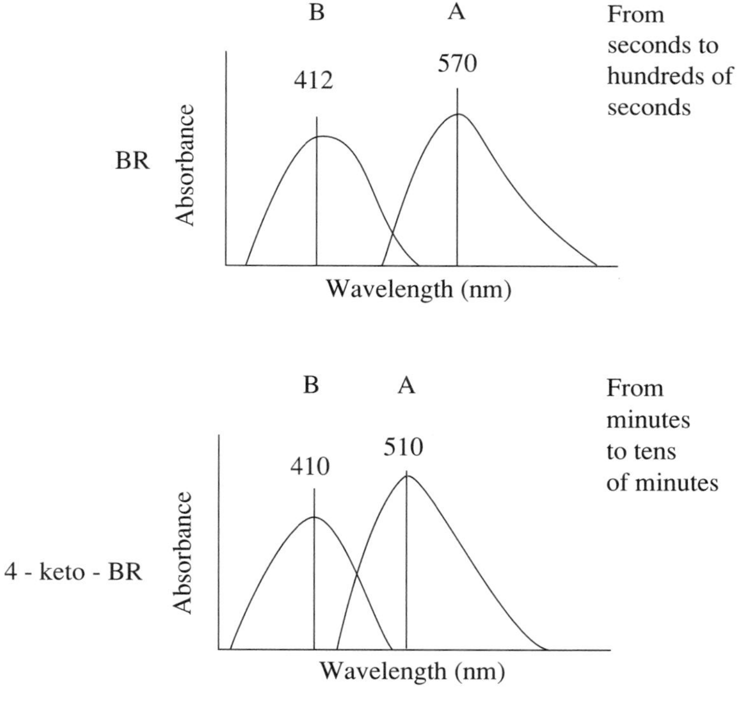

Figure 4.12. Comparison of the primary (A) and photoinduced (B) spectral forms, and the life-times of B into BR and 4-keto-BR in a gel matrix. Lifetimes depend on the degree of dehydration and the humidity.

shifted by 50 nm to the blue spectral region, however, the absorption maximum of a long-living 4K420 intermediate coincides with the M maximum in the BR photocycle. Contrast range in such films is, as a rule, slightly lower than in Biochrome-BR, whereas the time of optical information storage is by two orders greater [Druzhko et al. 1986] [Druzhko and Zharmukhamedov 1985] [Lukashev et al. 1992].

In some samples of polyvinyl films, the image was distinctive a week after the exposure. Figure 4.12 shows spectras of photochromic processes of BR and 4-keto-BR in gelatin films prepared from their suspension.

4E.3. 4-Keto-bR: A Potentially Useful Anomaly

The early research demonstrated a strong dependence of photoprocesses in the 4-keto-BR molecule on pH, and showed that they follow at least two separate photocycles—the fast (lasting seconds) and the slow (lasting minutes)—each defined by the isomeric state of a chromophore [Khitrita and Lazarova 1989]. Later studies revealed that the photoprocess in 4-keto-BR

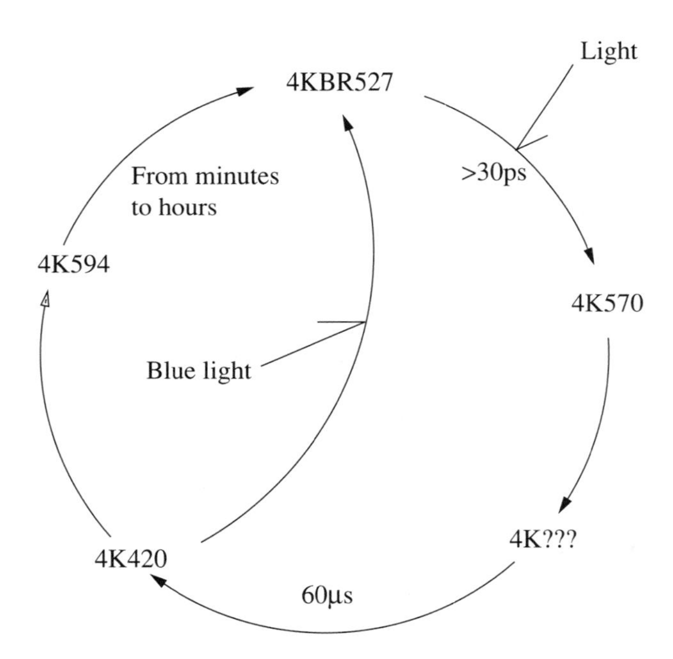

Figure 4.13. Photochemical reaction cycle 4-keto-BR incorporated into a polymeric matrix. Times of 4K420 transition into the primary state 4KBR527 depends on the type of matrix [Druzhko et al. 1986].

may follow three cyclic schemes, each having different M intermediate lifetimes and the absorption spectra of 390, 420, and 440 nm (see Section 3C.3). This peculiarity of 4-keto-BR may be important for its application in optoelectronics as a photochromic and electrochromic element.

Figure 4.13 shows the photoinduced and the electroinduced spectral shifts in 4-keto-BR immobilized in gelatin film with some chemical additions.

Optical information storage time in such films is two orders greater than in the analogous native BR-based film. Relative amplitudes of the electroinduced spectral shifts are also greater than in BR [Druzhko et al. 1995]. 4-keto-BR is easily reconstructed with nearly a hundred percent output using the microbiological, or the photoinduced hydroxylaminolysis method. Biochrome-4-keto may in the near future become the second most popular research object after Biochrome-BR.

4F. More Photosensitive Materials

It has long been noticed that if a plant leaf is partially covered from sunlight, in a short while it produces a print in the sun-protected area, since this area is not capable of photosynthesis. The print can be developed by soaking the leaf in a solution of potassium iodide. In the illuminated area of the leaf, glucose and starch are photosynthesized, respectively. Starch-containing areas darken after reaction with potassium iodide. The result is a negative image. Such "photofilm," however, has no practical value.

Images have been sometimes registered on the retina if the process of rhodopsin regeneration is suddenly stopped. These images appeared due to irreversible discoloration of rhodopsin in the retina. The irreversible discoloration of the isolated rhodopsin-containing segments is easy to observe: the bright-red segment suspension becomes pale-yellow in the light, and in the dark, its color does not return.

In 1878, Kune obtained the first optogram, or the photoimage on the retina. A rabbit was placed before a bright window so that it had to stare at it for 3 minutes. Then the animal was killed, and the retina was immediately fixated in alum. In 1937, J. Wold discovered that the photochemical processing of rhodopsin gelatin films is similar to the same in intensely cooled preparations of VR [Wald 1937]. After the image was projected on the film, and the film was soaked in hydroxylamine, it gave a negative image which could be fixated by drying (See Figure 4.14).

Beginning in 1968, the possibility of biological and enzymatic reactions applied as amplification mechanisms (like halogen-silver materials) to the new types of photoregistering materials has been investigated [Kaufman et al. 1968]. In the 1980's, the photomaterials with enzymes research gave the first inspiring results (see Section 4F.3)

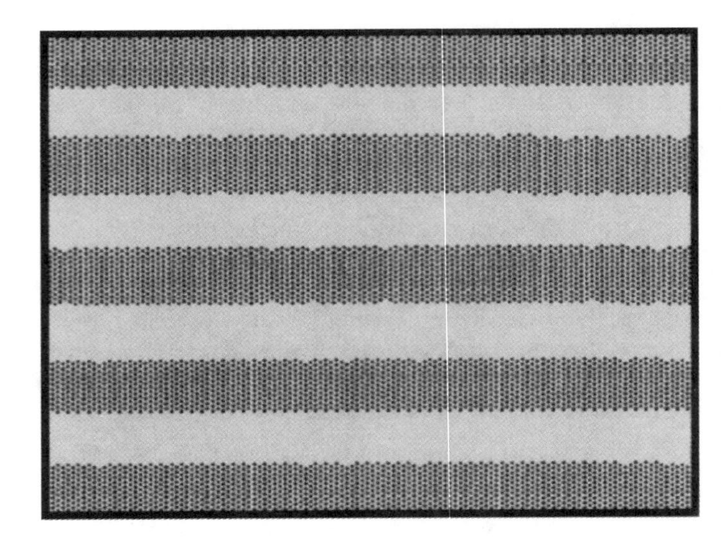

Figure 4.14. Photoimage obtained on visual rhodopsin gelatin film.

4F.1. Chlorophyll Photographic Film

Only a few attempts to create a photomaterial based on the leaves of green plants have been attempted. For example, it was suggested to use the ability of the dark-grown green leaf to change its absorption spectrum in the light. This change is due to protochlorophyll (absorption maximum 647 nm) reduction to chlorophyll (absorption maximum 676 nm) (See Figure 4.15).

The complexity of practical application of such film can be judged by the following description of preparation and application processes.

First, etiolated (dark-grown) plants are grown. Instead of chlorophyll, they produce protochlorophyll. The next step is to extract protochlorophyll-apoprotein complexes from the leaves of these plants. The process takes place in a dark cold room. A buffer maintains the pH in the extract between 7 and 10. To protect the apoprotein from oxidation and the loss of speed of the complex, it is mixed with glycerin, saccharin, polyvinylpyrolidone, etc. Everything is carefully mixed and grated. The resulting mass is filtered through a large cell filter, the filtrate is centrifuged for 30–60 minutes at 30,000 g. The supernatant contains photoactive protochlorophyll-protein complexes which is collected and used for the preparation of a three-compound mixture with the addition of a polymeric material. The mixture is then centrifuged and the sediment is collected.

The above protocol occurs at temperatures not over 50°C, and stored for a week in dark-green light. The triple complex can be stored at 0°C for several months. At room temperature, photoactivity disappears in 1–2 days. The film

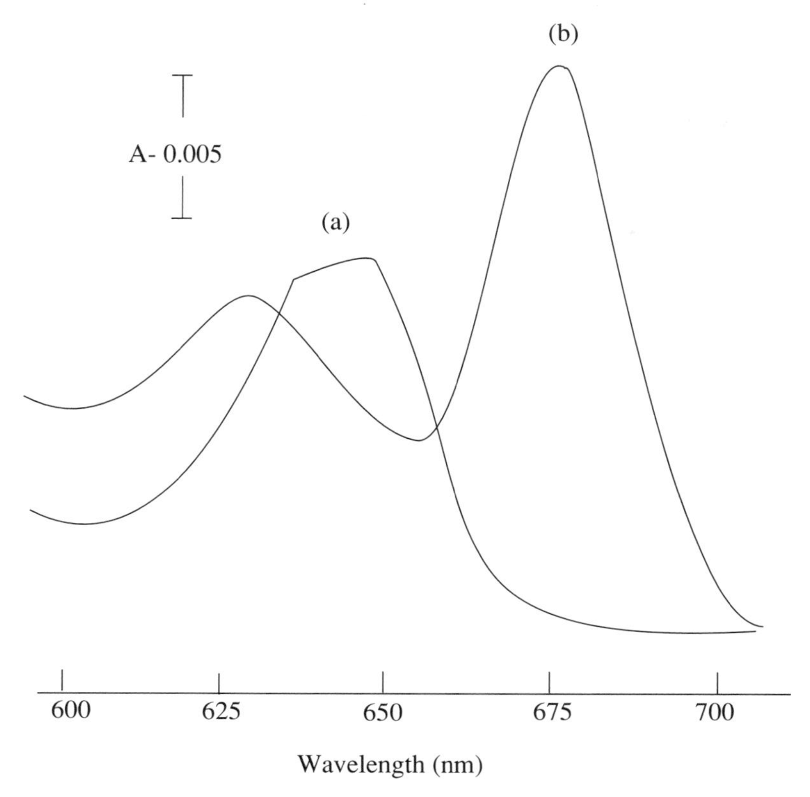

Figure 4.15. Absorption spectra of a protochlorophyll membrane recorded before (a) and after (b) illumination for 30 min from a 60 W tungsten lamp. Membranes were resuspended in a buffer at a protein concentration of 2.5 mg/ml [Bochove and Griffiths 1984].

must be protected from humidity by a water-resistant coating or a waterproof film. The formation of the photosensitive film itself is not a simple process. The mixture is applied to a copper plate and is dried at −130°C in vacuum for three days. Upon illumination of the film (with the focused light from the 250 W lamp) at 0°C through the mask, the absorption spectrum shifts from 640–675 nm. The obtained negative is brought into contact with the common halogen-silver film and makes a standard print. The photosensitivity of such films goes down by 50% in a month, even if it is stored at −10°C. We intentionally go into all these details to demonstrate that at present these films are not ready for application, not even under lab conditions [Dujardin et al. 1976] [Michel-Wolvertz et al. 1976].

Under certain conditions, chlorophyll behaves as a photochrome. It was in 1984 that the ability of light-excited chlorophyll and its analogs (pheophitin,

Table 4.8. Absorption maxima of chlorophylls

Pigments	Absorption maxima (nm)		Host
	Blue region	Red and IR region	
Chlorophylls			
a	430	670–700	Plants, green algae
b	470	645	Plants, green algae
c	460	630	Brown algae
d	450	685	Red algae
Bacteriochlorophylls			
a	375	800–900	Bacteria
b	400	1020	Bacteria
g		770	Bacteria

gematoporphyrin, phtalocyanine etc.) were found to undergo reversible photochemical interactions with molecular electron donors [Krasnovsky 1948]. Upon illumination, the reduced forms of the pigments are formed, and the position of the new absorption spectra depends on the nature of the donor-molecules. After the light is shut off, the spectrum relaxes to its initial position at a rate that depends on the properties and types of donor-molecules. Unfortunately, the quantum yield in such processes is extremely low. Table 4.8 lists the absorption maxima of chlorophylls.

4F.2. Visual Rhodopsin Photographic Film

An interesting idea to create an amplifying photofilm on the base of vertebrate VR was put forward in the U.S. Photomaterials consist of phospholipid vesicles from the surface of VR along with metal cations on the inside. Outside the vesicles, in the base matrix, metal chelates are added. Under the action of light, vesicles in the membrane open their channels and the chelates interact with the cations, forming colored complexes [Maillart 1978].

The amplifying effect is achieved through a single channel opened in the vesicle membrane by a photon. Many metal cations pass through the channel. By speed and resolution parameters, such films approach halogen-silver films.

Speed, of course, depends on the size of vesicles. In the above materials, their diameter is from 250 A to 10 μm. For optimal performance, the film must contain 5–30% water.

The spectral speed of the film lies within 350–600 nm. Upon drying, spectral speed decreases which is probably one of the ways to fixate the image. The second means of fixation is the application of another film to the exposed

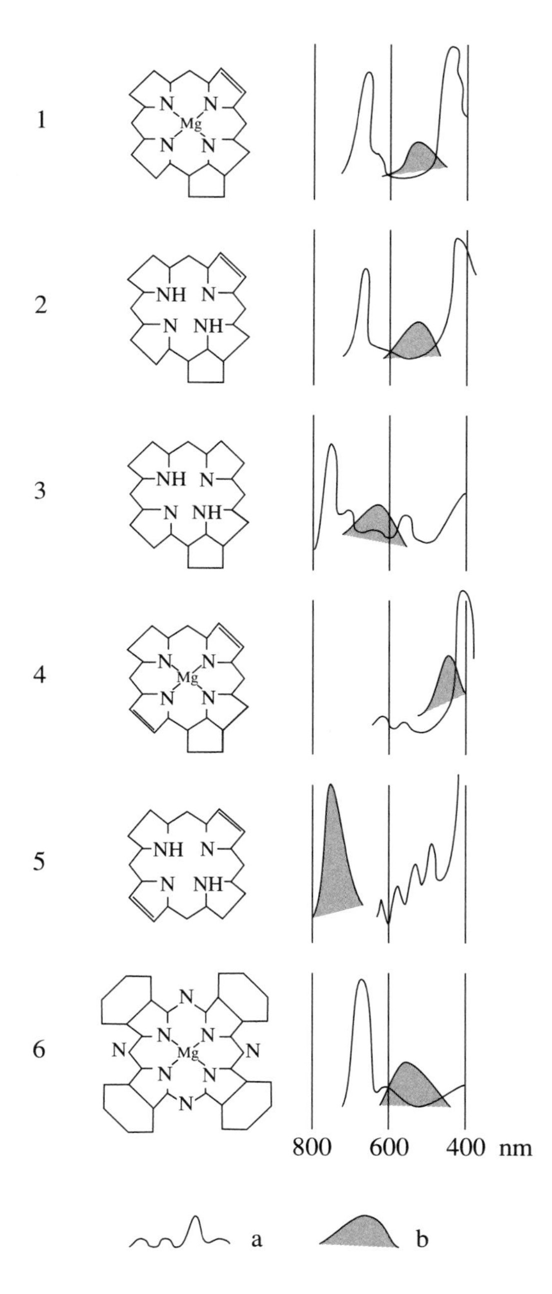

Figure 4.16. Photochromic absorption spectra shifts of a few chlorophyll types and their analogs upon photoreduction with ascorbic acid [Krasnovsky 1948](1). Chlorophyll a (2). Pheophitin a (3). Bacteriopheophitin. (4). Protochlorophyll (5). Hematoporphyrine (6). Pthalocyanine Mg. absorption spectra before (a) and after (b) photoreaction spectra of intermediates [Krasnovsky 1966].

one so that the formed colorimetric agent can pass onto that film. After the plates are separated, the print is there. Theoretically, it must be easier to regenerate the discolored rhodopsin by adding 11-*cis*-retinal than to restore the entire plate for a new exposure. However, it is not yet clear how many times this process can be accomplished, and how big is the percentage of regenerated molecules [O'Brien and Tyminski 1983].

The process of film preparation is no less complex compared to the one described in Section 4F.1 for chlorophyll films. The author has never heard of the preparation of VR or similar films being used even for laboratory usage.

4F.3. Photographic Films with Enzymes

The so-called fermentive photomaterials must also be regarded as biological materials [Martinek and Berezin 1979]. Many enzymes possess a high-specificity and act as biochemical reaction enhancers. This also means that when catalyzing one reaction, they do not affect the others, proceeding at the same volume. Speeds of such reactions depend on the structure of participating compounds. Some enzymes have different activities in the light and in the dark. It is possible to base a photosensitive system on these properties of fermentive reactions.

The first known patent for the production of enzyme photomaterials was taken by Kodak in 1968. A mixture of gelatin and enzyme (protease) was applied to the plate and dried up. UV light was used for the exposure. At the illuminated areas, the enzyme deactivated. The development was accomplished in a solution where nonactivated protease and dissociated gelatin were washed out. The result was a relief image. The exposure lasted for 30 minutes. The enzyme photo-inactivation quantum yield was less than 10^{-2}. To reach the speed of traditional photomaterials, it has to be about 10^4–10^6.

Russian scientists Berezin, Kazanskaya, and others, the authors of many species of enzyme photomaterials, think that attention must be drawn to the processes where enzyme activity is maximum in light, and is not present at all in the dark. In this case, the general quantum yield is at a maximum. The general quantum yield of a photoprocess in enzyme systems is determined by three parameters: the speed of substrate (the wash-out or the changing its optical density) conversion (the time of image development, i.e., the speed of wash-out or the change of color), and the catalyzer formation quantum yield. To visualize the image at every square cm of the photofilm, 10^{-7} M of color matter must form, at an extinction coefficient no less than 10^4 M^{-1} cm^{-1}.

For most enzyme reactions, there must form between 10–11 or 10–15 M/cm^2 active catalyst in the light for the process to flow. According to calculations, this requires a light flux of no more than 10^{-7} J/cm^2, i.e., the light flux must be comparable to the same used in photography. In reality, it is 2–3 orders greater. Modern enzyme photomaterials have the following parameter values: a resolution around 20 mm^{-1}, a speed of 3×10^{-5} J/cm^2 at density change by 0.2 D: exposure of about dozens of seconds at the wavelength of

313 nm and a development time (in common buffer solutions) of minutes [Kuan et al. 1980] [Dobrikov and Shishkin 1983].

4F.4. Other Types of Photochromic Materials in Nature

Apparently, scientists, being attracted to common chlorophylls and rhodopsins, have touched only the tip of the problem of applying a living systems photochromism to technology. Photochromism has to be the essential property of any live system, for it not to break after a few work cycles. We have not mentioned biological photopigments that do not lose these properties. Some of them already now could be of interest for bioelectronics. For example:

Phytochrome — this pigment in responsible for their periodicity, absorbing in the dark at 660 nm with an extinction of 76,000. Upon red light illumination, its maximum shifts to the IR spectrum region to 730 nm. In the dark, slow relaxation to the ground state occurs which is catalyzed in IR light [Bortwick et al. 1954]. Thus, phytochromes have the main properties of a photochromic material.

Phycochromes — the pigments of blue-green seaweeds. In the blue-green seaweeds, several types of phycochromes (a, b, c, d) with different absorption spectra have been discovered, all of them capable of photochromism. In a and b phycochromes, the process follows the standard photochromic scheme, and in c and d it follows the sequence of Diagram 4.2.

Diagram 4.2

$$\mathbf{A} \overset{h\nu}{\longleftrightarrow} \mathbf{B} \overset{kT}{\longleftrightarrow} \mathbf{C}$$

After the light-dependent conversion of A to B, the latter may convert into C forms in the dark, or also in the dark, return into B, or stabilize in the C form. Such a process allows one to control the time of information storage from the outside and is extremely interesting from the point of view of practical application [Dorion and Weibe 1970a] [Vernon and Ke 1971] [Bjorn 1979] [Krongauz et al. 1975].

Flavoproteines — A possible biomaterial for use with UV irradiation [Ninneman 1980].

P-coumaric protein — This photosensitive protein, discovered in 1984, was identified as a photochrome only in 1994 [Hoff 1994]. It may be responsible for a new type of negative phototaxis.

The reader is invited to refer to the surveys: [Vernon and Ke 1971] [Bjorn 1979] [Krongauz et al.1975].

5

Bacteriorhodopsins as Optoelectronic Materials

Due to their photochromic, electrochromic, photoelectric, and photochemical properties, along with their extraordinary chemical stability in PM, BR molecules are highly qualified materials for many opto- and bioelectronics' devices. It is not surprising that by 1997, over 150 patents for bacteriorhodopsin-related research had been issued. The initiative belongs to the Japanese scientists, or, to be precise, the Japanese companies that fund the patents. Unfortunately, neither American nor European businesses followed this example. The attempts of Birge, Oesterhelt, and Hampp with colleagues, to increase American and German patentability to the level of Japan were the struggle of solitary heroes against titans like Fuji, Canon, Sanyo, and Hitachi. By 1995, the patent distributions looked like this: Japan-90, U.S.-12, Germany-12, Russia-7.

Brief descriptions of some patents will give the reader a general picture of the diversity of the search for retinal-proteins application.

A. A color image sensor. This biosensor is based on the photoelectric properties of BR, HR, etc. In the authors opinion, a mosaic of differently colored retinal-proteins similar to a video screen will perceive the color image and transform it into electric current (or voltage) [Ogama 1987 1990] [Patent 05135013 Japan; 4896049 U.S.].

B. Recording mediums and color recording processes. A complex composition of BR, 3,4-dihydro- and 4,6-dihydro analogs of BR, and a color-forming agent, is incorporated into a polymer. Upon photoexcitation of BR and its analogs, the hydrogen ion concentration changes, causing the color-forming agent to change its color. The process is similar to modern color photography [Arai et al. 1986] [Patent 63092946 Japan].

C. Optical sensor. This patent describes a typical light-recorder whose principal scheme may be found in many scientific papers [Sora et al. 1987] [Patent 60203868 Japan]. A thin film of PM is pasted at the conductive glass plate. A second conductive layer is coated over the film. The conductive layers are connected to the complete circuit with a power source and a photo potential registering resistor.

D. Camouflage system. The authors suggest use of the electrochromic properties of BR and phycobiliproteins for camouflaging flexible cloths, tents, etc. [Cazeca et al. 1995] [Patent 5438192 U.S.]. The possibility of BR, and its color analogs, in an application for such purposes was investigated on one of the military grants in Russia in the 1980s (see Section 4D.1).

E. Photoactive membranes. Japanese investigators were not the first to suggest the application of the photodependent ion transport mechanism in liposomes with BR additions (see Chapter 5). They suggest using such liposomes for optical information processing [Saito et al. 1993] [Patent 05048176 Japan].

F. Optical random access memory. The patent describes the method, and the multi-beam laser system for information recording, processing and readout [Birge and Lawrence 1993] [Patent 5228001 U.S.] (see Section 6A).

G. Memory cells. The pn-junction used as a memory cell for a memory matrix with the control of the pn-junction realized by means of a BR-based film, and with optical data conversion into electronic data. This invention may be used as a graphical data converter into digital code [Belchinski et al. 1988] [Patent 1511765 USSR].

H. Micro-electrostatic actuator. As we know, the charges on PM fragments are re-distributed in light. The Japanese company Nipon used this property in its patent [Patent 05227765 Japan]. Upon light absorption, and monitoring the total induced charge, the film with PM moves toward the film with a permanent static charge. The authors think that such an "actuator" may be used to control the light in increasingly small object movements.

A large number of patents have been issued for the application of BR as a light-converter in photodiodes, solar elements, photoelectronic light energy detectors, optical detectors of various kinds and so on. Nevertheless, today all these schemes are much too ineffective and expensive for them to replace modern silicon-organic and other similar photoelements and photodetectors. In future biomolecular devices, these disadvantages may become their virtues, due to their affinity for the elementary base.

Many patents are based on the ability of BR for light-dependent and directed proton transport and/or for a pH change. Many new devices may be created on the basis of these properties by addition of chemical substances whose characteristics depend on pH. Such hybrid systems may be used in optical switches [Watanabe 1991] [Patent 02310538 Japan], in color image systems [Iwashita et al.1987] [Patent 01116536 Japan], and even for amplification of biochemical processes (see Section 5D)

As we said repeatedly in the preceding chapters, a single PM is a ready biochip, possessing photoelectric, electrochromic, photochromic, piezoelectric, ion-transporting and charge-separating properties. PM modification allows a programming of the parameters of these properties. PM may be regarded as a protoelement for many optoelectronic biomicrosystems of the future. There exist many constructive ways of biochip application: starting with the liposomes with one or two PMs incorporated in their surface, or a solution of vesicles or in the interface monolayer, and concluding with a two- and three-dimensional gel and polymer-based microsystems, with hundreds of thousands of PM fragments. Most patents use multi-layer systems of oriented PMs prepared in different ways.

The reader may refer to the surveys that most fully describe the possibilities of BR application for optical informational systems and optoelectronics [Birge 1990] [Oesterhelt et al. 1991] [Birge 1992] [Birgeand Gross 1995] [Hong 1996].

5A. Molecular Computing Elements

It has been on the discussion table for over 20 years that by the year 2000, modern silicon integrated circuits will reach their limit of single switch (i.e. transistors') density per volume unit. After that, quantum limiting will make it impossible to further improve the parameters (see Section 1C). The radical solution to this problem is to replace silicon-based computers with carbon-based ones [Conrad et al. 1982] [Tucker 1984]. The idea of rhodopsins application as a molecular element was around at the beginning of the 1980's. It was 1982, when Honig suggested at one of the conferences, to employ the ability of retinal-proteins to store energy by means of isomerization and charge separation for the creation of a molecular switching device [Honig 1982]. In that year, the author of this book already had a collection of biochromic films of different colors, suggested for use as macroelements of optoelectronic devices and biocomputer microelements [Vsevolodov 1984]. In October , 1985, at the session of the Japanese Chemical Society, the results of experimenting with the biophotoelement model based on the hybrid of an ion-selective field transistor and PM were reported (Figure 5.1).

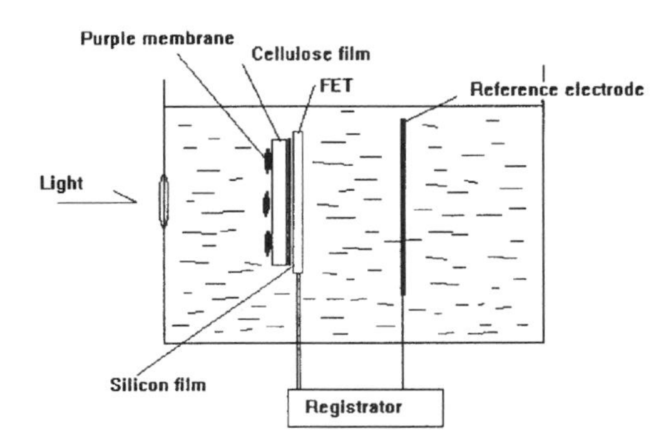

Figure 5.1. One of the first prototypes of a bioelectronic element on purple membranes (PM) and a ion-selecting field transistor (FET), applied as a biosensor for medicinal usage. Upon illumination, photo-dependent proton translocation occurs in the PM layer. The transistor reacts to the change of hydrogen ion concentration by changing the photopotential which is registered by a recorder.

In the same year, Hong suggested regarding BR as a prototype for a molecular computing element [Hong 1986]. Figure 5.2 shows the schemes of four methods for BR photocurrent measurement, offered by Hong as a

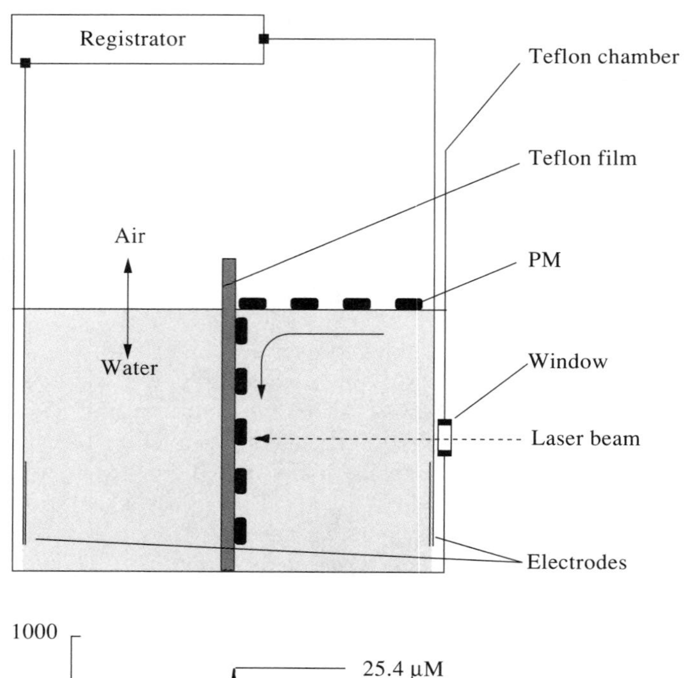

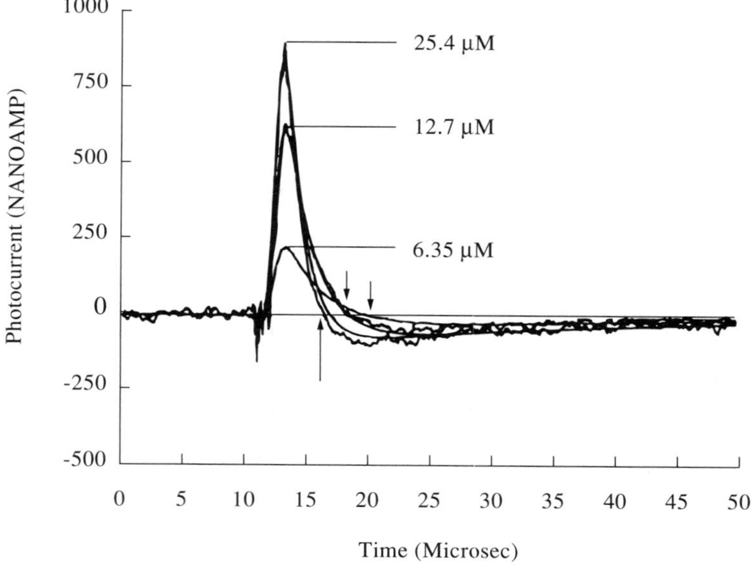

Figure 5.2(a)

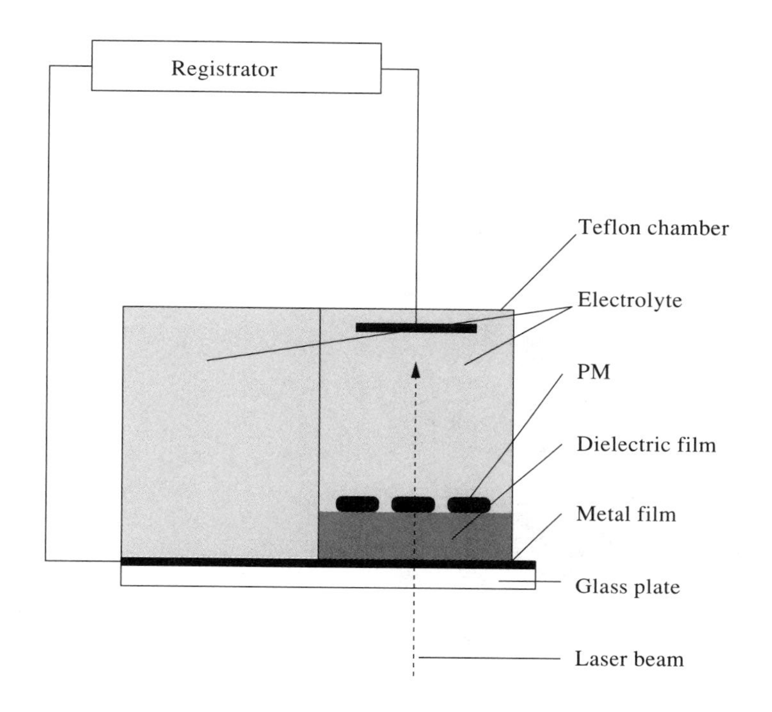
Registrator
Teflon chamber
Electrolyte
PM
Dielectric film
Metal film
Glass plate
Laser beam

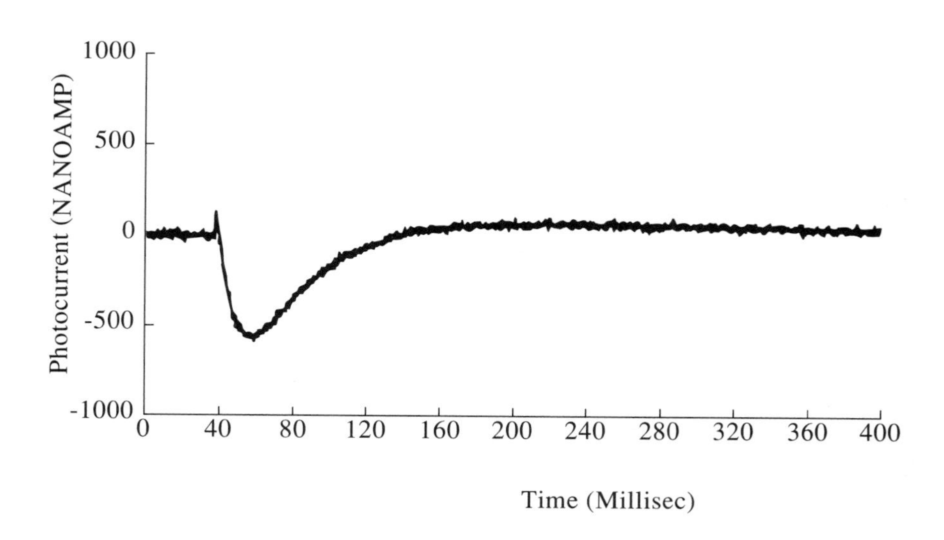
Photocurrent (NANOAMP)
1000
500
0
-500
-1000
0
40
80
120
160
200
240
280
320
360
400
Time (Millisec)

Figure 5.2(b)

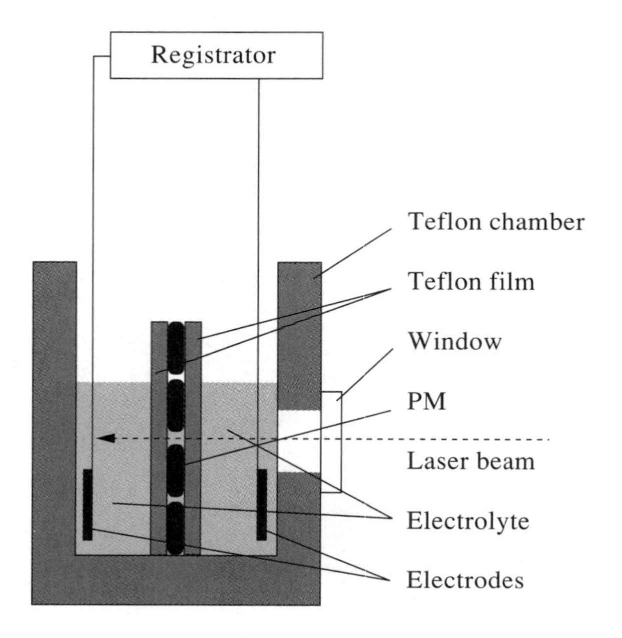
Registrator
Teflon chamber
Teflon film
Window
PM
Laser beam
Electrolyte
Electrodes

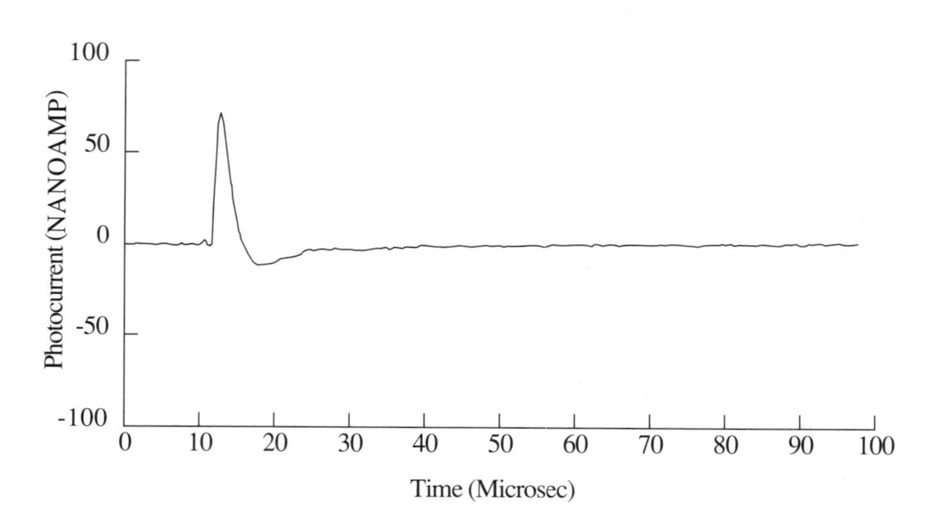
100
50
0
-50
-100
Photocurrent (NANOAMP)
0
10
20
30
40
50
60
70
80
90
100
Time (Microsec)

Figure 5.2(c)

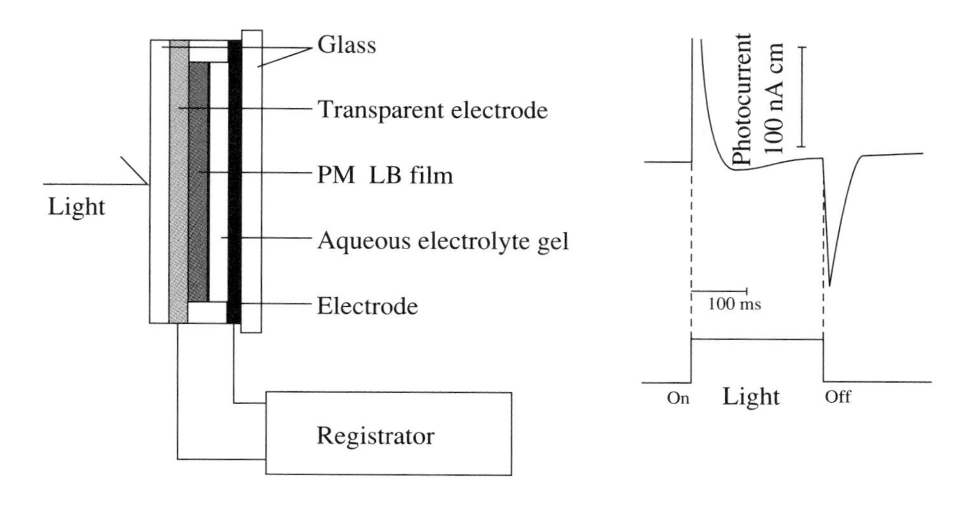

Figure 5.2(d)

Figure 5.2. Model membrane systems for use as prototypes for biomolecular computing components. (a) Model using Trissl-Montal method of model membrane formation. The interfacial layer of PM is formed first at the air-water interface with the water level below the aperture where the Teflon film is mounted. The water level is then raised so that the interfacial layer is attached to the Teflon film. The interfacial layer is probably a multilyer. (b) Model using extra-thin dielectric/metal films (transparent to visible light) assembled on a glass plate. The PM layer is then attached to the dielectric coating. Metal film serves as a conducting path. (c) Model using oriented PM (method described in a) sandwiched between two thin Teflon films (6.35 μm each). (d) Cross section of a thin sandwich-type photocell. PM LB film from 6 to 10 layers. Electrolyte gel layer 200 μm thick, comprised of 6% carboxymethylchitin and 1 M KCl. This photocell is used for a contrast, position-sensitive detector with 64 pixels [Modified from Hong 1986 (a,b,c)] [Miysaka et al. 1992 (d)].

prototype for the new biochip. The scheme of Figure 5.3 is obviously more stable and technology-friendly (Kononenko and Lukashev 1995). It can function on the basis of one PM fragment as a stable microbiochip with high technical characteristics and an unlimited warranty.

5.B. Ultrafast Electro-optical Detectors

During the formation of first photochemical intermediates, the photoinduced PM polarization occurs as a redistribution of charges inside BR at the speed of primary spectral transformations [Trissl and Montal 1977].

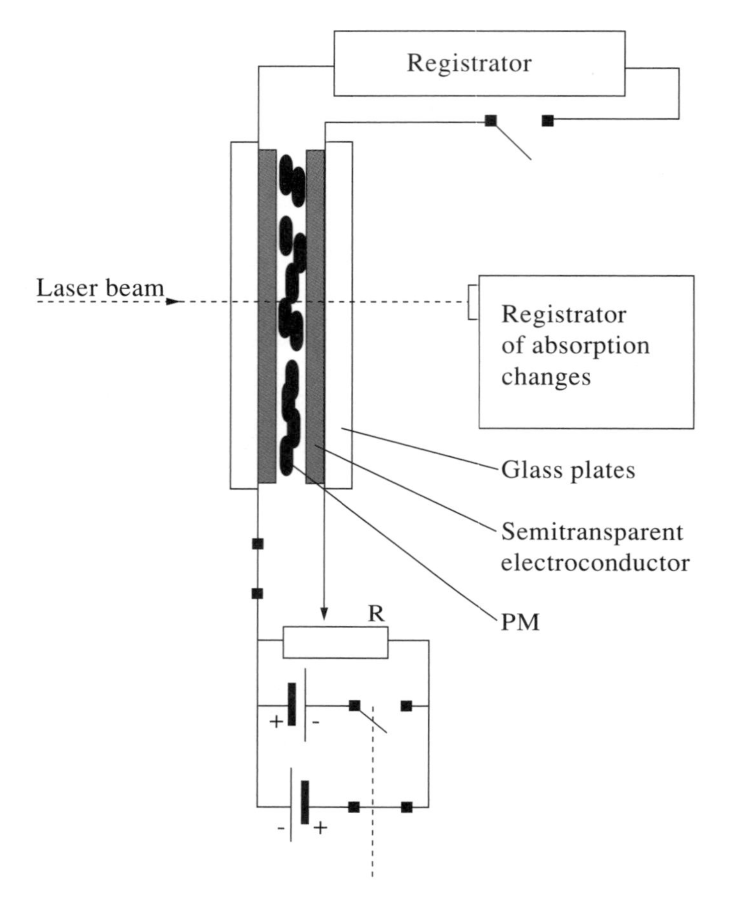

Figure 5.3. Prototype of a photobiosensor based on PM. Amplitude and kinetic characteristics for this biosensor see on Figure 5.4 [Kononenko and Lukashev 1995].

At the excitation of an BR artificial membrane, with a 15 nsec long laser impulse at 530 nm, the photoelectric response consists of three components, the first being formed in less than 1 ms [Drachev et al. 1981]. In the Trissl-Montall film [Montal and Mueller 1972], the photoelectric signal from HR lasts less than 10 mμ, and reverses its polarization with KCl substitution for sodium citrate [Michaile et al. 1990]. These and other results point at the possibility of creating a photodetector with high temporary resolution. How high will it be? The first spectral intermediate J in the BR photocycle forms in hundreds of femtoseconds (see Section 3A.3). What maximum speed of photoelectric response can be obtained by means of modern methods?

Upon illumination of a multilayer of dry, oriented PM, located between conductive plates with a supershort light impulse, a few picosecond-long electric impulse is registered on the plates [Rayfield 1988, 1989, 1994]. The electroinduced orientation of the purple membranes in a polyacrylamide gel can measure photocurrent impulses less than 30 nsec long [Liu and Ebrey 1988]. These reports demonstrate the theoretical possibility of creating a BR-based optical detector for the registration of ultrafast optical processes. The limitations lie, most likely, not in the PM or BR molecules, but in the registering devices.

5C. Biosensors

Biosensor design is the most attractive field of biotechnology. Large companies eagerly invest in the development of biosensors, hoping for later profits. The Japanese are developing biosensors to detect the freshness of fish and meat, the Israelis are trying to invent bio-sniffers to search for bombs by the smell of the explosive, the group led by Dr. Howard Weetall from NIST U.S. endeavors to create a biosensor to detect poisonous gas in a car, drug enforcement agencies seek a portable drug detector, and medical workers are waiting for a complete and separate organ biosensor to register any deviation from normal body functioning [Mathewson and Finley 1992]. However, the word 'biosensors' has a dual meaning. Initially, biosensors must have been the devices for biological application, i.e. the devices to measure the biological parameters of a functioning organism, or chemical and other characteristics of biological and organic objects and substances. Later or simultaneously, with this first meaning, another one emerged, defining a biosensor as a device constructed of biological elements, or having a biological element as one of its important functional parts.

Different BR-based biosensor constructions have been suggested. Such as multilayer systems, with varying BR-containing layers, and some cationic polymer. Such layers may serve as the basic element for many biosensors [Miyasaka 1990] [Miyasaka et al. 1992]. The BR-based spatial element immobilized in sol-gel glass looks promising in terms of reliability and longevity [Weetall et al. 1993].

One of the most recent surveys on BR-based biosensors and photosynthesizing reaction centers describes a technology-friendly, stable, and universal biosensor element, capable of light conversion into electric potential and, *vice versa*, electric potential conversion into optical changes [Kononenko and Lukahsev 1995]. Figure 5.4 presents amplitude and kinetic characteristics for this biosensor.

An analogous biosensor is easy to synthesize using 4-keto-BR (see Section 4E.2). The characteristics of some parameters should be greater than in native BR [Lukashev and Robertson 1995]. Lewis and colleagues offered an interesting idea for such biosensors. The layer that contains areas with cross-

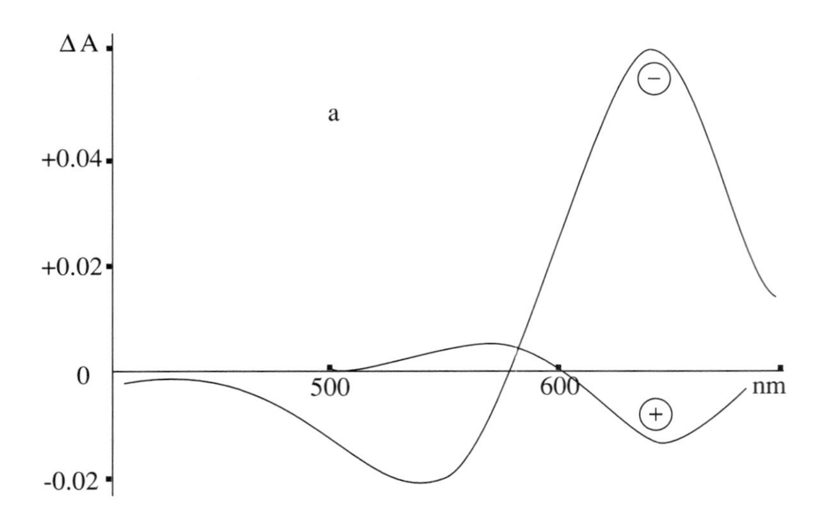

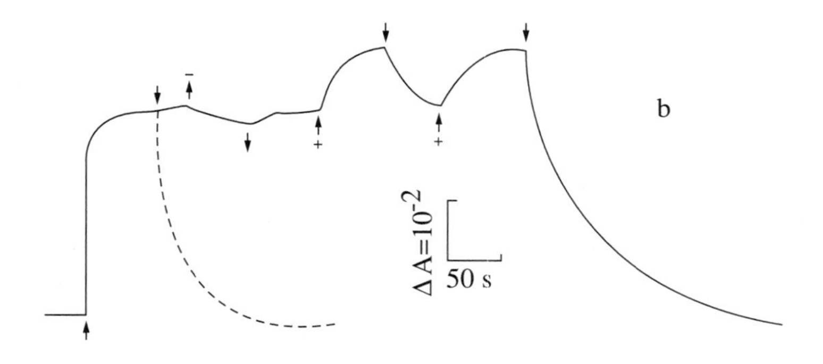

Figure 5.4. Optoelectric and electrooptical potentials induced on electrophoretically oriented PM film (see Figure 5.3 for the construction of the experimental cell). a) Electrically induced absorption changes depending on the direction of an applied electric field (10^3 V/cm). b) Light and electricity induced kinetics cause absorption changes in the PM film at 412 nm (M intermediate of the BR cycle). ↑ light switch on; ↑ electric field switch on; ↓ light switch off; ↓ electric field off; + refers to the direction of an applied electric field from the cytoplasmic to the periplasmic side of the PM, and a minus sign means the electric field is applied in the reverse direction. The dashed line is if the light was switched off at this point [Konenko and Lukashev 1995].

oriented PMs is placed between the two electrodes. Thus, a biosensor with an edge effect is obtained: at the scanning of its exit window with a focused light beam, the entrance potential changes to the opposite while crossing the boundary between the two areas [Takei et al. 1991].

5.D. Photosensitive Hybrids of Purple Membrane with Mitochondria and Nonphotosynthetic Bacteria

It may appear that this section is not in the same spirit as the rest of the book. However, as was mentioned in the Introduction, the application of biological components in future biocomputers may not be restricted to molecules and their complexes. It is possible that cells, both natural and artificial will also be used as elements to supply energy and/or to process information.

It is tempting to use the PM as a photoactivator of biological activity in different nonphotosynthesizing cells or unicellular microorganisms. This could save part of the substrate for biomass production using sunlight as the driving force.

As is well-known, PM fragments orient in a particular direction when incorporated into vesicles, allowing for photodependent proton transport inside and outside. Simultaneously, a electrochemical potential of hydrogen ions appears, serving as a unified energy form, similar to ATPase [Skulachev 1984]. The presence of ATPase and ATP in this system creates a possibility for the synthesis of ADP from ATP. There is the evidence that a BR-ADP-ATPase system under certain conditions works as an ATP-synthesizing bioreactor [Inatomi 1986].

The difference of some unicellular bacteria from vesicles is that their plasma membrane is protected from osmotic shock by the cell wall, penetrable only for certain ions and nutrients. Some of them have no organelles in their cytoplasm. However, they have enzymes. It is possible that during the contact between the bacterial surface and the PM, the energy of the hydrogen electropotential is transferred onto bacterial surface. Experiments have been conducted on two species of the most primitive, nucleus-depleted heterotrophic bacteria *E. Coli* and *Bac. subtilus*.

The results showed that the final concentration of bacteria, grown in the PM-containing mineral medium without nutrients is greater in the light than in the dark [Vsevolodov et al. 1986]. Electron microscopy showed that PM fragments come into contact with the cell wall and attach to it, possibly, with the help of phimbrias (plasma protein prominences penetrating the cell wall). The potential change on the PM may spread across the cell membrane, utilizing it as an energy substrate instead of ATPase. The transportation of a proton across the bacterial membrane with the help of PM is unlikely but not impossible.

Table 5.1. Changes in concentration of *Bac. subtilus* cells grown in a mineral medium without organic additions or purple membranes*

Initial concentration	End concentration	Ilumination by white light	Available PM in media
1.5×10^6	5×10^7	Yes	Yes
1.5×10^6	2×10^6	No	Yes
1.5×10^6	1×10^6	Yes	No

* The number of purple membranes per one bacterial cell was 10–20.

It would be interesting to investigate the effect of the purple membrane on respiration and glucose transport in bacteria, since these processes directly depend on the electrochemical potential. Such investigations have been conducted on the heterotrophic aerobic bacterium *Rhodococcus minimus* [Baryshnikova et al. 1988]. The investment of energy into the speed of light-driven glucose transport was different with and without PM. In the hungry strain, the rate of growth in the light was 2–3 times greater with the presence of PM. Conversely, the presence of PM in the medium had zero effect on respiration. In all of the above experiments the PM had maximal effectiveness at the concentration of no more than 100 fragments per one bacterial cell. Higher concentrations caused lysis and the cell deteriorated.

At the illumination of the PM suspension, and the suspension of freeze-thawed mitochondria, it was found that light exposure prevented the inhibition of succinate oxidation. The illumination also led to the decrease in the inhibition of the rate of ferricyanide reduction by mitochondria in the presence of succinate. Elimination of the inhibition of succinate oxidation is known to be the result of excitation of the mitochondria. Light-induced excitation of the mitochondrial preparation in the presence of PM may be explained by certain interactions between PM and the mitochondrial membrane. Therefore, PM participates in the formation of the electrochemical proton gradient on the membranes of mitochondria at the points of contact with PM [Okon and Vsevolodov 1987]. Unfortunately, these three works hardly provide enough material to make conclusions about the mechanism of PM interaction with the membranes of unicellular organisms, or its effectiveness as an energy converter in hybrid systems.

5E. Position-sensitive Detectors

Position sensitive detectors might be used in technology associated with optoelectronics, particularly, as a recognition device, and a control measur-

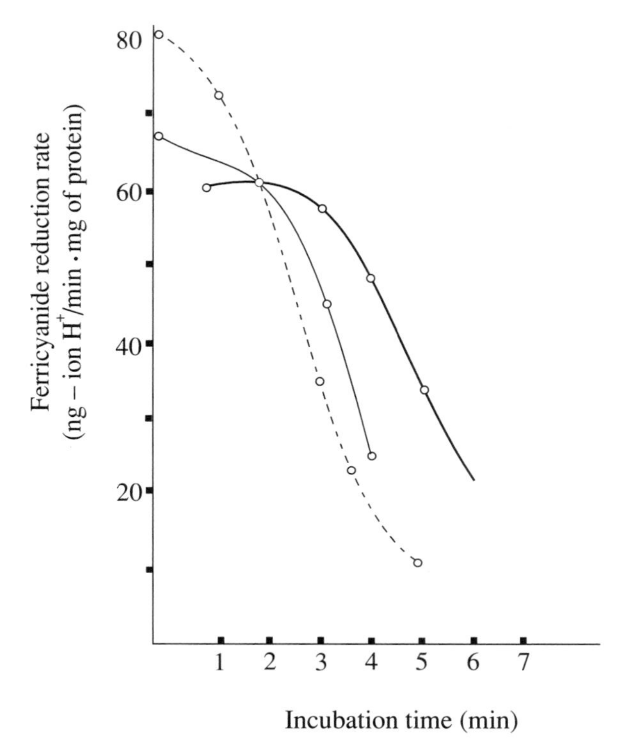

Figure 5.5. Time dependence of the ferricyanide reduction rate by aged mitochondria. Purple membrane absent (dashed line) and present in the dark (thin line) and under light exposure (thick line) [Okon and Vsevolodov 1987].

ing system. The detector is based on a highly ordered multilayer film of PM. The PM suspension-based film is poured at the conductive transparent base by means of electrophoretic sedimentation (see Section 4C.2). It constitutes one of the detectors electrodes. The film is covered by a dielectric plate containing a series of electrodes (pixels) of the assigned size, form and quantity. Figure 5.6 presents the simplest model of a position-sensitive matrix photodetector.

The typical kinetics of the photopotential and photocurrent increase and decrease is presented in Figure 5.7. Electrocurrent is measured on the 50 Ohm resistor, and photopotential is measured on a load resistor of 10^{15} Ohm. The intensity of permanent excitation by green light did not exceed 20 mV/cm^2. At such values of light intensity and load resistance, the photopotential on one pixel reached 0.7 V. A solo 7 ns laser impulse excited at a separate pixel generated an amplitude up to 3–4 μA [Kononenko et al. 1993]. A similar detector with 64 2×2 mm pixels based on 6–10 bacteriorhodopsin layers

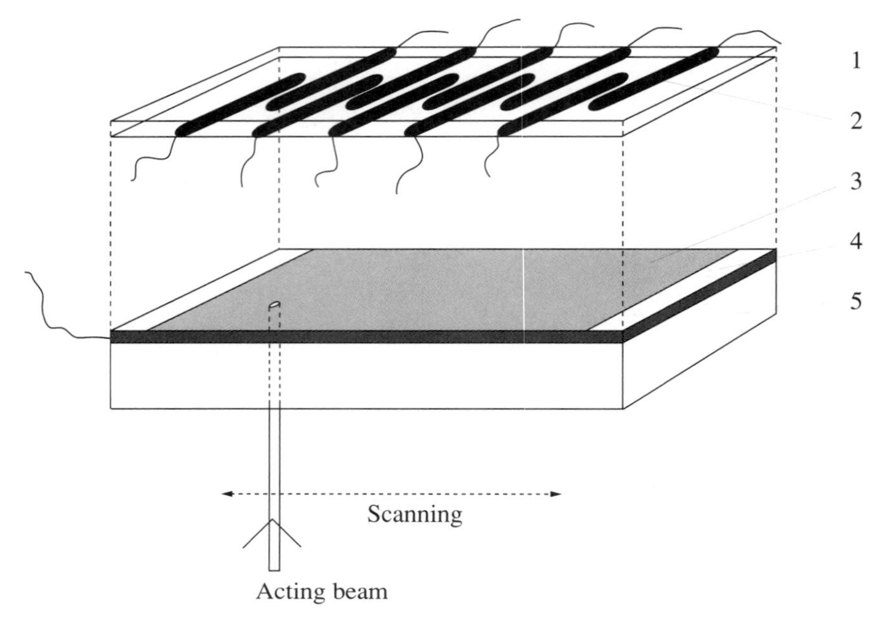

Figure 5.6. Diagram of a position-sensitive matrix photodetector on bacteriorhodopsin. (1). Dielectric plate (2). Metallic electrode-pixels (3). Electrophoretically oriented purple membranes (4). Transparent electroconducting coat (5). Glass support.

coated by the Langmuir-Blodgett method, allowed registration of the photocurrent impulse at 0.1–0.2 μA [Myasaka 1992].

These models are not made for commercial application. They demonstrate yet another possibility of retinal-proteins application in technical components. Naturally, this idea was patented in Japan [Fukumaza 1993] [Patent 06253606 Japan].

5F. Photovoltaic Devices

Photovoltaic devices (PD) can be applied to BR technology using the photoelectric properties of BR [Skulachev 1982]. These properties are described in Section 3A.6. One would think that measurement of the photocurrent would provide a voltage value, and vice versa. However, real solutions to such cross-measurement are not simple, and the advantage of current measurement before voltage, and vice versa, depends on specific experimental requirements. Experience shows that photovoltage measurements are easier, and do not

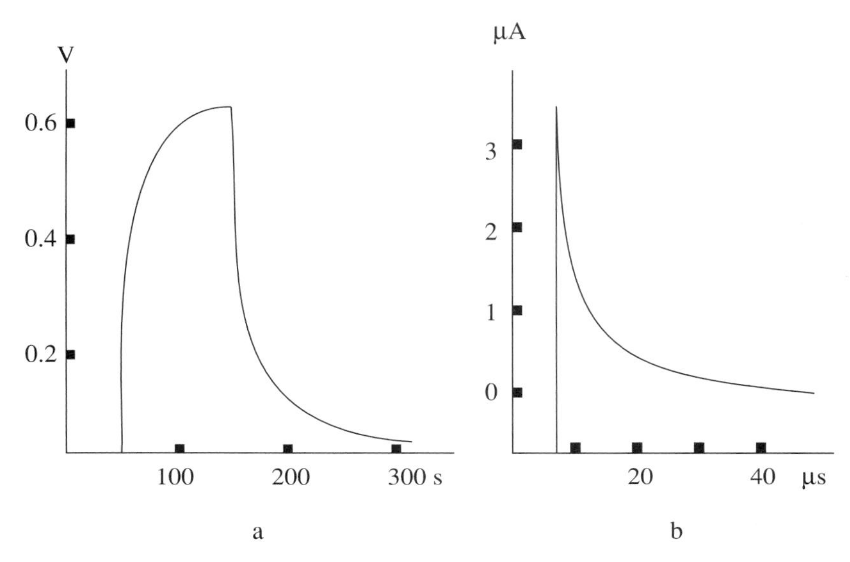

Figure 5.7. (a) Photoelectric signal induced by constant light upon a single pixel. (b) Photoelectric signal induced by a single laser flash upon a single pixel.

require complex registration schemes, since they are less sensitive to the preparation parameters and, more importantly, allow measurement of ultrafast elecroresponses with revolution times reaching 100 ps [Groma et al. 1984] [Trissl 1985]. Conversely, current measurements provide more information on the number and distance of charge motions in the PM, and the kinetics of the photocurrent is related to the kinetics of the photocycle intermediates [Fahr et al. 1981]. There are different ways to avoid experimental obstacles in order to obtain undistorted information on the charge processes in BR. For example, electroinduced PM orientation in polyacrylamide gel allowed measurement of photocurrent responses in the range from 70 to 30 nsec [Liu and Ebrey 1988].

The most popular PD construction is a combination of transparent conductive layers, gels or electrolytes serving as electrodes, and oriented PM in between the electrodes. The measurement of the potential or current is made according to standard physical methods. Theoretical and practical possibilities of the creation of such devices, based on photoelectric and photovoltaic properties of BR, have been discussed by Felix Hong and colleagues [Hong 1994] [Fuller et al. 1995]. Four most typical cells for the measurement of photovoltaic properties of BR were shown in Figure 5.2. Photoelectric responses normally consist of three components B1, B2, and B2'. B1 is a fast component, corresponding to the redistribution and/or

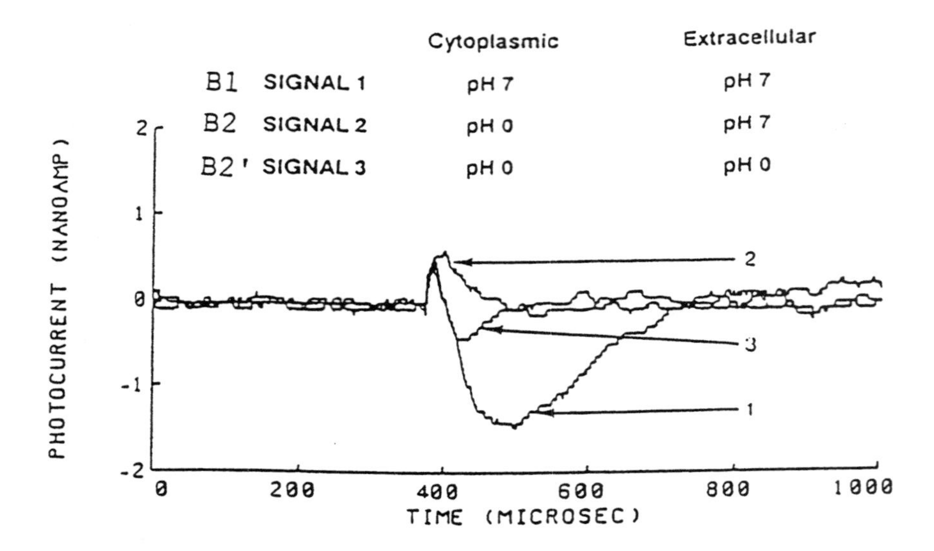

Figure 5.8. EPR-like signal in PM with three components. The B1 component has positive polarity and pH independence, whereas B2 and B2' components have negative polarity and pH dependence. PM was incorporated into a collodium membrane according to the method developed by Drachev et al. [1978]. When both sides of the membrane face neutral pH, both B1 and B2 are observable. When both sides are low, negative component B2' appears. This component is identified as hypothetical because it has a different relaxation time and its pH dependence is opposite to that of the B2 component—B2' is enhanced by low pH [from Hong and Okajama 1986].

separation of charges in the BR molecule, and is not sensitive to thermal and pH changes. B2 and B2' are the slow components which correspond to proton uptake and discharge, and are pH dependent. Depending on the type of retinal-protein, on the construction of cells, and measuring schemes etc., the number of B components may change.

PD is a general name, defining a device as a converter of light into electric energy. As the reader might have noticed, depending on their application and sometimes on the whim of the researcher, these devices may be called biosensors, photosensors, photocells, photodetectors, biochips and so on. Nomenclature does not alter the principles and the mechanisms at work. Which of the PD constructions will prove technologically convenient for the producer and appealing for the customer, is not yet clear. Perhaps, these will be the already patented constructions [Miyasaka 1990] [Patent 04006420 Japan] [Oyama et al. 1990] [Patent 03295278 Japan], or perhaps, new ones will appear.

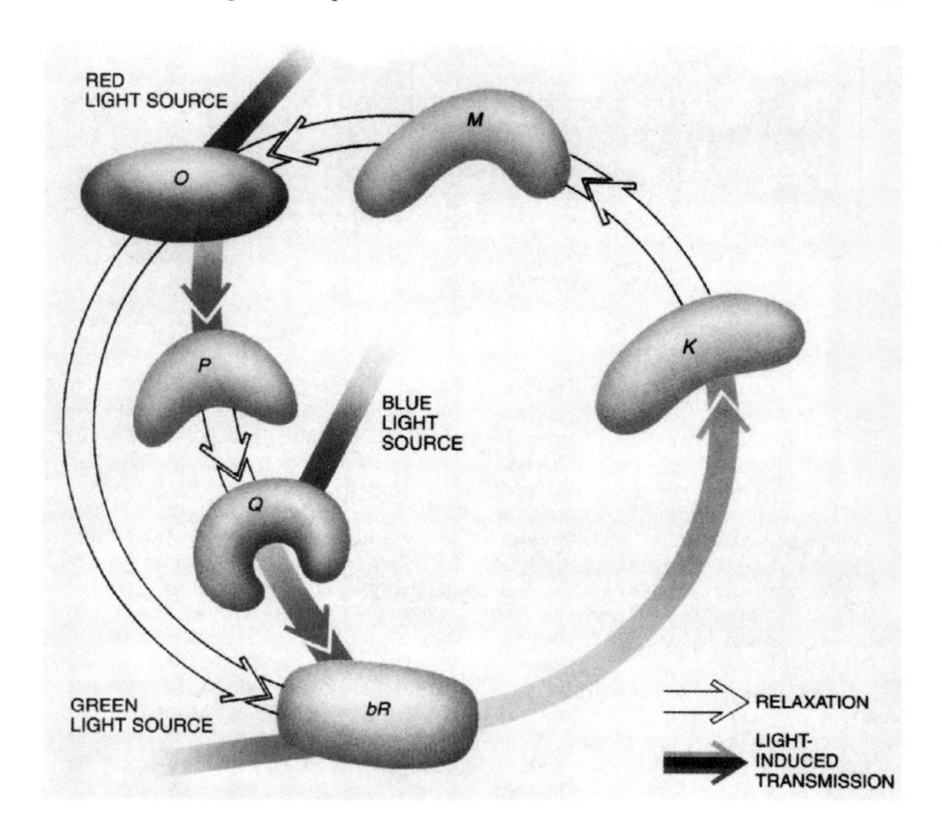

Figure 5.9. Photocycle of bacteriorhodopsin permits data storage. If the O intermediate is exposed to red light, a so-called branching reaction occurs. O converts to the P state, which quickly relaxes to the Q state. The Q state remains stable for a very long time. Blue light will convert Q back to ground state BR [Birge R 1995].

5G. 3-D Memory in Bacteriorhodopsin

The idea of creating a BR-based 3-D memory element was suggested in a survey-forecast [Vsevolodov 1984]. The idea was realized by Birge [Birge, Govender, 1991] [Birge 1995] [Chen et al. 1995]. For optical recording, he used his original idea of affecting BR with a sequence of impulses of different frequencies: first BR is excited by a green light impulse, and at the moment of the appearance of O intermediate, a red impulse arrives. Having absorbed the red impulse, O intermediate transforms into the short-lived P which, in turn, converts into long-lived Q. According to the estimates, Q lifetimes may last for years, if not for the blue light impulse which returns it to the primary state

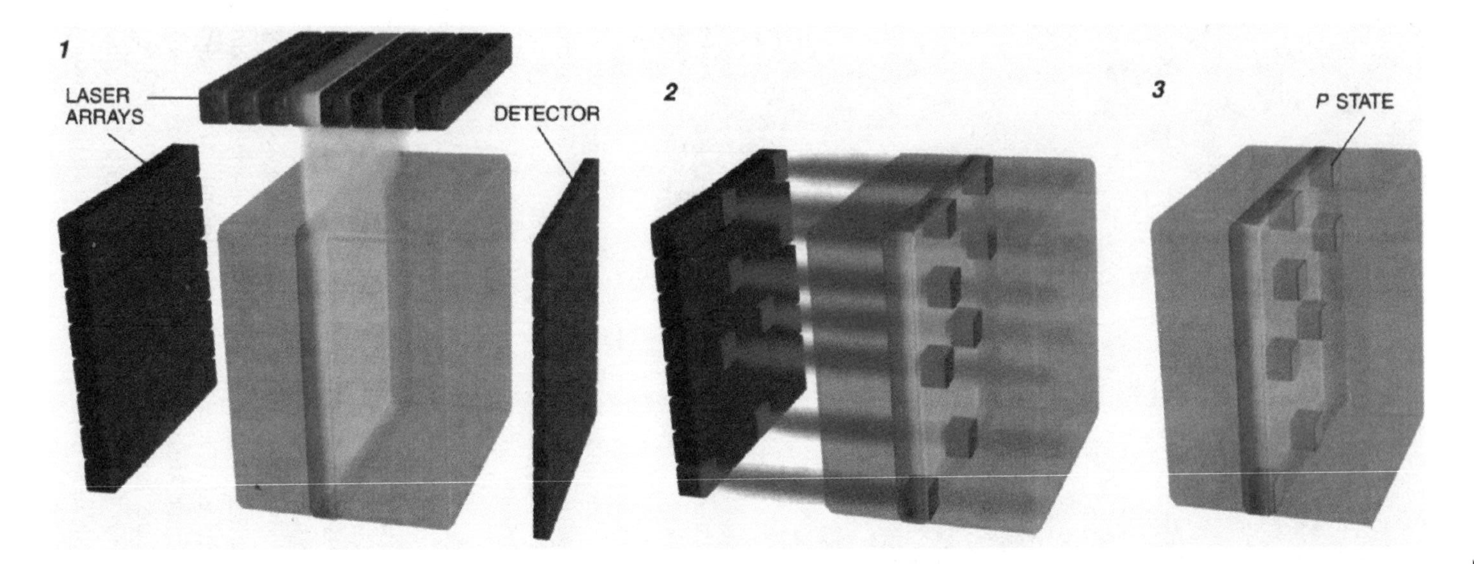

Figure 5.10. Sequence: (1). The first laser I impulse (green light) converts the BR molecules in the beam plane to the O state. (2). The second laser 2 impulse (red light) converts the programmed part of O -molecules to the state P. The rest remain in O.(3). P quickly converts to Q. (4). The read-out occurs at the switch-on of laser I (green light) and at the activation of the Q state. (5). Laser 2 (red light) is then turned on, however, in a low power regime. With the help of this weak red light the information, recorded by molecules in O and Q states, is passed onto the diode grating.

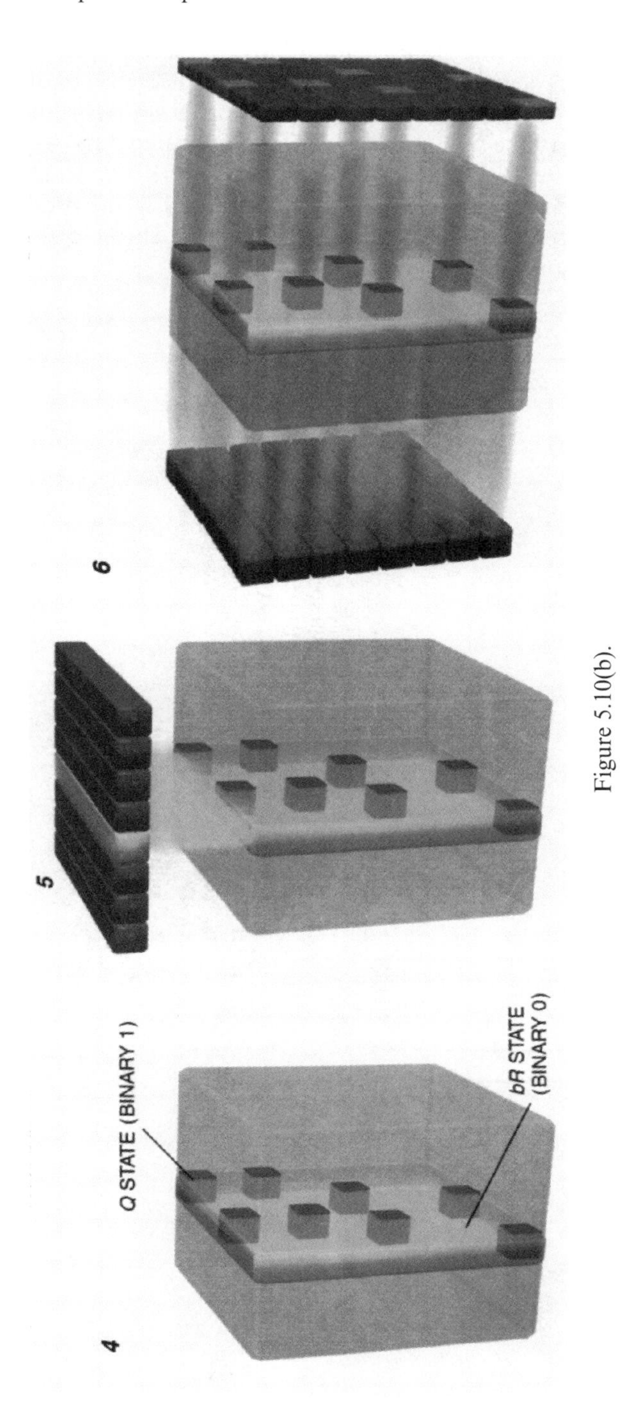

Figure 5.10(b).

of BR. On the basis of this photochemical process, parallel processors can be created. Figure 5.9 demonstrates the scheme of the described photoprocesses. Figure 5.10 shows the sequence of steps of information recording and read-out in a polymeric cube with oriented BR, incorporated into the cube [Birge and Govender 1991]. Technical characteristics of this element and the complexity of the assembly scheme are not essential at the present stage of research. At present, it is significant that a photochromic material with controllable memory time can be created.

6

Applications of Biochromes and Similar Films

By the end of 1985, it had become clear that biological biochromic films may find wide application in those fields of science and technology where the application of traditional organic and inorganic photochromic materials had been long suggested, but had to be postponed because of the low speed and modest cyclicity of the traditional photochromes (see Section 4B.). The results of Biochrome-BR research which had accumulated during the period from 1978–88, revealed their advantages against the traditional photochromes [Ivanitsky and Vsevolodov 1985] [Vsevolodov 1988]. In one of the referred works, the results obtained by Russian scientists (mainly physicists) were published, showing the following multiple capacities of biochromic films:

* Photomodulation of the diffraction index of BR during a BR photocycle—Photoanisotropy of the BR molecules
* Four-wave interaction in Biochrome films
* Amplitude-polarizational interaction in the Fabri-Perro nonlinear interferometer
* Polychromatic recording of dynamic holograms
* Anisotropy of electroinduced spectral changes in the oriented PM films
* Reversible two-beam-diffraction in Biochrome films

Researchers could not be indifferent about these encouraging results. It was also obvious that Biochrome and similar films based on BR analogs and variants could make a contribution to optoelectronics. The following years were marked by progress in the development of photosensitive films based on native BR, its analogs and genetic variants. Such research constituted a large contribution into the research of the new photomaterial and its properties [Birge 1990a] [Hampp et al. 1990 1992a] [Oesterhelt et al. 1991] [Haronian and Lewis 1991] [Werner et al. 1992] [Kononenko and Lukashev 1995].

Table 6.1. Comparison of traditional 2D photomaterial for optical memory

	Photographic plate	Optical disk	Photochromic films	Ideal
Cyclicity	No	No	10^3	$> 10^6$
Resolution (lines/mm)	1000	1000	> 5000	> 5000
Density (bytes/mm^2)	10^5	10^5	10^6	$> 10^6$
Sensitivity (J/cm^2)	10^{-8}	5×10^{-2}	10^{-2}	5.10^{-4}
Storage information:				
a. temporary	None	None	ms-hours	fs-days
b. archive (years)	10–50	50	None	50–100
Readout speed	None	None	0.01 ms	ps-fs

6A. Optical Memory

As was repeatedly remarked earlier, optical memory has greater capacity and reliability than magnetic memory. Table 6.1 shows comparative characteristics of optical, magnetic, and ideal memorizing mediums. The table data refers to the materials and optical memory devices (such as reference tapes and magnetic and optical disks) that are widely used in industrial and consumer technology. The modern laboratory research is concentrated on the development of new materials, components, systems and theories for the close and distant future. Some of the directions of this research are listed below:

* 3-D holographic memory
* associative memory and memory systems
* autowave and photosensitive biostructures
* self-training methods and self-organization optical-neural networks
* pattern recognition and machine vision
* optical neural computing
* optical interconnections
* pure optical readout
* stand-by photochromism (monitoring of the life-time)
* bi-stable and nonlinear devices
* artificial intelligence systems

1992 marked the founding of a new international journal called *Optical Memory and Neural Networks*. It is entirely dedicated to the above research directions. The terms "optical" and "neural," united under one title, speak in favor of the creation of a complex compact associative operative memory based on photosensitive proteins and complexes.

During the last decade, many technical systems for information recording, storage and readout have been developed and constructed for modern film-like photomaterials, including photochromes. Many of them have not gone beyond the lab and research center boundaries. It is natural that BR-based films (such as a photocarrier for optical disks, or a microfiche contrast range increasing photomaterial) have been tested in such systems in the first place. The fine sensometric and photochromic characteristics of biochrome and biochrome-like photomaterials allowed advancing them for application in prospective technologies and projects. Many patent holders offer to use PM as associative memory chips in a computer (not necessarily a biocomputer). An optical switch based on oriented BR layers, prepared by the Langmuir-Blodgett method is such an example [Schulz 1992.] [Patent 4241871 Germany] [Chen et al. 1994] [Patent 5346789 U.S.]. Another example is based on nonoriented PM multilayers, incorporated into polyvinyl alcohol at high pH, which notoriously prolongs the BR photocycle times and thus, the BR film is able to store the image longer than at normal pH [Chen et al. 1991].

The use of BR as memory cells in neural network architecture has been seriously discussed in literature for several years [Haronian and Lewis 1991, 1991a]. Section 6C describes the exotic way of biochromic film-based memorizing by means of generating the second harmonic of laser irradiation in the film. Nowadays, BR-based 3-D optical memory chips are not fantasy, but reality (see Section 5G). Rapid development of BR analogs, the use of gene engineering for the creation of new BR variants, the successful search for new natural phototransforming proteins are all factors stimulating the production of new biochromic films.

6B. Can Bacteriorhodopsin-based Compact Disks Exist?

Compact disks for laser phonographs appeared 10 years ago along with the dreams to use them in personal computers. Many specialists doubted the possibility of their introduction into computer marketing. Today compact disks are a common attribute of the personal computer, and it costs 10 times less than 10 years ago. However, large information systems require disks with larger memory. Despite their appearance even before compact disks, optical disks still encounter different problems.

Optical disks may be of three types:

* for the permanent storage of information recorded by the manufacturer. Such disks are designed for singular information recording and multiple read-outs. They are convenient for data and program storage, etc.;

* for permanent recording and additional recordings by the user;
* for multiple re-recording and temporal storage of information. We understand temporal storage as a time period from 8 hours (operative memory during the working day) to several years (the time of work on a project, etc.). Modern disks have diameters between 50 and 405 mm. Semiconductor lasers are used for the recording. The ways of recording may vary from bitmapping (the sequence of points and dashes) to holographic recording. Bitmapping is used for CD systems, and is the only recording method presently used in commercial production.

Recording media may be of different types such as thin nontransparent metal layers in which dots and dashes of the micron scale are burnt; amorphous crystals, transforming into crystals under the laser beam; thermoplastic materials, photochromic materials, etc. Only the first type, as the most cheap and reliable, is involved in industrial production. Compact disks are produced and sold by the millions. A compact disk is a transparent plastic disk coated with a few micron thick aluminum layer. These disks are used for high-quality recording of hour-long music and information programs, PC application packages, encyclopedias, study programs, etc. The big, so-called laser disks and special disks for professional use, may have up to 4Mbyte disk space. One such disk replaces 40 reels of magnetic tape or 50.000 microfiche. Information searches on a disk are infinitely faster than on a magnetic reel or on a microfiche catalogue. Optical disks are cheaper than magnetic disks of the same size. Big optical disks are convenient for the storage of archive files since no external factors can distort or erase the information except by physical destruction. However, optical disks for multiple recording are still in the process of development.

It is no small wonder, that several attempts have been made to create a BR-based optical disk for multiple recording and re-recording of information. According to the calculations, 10 times more information can be recorded on one unit of square surface of the BR layer than on the same square surface of a metal layer. The "write-read-out" speed at low temperatures reaches 10 psec. In 1985, our group received a 5-year grant from the Ukrainian Academy of Sciences for the creation of such a disk. The disk with 405 mm diameter was taken as a base. The results were inspiring. However, the work had to be stopped halfway for the reasons described in Section 4D.1. A different group took the base of a standard CD, and coated one of the side surfaces with the mixture of PM and gelatin by means of a spin coating technique. The result was a homogenous photochromic layer of even thickness. The disk provides spatial resolution of about 10 lines/mm; dynamic range of 60 dB in the mode of irreversible recording, and 34 dB in the reversible mode at a temperature between 210–215 K, and sensitivity around dozens of mJ/cm^2 [Belchinsky et al. 1991]. The same disk was created by Robert Birge with coworkers [Lawrence and Birge1984] [Birge et al. 1989] [Birge and Lawrence 1993].

6C. Generation of the Second Harmonic of Laser Radiation

The generation of the second harmonic of laser radiation is the most obvious manifestation of optical properties in BR [Aktsipetrov et al. 1987]. Retinal has a large (about 10^{-28} esu) nonlinear polarizability. Nonlinear polarizability of the retinal chromophore in molecular BR is 10 times larger than in free retinal [Huang et al. 1989]. Optical nonlinearity is the important and necessary characteristic of optoelectronic materials. Effectiveness of the material application for information storage and computation depends on the value of nonlinearity.

At the read-out of information from any photochromic material, this information is being erased, since the photoinduced form is photoactive. Upon IR radiation (1062 nm) of a thin layer of nonoriented PM, the radiation partially transforms into the second harmonic. The second harmonic irradiation maximum is 532 nm. This radiation excites only the ground state molecular BR, not interacting with the intermediate M. This permits avoiding the erasing of optical information by the readout beam in the process of recording [Aktsipetrov et al. 1987, 1987a] [Chen et al. 1991]. Simultaneously, it is possible to modify the intensity of the harmonic by means of additional lighting of the film from an external controlling light source. Figure 6.1

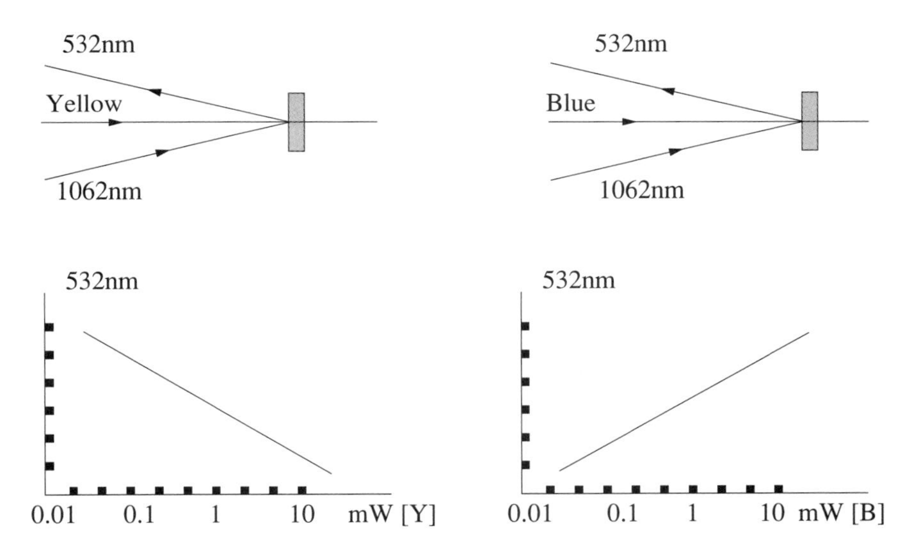

Figure 6.1. Generation of second harmonics of laser radiation by Biochrome-BR film. IR-laser radiation (1064 nm) incident on the surface film is doubled in frequency (523 nm) after it interacts with molecular bacteriorhodopsin. The intensity of the radiation emitted can be modified by illumination with light of yellow (Y) or blue (B) wavelength.

presents the schemes of second harmonic generation, and its experimental dependence on the intensities of the yellow and blue light.

Radiation of BR film with a 1064 nm laser results in outgoing radiation of half the wavelength (532 nm) and doubles the frequency. The intensity of the emitted radiation may be modified by illumination with either yellow or blue light. Since all the photocycle intermediates are photoactive, the ratio of the BR ground state to the photoinduced state may be changed by illumination; by changing the proportion of ground state BR molecules. The intensity of the 532 nm harmonics can also be changed. It is also worth noticing that air-dried films based on electrophoretically-oriented purple membranes represent the best existing model of an electret—a permanently polarized dielectric material.

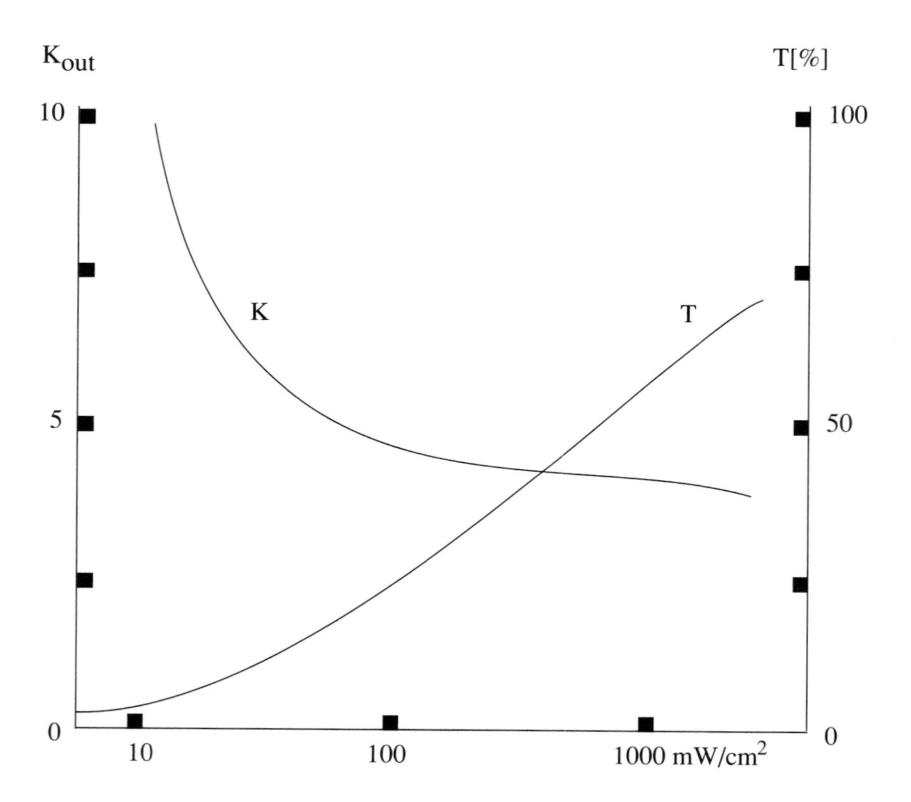

Figure 6.2. Dependence of contrast value (K) and optical transmission (T) of Biochrome-BR film on the intensity of the excitation light. Initial contrast—3.5; initial optical density of the film—3.0 in the range of measuring; the measuring wavelength—514 nm.

6D. Increasing the Contrast of a Photostat Copy

It was found that biochrome can be used to enhance the contrast of the unfocused He-Ne laser beam with an output power, varied continuously between μW and 10 mW. Filtering in the course of nonlinear absorption in Biochrome film enhanced the input signal contrast by over 100 times. An experimental investigation is represented in Figure 6.2 [Korchemskaya et al. 1993]. These properties can be used in the optical processing of images in the Fourier plane [Korchemskaya et al. 1990].

Biochrome films have been tested in microphotographic systems for reprography. Such diagrams are simple and are based on conventional reprographic configurations. The low-contrast (for instance, old microfiche) transparency may be enhanced using Biochrome-BR film.

This procedure is based on the bleaching of biochromes upon illumination of the transparent area of the transparency. This method is well-known under the name of masking. Halogenyde-silver photomaterial, normally used in this procedure, needs to be developed after the exposure and then brought into more precise coincidence with the original. Upon obtaining the higher-contrast copy, the photomaterial is trashed. Biochrome requires no developing or matching, and can be used indefinitely without need for replacement. In the course of a few years, Biochrome-BR testing for such purposes was conducted on one of the Russian military grants. Old microfiche was copied on photosensitive paper with or without Biochrome-BR coating. At other equal conditions, the legibility of signs was 50 and 100%, respectively [Vsevolodov and Filin 1985, unpublished].

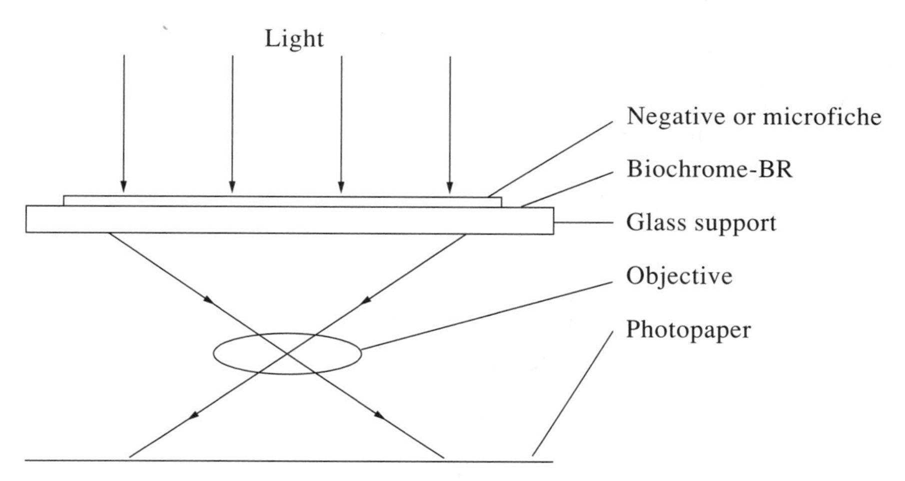

Figure 6.3. The method for Biochrome-BR film application in contrast enhancement of low-contrast negatives and microfiches. It is based on the ability of Biochrome-BR to bleach itself in the transparent areas of negatives, even when exposed to weak light.

6E. Polarization Holography

The birth of holography permitted dealing with the problems of information storage and processing in 3-D space which before had been regarded as science fiction. Virtually every operation on light waves became possible—transformation of wave fields, their comparison in thermal and spatial intervals (filtration and processing of optical information, interferometry and so on). In all holographic methods, three characteristics of the light wave are normally used. These are: length, amplitude, and phase. The fourth characteristic (polarization) is not recorded or reproduced, however, it carries the additional, and often very important, information on the object. The attempts to write and reproduce the fourth component in common photomedia encountered essential difficulties related to high precision requirements to the geometrical parameters of the scheme during 3-D image recording and reproduction [Kakichashvili 1972, 1982]. These difficulties can be avoided with the help of photoanisotropic photomedia, which can write and uniquely restore the direction of beam polarization. Such materials have been known since 1919 [Veiger 1919] but there are not too many of them, and they possess low sensitivity. The most popular one is methyl orange [Todorov et al. 1984, 1984a]. The search for a photomedium turned out to be a central problem. Biochrome-BR, having decent holographic effectiveness (see Section 4D.3) and capable of photoinduced anisotropy [Burykin et al. 1985, and references therein], can easily write and reproduce polarization holograms. Table 6.2 a, b presents proportional relations of diffraction effectiveness values of polarization holograms, depending on the states of polarization of write and read-out beams [reprinted fromVsevolodov et al. 1986; Hampp et al. 1992a]. Both tables demonstrate that diffraction effectiveness values of different polarization gratings are close to each other (only in one combination do they have a three-fold difference).

This is the important advantage of the BR-based holographic media, since

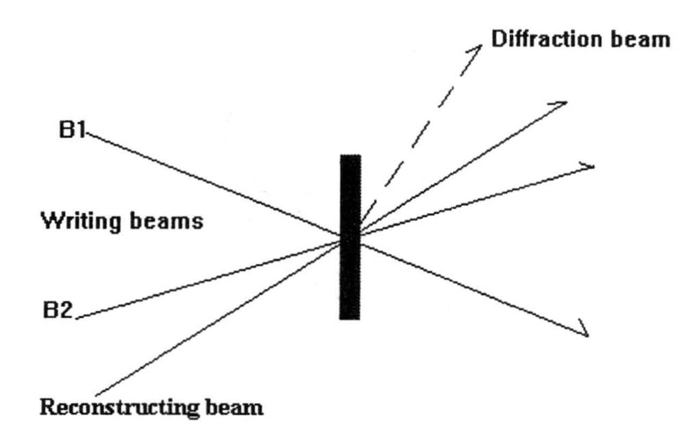

Table 6.2a. Polarization recording properties of Biochrom-BR films

No	Writing beams		Readouting beams		η
	B1	B2	Reconstracting	Diffracted	(relative)
1	↻	↻	↻	↻	2,5
2			↺	↺	2,5
3	↻	↺	↻	↺	1,5
4			↺		0
5	↕	↕	↕	↕	2,7
6			↔	↔	1,1
7	↕	↔	↕	↔	1,0
8			↔	↕	1.0

Table 6.2b. Polarization recording properties of Biochrome-BR films*

No	Writing beams		Readout beams		η
	B1	B2	Reconstructing	Diffraction	(relative)
1	↻	↻	↕	↕	0.8
2			↔	↔	0.8
3	↻	↺	↕	⬭	0.9
4			↔	⬯	0.9
5	↕	↕	↕	↕	0.9
6	↕	↔	↕	↔	0.3

*Modified from Hampp et al., 1990

in the common media (methyl orange, criptocyanide in solution, some mono- and diazopigments, etc.) the difference is more than 30-fold [Todorov et al. 1984].

Wavefront reversal is one of the most important research directions in nonlinear optics. It is also related to photoanisotropic abilities of photomedia [Zeldovich and Shkunov 1979]. Due to its high spatial resolution and the ability for polarization holograms recording, a high quality wavefront reversal with nearly 100% effectiveness has been obtained on biochrome. Figure 4.9 presented the experimental dependence of the reversed wave on the ellipticity of the signal beam [Vsevolodov et al. 1986].

6F. Real-time Holography

Because of its real-time write, erase and read-out characteristics, biochromic film is suitable for application with varying time scales. Holographic interferometry is an important tool for nondestructive testing. However, high-frequency signal registration is always hindered by harmful low-frequency oscillations of interferometer blocks and of the subject of investigation, and by alterations in the optical characteristics of the medium, penetrated by the probing beams. Commonly, in order to stabilize the working point of the interferometer complex, electromechanical stabilizers are applied. There is, however, a simpler and more effective way, called time-averaging interferometry [Powell and Stetson 1965].

This method requires a highly sensitive reversible holographic medium with the relaxation time coincident to the time of parasitic low-frequency oscillations (around hundreds of ms).

Biochromic film has shown good results when tried as a reversible holographic medium [Zaitsev et al. 1992a] [Kozhevnikov et al. 1990] [Hampp et al. 1990, 1990a]. For dynamic hologram recording, Kozhevnikov used only the direct transformation (BR $\longrightarrow$ M) in the BR photocycle, while Hampp used both the direct and the reverse (M $\longrightarrow$ BR) transformations (see Section 3A.5), and called the recorded holograms B-type and M-type holograms respectively. Investigation on time-averaging interferograms of a piezo-buzzer gave good results and revealed advantages and limitations of BR films for interferometry [Renner and Hampp 1993]. The main limitation is a low sensitivity value (around 30 mJ/cm^2 for M-type holograms) of the experimental film. The desired value is 0.5–1.0 mJ/cm^2. The sensitivity value for B-type reaches 1 mJ/cm^2, however, application of this type of hologram requires powerful blue light lasers. The film was based on D96N variant with 25 μm thickness and optical density of 0.7 around 570 nm.

Other experiments on optical hologram recording involved BR films based on methylcellulose. Despite its modest sensitivity, the methylcellulose based biochromic film [Zubov et al. 1987] showed decent results when applied as an adaptive holographic medium [Kozhevnikov and Lipovskaya 1990].

The important advantage of biochromic films in such systems is that the lifetime of holograms can be adjusted in a wide time range, and interferograms in biochromes can be tuned to the desired time scale. The use of Biochromes for various systems based on properties and peculiarities of dynamic holography must be one of the most important directions of biochromic application in the nearest future [Vsevolodov et al. 1992] [Downie 1994].

6G. Holographic Connections for Optical Communication

The appearance of optical fibers connecting networks raised the problem of switching a great number of channels without amplitude losses, or the narrowing of signal bandwidth. It still remains a problem, despite the fact that the schematic solution for the switch has been around since the beginning of the 1980s. The switch is a photochromic (for example, bismuth-silicon oxide) rectangular crystal with photorefractive performance characteristics. The beam from one of the fibers of the input channel, having passed through the crystal, deflects in the desired direction and is incident on the requested output channel. Deflection occurs on the diffraction grating recorded in the crystal by a special laser system exclusively at the spot of beam passage. Spatial frequency of the grating changes depending on the subscribers request and the beam, as a result of diffraction, deflects to the needed angle. Thus, the beam can be projected to every requested channel. However, the crystals are low-sensitive (0.1–1.0 J/cm^2) and not sensitive at all to red light, therefore, expensive and complex blue and green light lasers have to be used. Furthermore, losses in the switch depend on the value of holographic effectiveness which in photochromic crystals does not exceed 1%. In a working model, the effectiveness must be a few times greater. Spiroperans have sufficient effectiveness, but lose photosensitivity after a few hundred cycles whereas the switch in the telephone network system must have a cyclicity reaching millions.

Biochrome application for such schemes is more convenient. Their sensitivity, even in the red optical region, is by two orders higher allowing use of cheap semi-conductive red light lasers. As for holographic sensitivity, it depends on optical density and currently reaches 7%. First results obtained from experimenting with Biochrome-BR in such switches, are inspiring [Mikaelian et al. 1991] [Mikaelian and Salakhutdinov 1992]. Figure 6.4 demonstrates the scheme of a functioning model of an optical switch. It depicts an optical switch located between input and output clusters of optical fibers. Such clusters transmit large, complex information flows, and optical switches provide controllable and independent switching among a large number ($>$ 10,000) of channels.

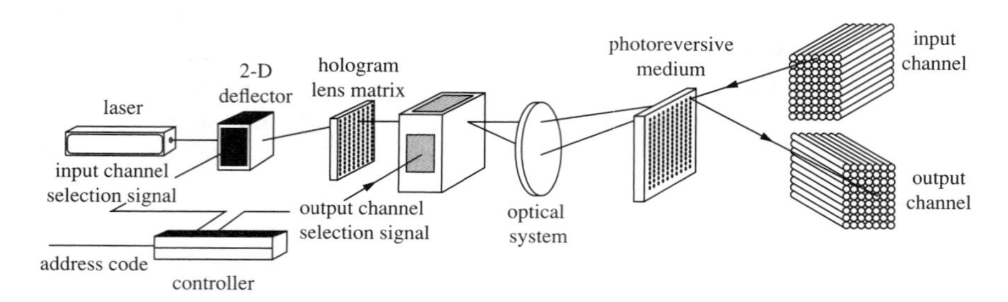

Figure 6.4. The use of Biochrome-BR film for optoelectronic switching of fiber nets. The controller splitter divides the laser beam into two beams and the deflector sends it to an address-defined site on the film to be recorded on a microhologram. One of the input beams directed onto the microhologram is reflected at a predetermined angle and falls onto the required output optic fiber. Depending on the address, microholograms can be recorded at different sites of the film, enabling the input and output channels to be switched selectively.

6H. Biomedical Applications

Nowadays, dynamic holography methods are more and more frequently used to investigate biological objects *in vivo*. In some cases, registration and processing of information, obtained from these experiments, requires reversible photoregistering mediums. However, in medical practices, only low-intensity light sources are used for safety purposes. Most traditional photochromic mediums do not react to light intensity less than 10^{-2} J/cm^2. The energetic sensitivity of Biochrome-BR is a few orders higher. Furthermore, it has similar reaction to light, ranging from green to red, and can be used in devices, using low-power and harmless helium-neon lasers and semiconductor lasers.

Korchemskaya and colleagues suggested using Biochrome-BR in order to determine the state of tissue *in vivo*. The polarization of the light wave, scattered from the tissue, depends on the subtle difference between the states of the tissue. This scattered wave serves as an object wave for dynamic holographic recording. Biochromic film allows registering the object wave polarization. In such a technique, the slightest difference between tissue states may be determined from the amplitude and polarization analysis of the diffractive wave [Korchemskaya et al. 1994].

Analogous methods may be suggested for the creation of high-sensitive devices for the investigation of the homogeneity of transparent tissues, and for diagnosing early pathological changes in the cornea and crystal at the early stages of cataract formation. Biochromic films have sufficient sensitivity to be utilized in simple individual indicators to measure working place illuminances or UV radiation levels (see Section 6J). Such indicators may be in the form of a badge, glued on a wristwatch belt, on the arm (for instance, during a stay at the beach) or on a transparent plastic card. More complex and precise indicators will be based on the array of biochromic film areas of different sensitivity, calibrated according to a specific illuminance scale. In order to construct a UV-tester, it will be necessary to investigate photochromic characteristics of those retinal-proteins and their analogs whose ground form absorption lies in the UVA and UVB optical regions. It is possible that in the nearest future, new natural proteins with photochromic properties similar to PYP (see Section 3H) will be discovered.

6.I. Military Use

Light radiation from a nuclear explosion spans the range from far UV to IR. It is so powerful that in parts of a second it burns eyes, skin, and clothes if they are not made out of a special white material. Common dark eyeglasses will not help because their optical density will have to be too high especially at night. The blinking reflex period is 0.15 seconds, and even during this small time the eye gets irreversibly damaged and temporarily blind. Photochromic eye glasses are the simplest device to protect eyes from the powerful light flux. Photochromic sunglasses can be found in stores. Photochromic eyeglasses designed for military purposes must have the following characteristics: an initial light filtration of no less than 50%; optical density in the "closed" state no less than 30 D; the speed of "closing" less than 0.1 mseconds, return time to the initial filtration level of 2–10 seconds, a temperature range of −40 to +50°C, a number of cycles over 1000; minimal distortion of color [Robinson and Gerlach 1969].

The main limitation of the traditional photochromes is their fast photodissociation, i.e., low light resistance. After a few photocycles, their filtration alters by 50%; after 50–100 cycles their photochromic properties completely vanish. Biochrome meets only part of the demand, and can be used as an addition to another photomaterial. For instance, traditional photochromes have the initial band in the blue spectral range, which in light, moves into green and red. In Biochrome-BR, this process is the opposite. Application of a BR analog or BR variant mixture is also possible.

Protection from laser radiation (from laser weapons or "peaceful" lasers in the labs) with the help of photochromes is even more complicated. The data show that, depending on the type of laser, eye adaptation, direction, focusing, etc., the photoinduced change of optical density must occur in nanoseconds and must reach an optical density of 7–9 units in the range of irradiation during the same time period [Brown 1971]. As far as the author knows, photochromic materials with such characteristics do not exist yet. Time will show if there are any among the biological photochromes.

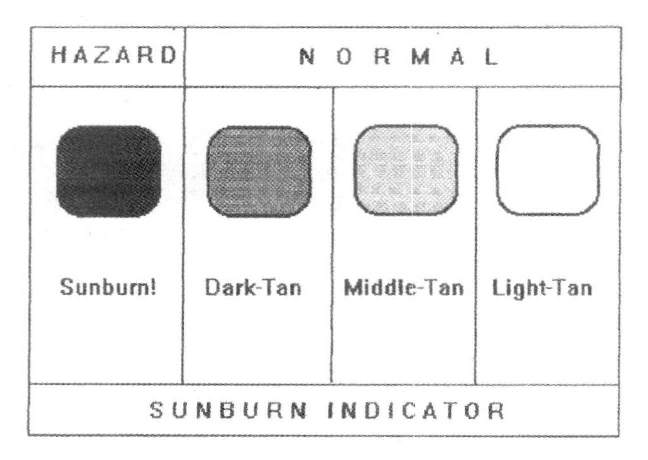

Figure 6.5. One of the possible types of a detector of dangerous levels of sun radiation. Such a tester is reliable, long-lasting, and sufficiently precise.

In the 1980's, the author received a grant to research the possibilities of BR and its analogs application for the camouflage of military technology, and the uniforms of military personnel under the conditions of nuclear war. Technology with reflective coating, and uniforms made of white material were covered and saturated with a solution of polyvinyl spirit mixed with PM suspension and a few BR analogs. At normal illuminations, the coating, depending on the analog composition, made everything look as a normal uniform. Upon a powerful light flash simulating a nuclear blast, the coating lightened and the white base reflected the light. In parts of a second, the disguising color was restored. We achieved certain results, but as the cost of the coating was high with no prospects to make it cheaper, the work was stopped.

6J. Light Testers

Approximately a billion people experience health problems related to sunlight exposure. Numerous epidemiological studies have shown a direct correlation between light exposure and certain types of cataracts. Much of the evidence points to UV light playing a particularly important role in the development of many eye diseases and age- related muscular degenerations [Hightower and McCready 1992].

Today there are plenty of electrical and electronic radiometers for the measurement of visual light and UV-irradiation. The Robertson-Berger

Table 6.3a. Radiation emission values of some sources.

Sources (types)	Visible radiation	UV radiation emission (mJ/cm^2)		
	Red-Violet 700–450 nm	UVA 400–320 nm	UVB 320–280 nm	UVC 280–200 nm
Sun (40° latitude)	100–70	10–5	0.4–0.1	0
Sunlight lamps		15–7	0.1–0.01	0
Arc lamp with filter		50–40	0.3–0.2	0
UVB lamps		0.3–0.4	0.2–0.05	0.0003

* Radiation emission values based on data reported in reference: Passchier and Bonjakovic 1987

Table 6.3b. Damage threshold for eye tissue

Wavebands (nm)	Damage thresholds (mJ/cm^2)*		
	Cornea	Lens	Retina
305–315	10–15		
310–320		2–15	
400–500			20–40

* Damage thresholds data was composed on the basis of numerous data reported during 1978–1994.

meter, and others, are currently in use. These radiometers are complex in their design, expensive and inconvenient for regular customers.

The energetic photosensitivity of biochromic films exceeds the energetic thresholds of hazardous radiation levels for eye and skin (see Table 6.3a, b).

The basis of the indicator is a photochromic film based on BR and analogs of BR (ABR). At present, several types of ABR with blue and UV absorption are available. However, only several blue ABRs are described in the literature and their initial absorption covers only a part of the UVA range [Mao et al. 1981]. Investigations of the present BR analogs with initial absorption in blue and near-UV, as well as the search for the new ones, are yet to continue.

Indicators could well turn out to be a transparent polymeric film (hard or flexible), of 0.2–0.5 mm of thickness, with one or a few 1 × 1 cm or more PF patches. They will have different energetic sensitivities. The number of PF patches, sensitivity, and types (BR or ABR) will depend on the type of the indicators. Every indicator type will be calibrated according to hazardous threshold levels for eye tissue and for different types of skin. More accurate indicators have to be calibrated against skin types.

Today only dihydrorhodopsins with short-wave initial absorption maximums have been investigated. However many analogs of chlamyrhodopsins and sensory rhodopsins have a maximum of absorption of the ground state form in UV of the area of a spectrum (see Sections 2B.5 and 3C.5).

6K. From Toys to Computers

The high sensitivity of Biochrome-BR in the visual range may be used for the invention of many games, scientific experiments, and toys. If insulated between two transparent polymeric layers (like in ID cards, driver's license cards, etc.), biochromic film preserves its characteristics for unlimited time,

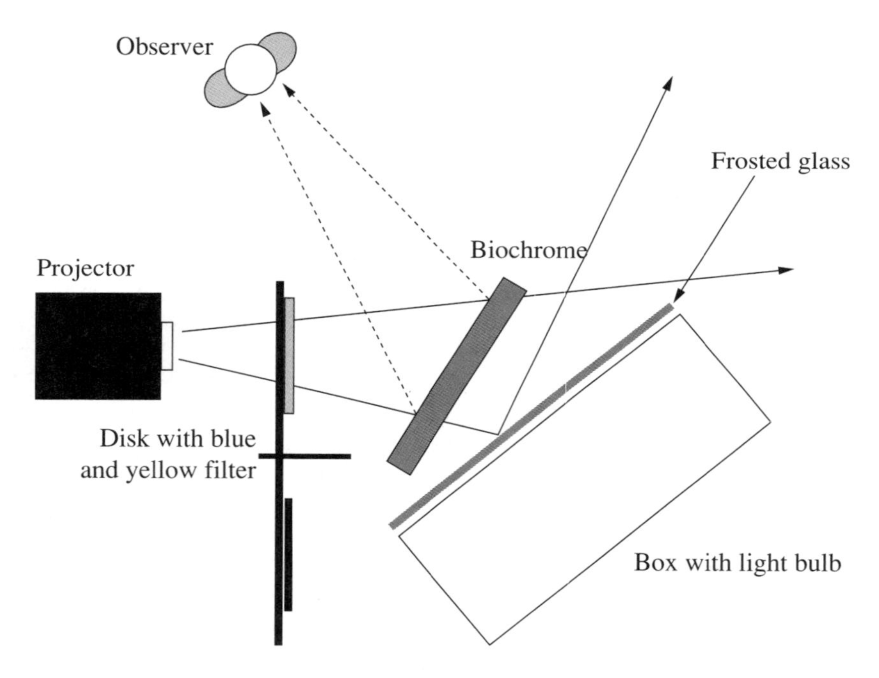

Figure 6.6. A demonstration of optical information processing (image optical conversion). The frosted-glass light source exposes Biochrome-BR film, keeping the initial (BR) and the photoinduced (M) forms in equilibrium. The ordinary negative or positive film or slide is placed into the projector. It is being projected on Biochrome-BR film. Upon projection through the yellow filter, the equilibrium shifts towards the photoinduced form and the negative image appears. Upon projection through the blue filter, the equilibrium shifts toward the initial form and the negative image transforms into a positive one. For the demonstration of information storage (optical memory) one can block the projector beam after a clear image appears on the film.

and may serve as a multi-purpose toy, good for both entertainment and learning. Such photochromic plates (PP) can be used in the following ways:

1. For writing or drawing with a laser pen-pointer,
2. Since the image on PP disappears in a minute or less, it is good for computing: how many words can you write before they disappear?
3. PP can be used to demonstrate temporal optical memory. For example, the projectional copying of slides and/or negatives in a photographic enlarger,
4. With the help of a photoflash, you may copy slides, negatives, photographs, posters on PP, and then copy the temporary slide onto photopaper,
5. PP is good for picture-puzzle games (finding differences between the two seemingly similar pictures). New interesting effects may appear, if two pictures are subsequently copied on PP. At the same time, it is one of the simplest illustrations of optical information-processing methods,
6. Rotation of PP between crossed Polaroids gives the effect of a strangely colored kaleidoscope which serves as an elegant demonstration of the photoanisotropic properties of molecular BR (see Section 6E)
7. Finally, it is worth mentioning that amateur photographers may use PP to improve the contrast of old negatives and photoplates (described in section 6D).

An impressive experiment can be staged at home if you have a blue light filter and a slideprojector. The outline and description of the experiment is shown in Figure 6.6. It is perhaps, the simplest demonstration of one of the elements of the optical image-processing computer—the transformation of the negative image into a positive, and *vice versa*. Take a transparent plate instead of a slide and make a drawing on it using blue and black markers. On the biochromic plate, blue details will disappear exactly at the moment of the appearance of the black ones, and the reverse is also true. With a minimum of imagination, most anyone can invent a series of "magical drawings".

A PM suspension with gelatin may be added to children's chemical reagents' sets for the domestic production of films. The materials are absolutely harmless. Addition of BR analogs and variants to the suspension, will make the experiments more exciting. All these games are related to optical information processing by optical means. Naive as they are, they will aquaint your child with mysterious scientific words- biotechnology, biocomputer, optical transformations, optical memory...

7

Conclusion

Having completed the book, the author admits that he could not possibly review all of the work in this area. He is tempted to write more, make changes, or even rewrite the book anew. Moreover, in less than a year of writing, so much new has appeared. However, as Kozma Prutkov, a Russian writer that never existed used to say, "Nobody can embrace the unembraceable."

Apparently, other books will tell of photoelectric cells, using only retinal molecules [Li et al., 1994] of a hydrogen photogenerator based on the hybrid of a modern semiconductor and the bacterium *Clostridium butiricum* [Krasnovsky et al., 1985], of piezoelectric and electret properties of BR [Ketis 1981], of holographic investigation of the molecules of bacteriorhodopsin [Vsevolodov et al., 1996], of the similarity between PM and the domains of liquid crystals, of color holography on biochromes, and so on. Yellow proteins warrant more attention than we could give in the book because interest has only recently avalanched. It appears that more attention should have been paid to photovoltaic systems since it is known that the U.S department of energy has spent millions of dollars in this area. Chlorophylls have had but a brief description, and nothing has been said of analogs, whereas the history of their investigation is no less exciting than that of retinal–proteins, and their future also belongs to bioelectronics. The curious reader will certainly find this material in the References section and reviews [Petty et al., 1995] [Birge, 1994]. In the next few sections we will render a conclusion, and touch upon the natural, nonretinal-protein complexes also regarded as ready bioelectronics elements.

7A. Proteins in Bioelectronics

Photoreaction centers were described in Section 1A. It is known that primary photosynthesizing processes are characterized by a high quantum yield, reaching 80%. Investigations of photoelectric characteristics of bacterial photosynthetic membranes and isolated reaction centers in mono- or/and multilayers have been active since the beginning of the 1980s [Tiede et al., 1982] [Erokhin et al., 1987] [Zaitsev et al., 1992]. The results stimulate attempts to use different vegetable proteins in artificial systems of light conversion into electric energy [Allen 1977]. As is well-known, under certain conditions, certain phenomena are observed in chlorophyll solutions: the photochromic effect, the electrochromic effect, photodependent conductivity, pH changes in the light [Kiselev and Kozlov 1980] [Gudkov et al., 1975]. Chlorophyll monolayers coated onto electrodes of different types, give a high amplitude photoresponse—up to 50 mV [Kiselev and Kozlov 1980]. The response time being measured in minutes, these systems are more convenient for biosensor systems than for informational systems.

RC may be regarded as a potential photomaterial for the registration of IR radiation since they have a high molar extinction coefficient (300000 M^{-1} cm^{-1} around 800 nm) and a quantum yield close to one. According to calculations, the PM suspension film with optical density close to one will have sensitivity close to 0.1 mJ/cm^2 [Shuvalov 1984]. Apparently, plant photosensitive pigments, as well as retinal-proteins, can claim the role of biomolecular elements for optoelectronics and, possibly, for biocomputers.

VR demonstrates BR-like photoelectric properties and, under certain activity- preserving conditions, can act as a phototransducer with a temporal resolution of less than 1 msec. Investigations of VR-ERP, incorporated into an artificial collodium membrane, confirm this hypothesis [Drachev et al., 1981]. HR may find more universal applications in bioelectronics than VR and BR, due to the strong dependence of their photocycle on the concentration of Cl^- ions in the medium (see Section 3E). The polarity of the HR photoelectric signal in Trissl-Montall films reverses upon the replacement of KCl with sodium citrate [Michaile et al., 1990].

As for photosensitive proteins, we made it clear that they can be used in the major systems—of information perception, primary recognition and sorting, processing of the necessary information, and memorizing of the result. They can participate in the systems of solar energy conversion, and its storage as an electrochemical potential. Auxiliary systems controlling movement, mechanical actions, self-repair etc., are subjects for bionics and biomechanics.

There is however, a topic that we have not mentioned in this book—not only because it goes beyond its scope, but also because it is still difficult to pronounce something distinctive on its behalf. This is the problem of proteins and protein systems responsible for other senses—hearing, taste of smell, sense of touch. It is possible that the same proteins and protein systems that are found in the primary receptors of sense control organs will be used for the creation of artificial biosensors.

In every photochemically-active protein complex, the light impulse is memorized in the form of reaction products. If at least one reaction product is stable, the light impulse gets "memorized" for the lifetime of this product without changes. If the product of the photoreaction depends on an external electric field, the result is a functional element with two input controllers. All experiments, so far, as it follows from the material of this book, are staged on artificially constructed microsystems. Further development of nanobiology will enhance miniaturization [Nagayama 1992]. At what stage of biocomputer technology development will preprogrammed self-assembly of proteins will acquire major importance? Today this question refers more to theory than to practice.

7.B. The Truth and Science Fiction of Biocomputers

Even the most stubborn electronics specialist will not argue that the best existing electronic device based on the Josephson effect, with a propagation delay of 10^{-12} s, and power dissipation of about 10^{-6} W, cannot compete with molecular devices with a power dissipation of less than 10^{-8} W, and the speed of primary processes close to less than a picosecond at room temperatures. Biomolecules win in the size parameter also. This is the undisputed reality. However, electronics and bioelectronics are destined to develop parallel to each other as two spirals of DNA, for a long time, creating more and more hybrid children of human progress. What is waiting at the end of this spiral? Nuclear electronics, i.e., information processing and storage at the atomic level? Possibly, but this is not the final goal, only a rite of passage... to what? The atom is a giant compared to many physical particles actively and independently interacting with each other. Which generation of scientists will take control over and create a femtocalculator to benefit the impossible, to fill the belly of scientific achievements? And bioelectronics will, in this vision of progress, be forgotten...

When should we expect bioelectronics to enter our daily lives? The first patents for liquid crystals were issued half a century ago. The first devices, clocks, indicators, etc., on liquid crystals appeared in stores 20 years ago. It took 30 years to create thermally stable liquid crystals. Judging by this example, the devices based on bioelectronic materials are not to be expected before 2010. This simple calculation coincides with the chart presented in Figure 7.1 (reprinted from the survey of David Bloor [Bloor 1991].

Having read the Introduction, the reader might have acquired an optimistic conception about the development of biomolecular computing technology as the final goal. However, on the temporal scale of human progress, the biocomputer is nothing but the nearest future, a single step forward... To create the model of the human brain, with nothing but the now common notions of information input-processing-output, is the same as to sit for a second at the start of a long-distance race. Perhaps future scientists will perceive this differently. Nowadays, we do not have even a name for a science that will study intelligence, and not the brain. Today, we can make only a

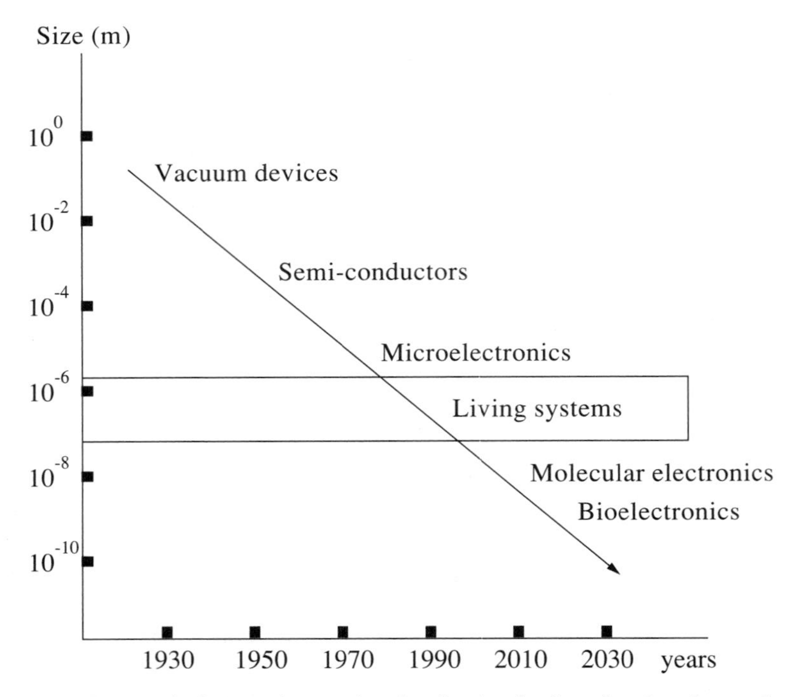

Figure 7.1. Extrapolation on the continual reduction in size of active electronic and bioelectronics devices [modified from Bloor 1991].

weak attempt to speculate about this future science, based on the rare body of original thinking.

It is possible that the difference between information processing in biological and organic molecules will soon be determined [Hong 1992]. However, the more complex mechanisms humanity creates, the more poignant is the question: what is the difference between live and inanimate? This is especially important to comparing the work of macromolecules and protein complexes with artificial systems of similar sizes. There exist a few theories, concepts and methods of research on this subject: The Theory of Coherent Excitation [Frohlich 1983], The Theory of Limited Diffusion [Levitt et al., 1985], the hypothesis of a "Protein Machine" [Chernavsky et al., 1987]. The diversity of these concepts and the absence of a general one, tells us that the problem has just been approached. The authors of the latter theory, think that it is different from the others because it takes into account the information that determines molecular function. In their opinion, therein lies the difference between the biomolecule and inanimate objects of comparable sizes. They regard a molecule as a machine of molecular size with typical characteristics normally assigned to micromachines. The work of the protein organism is united by the interrelated work of a certain array of such singular molecular machines—proteins, enzymes, and other molecules. Each of the protein-machines may be regarded as an electromechanical machine since electrical and mechanical forms of energy are closely interrelated at the molecular scale; which constitutes the significant difference between micro- and macrosystems. The future will show which concept will triumph at creating a biocomputer.

Attempts are underway in determining the molecular dynamics of proteins using modern computers [Karplus and McCammon 1983]. This approach to the life-death problem is considered promising, however, the enormous amount of parameters entered demands extra powerful computers and consumes a lot of time and has not yielded any unique results.

The self-assembly of biocomputer elements, blocks, and schemes, remains the most attractive, though, the most fantastic problem. The theoretical foundation for it is being created right now [Conrad 1990, 1992].

We repeatedly mentioned vision as the most informative way of information input. But is it the primary one? Psychologists have shown experimentally that with the absence of all information perceiving organs, excepting tactile (skin) organs, still allows a full life to be led, whereas the absence of tactile transducers, with intact vision and hearing, fully distorts brain functioning [Vekker 1997]. The conclusion is obvious: the modern understanding of a biocomputer as an imitation of the human brain is incorrect. More likely, a biocomputer is the name of the device invented by humans as an imitation of a part of the brain available for modern knowledge. We cannot see beyond the edge of knowledge.

It is also known that every intelligence—natural or artificial—is only part of the accessible physical world. If this physical world abides by laws—and only in this case—can we count on the connection between physical and biophysical constants and, thus, create a theory for the optimal working of a biocomputer. Optimally functioning biocomputers should not distort the system which is the object of its study. Its elementary microcomputers must then consist of superelementary microcomputers, built on elementary particles. The author of this theory, Efim Liberman, called them molecular particles [Liberman 1978]. Molecular particles may long have been at work in our brain which we do not know of...

Let us not wander into the jungle of scientific thought that grows around every new achievement, generated by somebody's personal natural computer. Let us leave these problems for our successors, and may curiosity never abandon humanity!

8

References

Abdulaev NG, Kiselev AV, Ovchinnikov Yu A (1976): Effect of proteolic enzymes on purple membranes of *Halobacterium halobium*. *Bioorganicheskaya Khimia* **2**: 1148–1150.

Abdulaev NG, Feigina M Yu, Kiselev AV, Ovchinnikov Yu A, Drachev LA, Kaulen AD, Khitrina LV, Skulachev VP (1978): Products of limited proteolysis of bacteriorhodopsin generate a membrane potential. *FEBS Lett.* **90**: 190–194.

Adamus G, McDowell JH, Arendt A, Hargrave PA, Smyk-Randall E, Sheehan J (1987): *Biophys. Struct. Mech.* **9**: 86–94.

Adey WR, Lawerence AF, eds (1984): Nonlinear Electrodynamics and Biological Systems: New York: Plenum.

Aktsipetrov OA, Akhmetdiev NN, Vsevolodov NN, Esikov DA, Shutov DA (1987): Generation of second harmonics of laser irradiation by bacteriorhodopsin molecules. *Sov. Phys. Dokl.* **32**: 219–220.

Aktsipetrov AO, Akhmetdiev, NN, Vsevolodov NN, Esikov DA, Shutov DA (1987a): Methods of information readout by an optical memory system. Patent 4,251,628, USSR.

Alexiev U, Scherrer P, Mollaghababa R, Khorana HG, Heyn MP (1994): The proton released during the photocycle of bacteriorhodopsin migration along the surface of the purple membrane. *Abstr of VI Intern. Conference of Retinal Protein.* **19–24** June, 1994, Leiden, NL: 65.

Alexiev U, Marti T, Heyn MP, Khorana HG, Scherrer P (1994a): Surface charge of BR detected with covalently bound pH indicators as selected extracellular and cytoplasmic sites. *Biochemistry* **33**: 298–306.

Allen MJ (1977): *Direct conversion of radiation into electrical energy using plant system. Living system as energy converters* (Buvet R. et al. Eds) North-Holland: 271–274.

Arai R, Haruta M, Yamamoto N, Yano T, Kishi H, Sakuranaga M (1986): Recording material and color recording process using this material. Patent 63,092,946 Japan.

Arnold FN (1992): Engineering enzymes for non-aqueous solvents. *Trends Biotechnol.* **8**: 244–249.

Arnaboldi M, Motto MG, Tsujimoto K, Balogh-Nair V, Nakanishi K (1979): Hydroretinals and hydrorhodopsins. *J. Am. Chem. Soc.* **101**: 7082–7084.

Asato AE, Li X-Y, Mead D, Patterson GML, Liu RSH (1990): Azulenic retinoids and the corresponding bacteriorhodopsin analogs. Unusualy red-shifted pigments. *J. Amer. Chem.* **112**: 7398–7399.

Avi-Dor Y, Rott R, Schnaiderman R (1979): The effect of antibiotics on the photocycle and protoncycle of purple membrane suspension. *Biochim. et Biophys. Acta* **545**: 15–23.

Bakker EP, Caplan SR (1982): Phospholipid substitution of the purple membrane. *Meth. Enzymol.* **88**: 26–30.

Balashov SP, Litvin FF (1976): Primary photochemical reactions of bacteriorhodopsin in purple membranes and cells of Halobacteria at 4K. *Bio-organic Chemistry* **2**: 565–566.

Balashov SP, Litvin FF (1981): Phototransformation of metabacteriorhodopsin. *Photobiochem. Photobiophys.* **2**: 111–114 and reference therein.

Balashov SP, Karneeva NV, Litvin FF, Sineshchekov MV (1987): Primary photoreactions of *trans-* and 13-*cis*-bacteriorhodopsin. In: *Retinal proteins* (Ovchinnikov YuA, Ed) Utrecht, Netherlands VNU Science Press: 505–517.

Balashov SP, Litvin FF, Sineshchekov VA (1988): Photochemical processes of light energy transformation in bacteriorhodopsin. In: *Sov. Sci. Rev.*, Section D: *Physicochem. Biol.* (Skulachev VP, Ed) **8**: 1–61.

Balashov SP (1995): Photoreactions of the photointermediates of bacteriorhodopsin. *Isr. J. Chem.* **35**: 415–428.

Bamberg E, Hedemann P, Oesterhelt D (1984): The chromoprotein of halorhodopsin is the light-driven electrogenic chloride pump in *Halobacterium halobium*. *Biochemistry* **23**: 6216–6121.

Bamberg E, Oesterhelt D, Tittor J (1994): Function of halorhodopsin as a light-driven proton pump. *Ren. Physiol. Biochem.* **17**: 194–197.

Baofu Ni, Chang M, Duschl A, Lanyi J, Needlman R (1990): An efficient system for the synthesis of bacteriorhodopsin in *Halobacterium halobium*. *Gene*. **90**: 169–172.

Barabas K, Der A, Dancshazy Zs, Ormos P, Marden M, Keszthelyi (1983): Electro-optical measurements on aqueous suspension of purple membrane from *Halobacterium halobium*. *Biophys. J.* **43**: 5–11.

Baryshnikova LM, Shakhbazyan V Yu, Sokolova LM, Vsevolodov NN (1988): Effect of the purple membranes on glucose transport and respiration in *Rhodococcus minimus* B239. *Appl. Biochem. and Microbio.* **24**: 529–534.

Barlow HB (1988): The thermal limit to seeing. *Nature* **334**: 296–350.

Baselt DR, Fodor SP, Van der Steen R, Lugtenburg J, Bogomolni RA, Mathies RA (1989): Halorhodopsin and sensory rhodopsin contain a C6–C7 *-trans* retinal chromophore. *Biophys. J.* **55**: 193–196.

Bayley ST (1979): Halobacteria—a problem in biochemical evolution. *Trend Biol. Sci.* **4**: 223–225.

Bayley H, Hojeberg Bo, Huang K-S, Khorana HG, Liao M-J, Lind Ch, London E (1983): Delipidation, renaturation, amd reconstitution of bacteriorhodopsin. *Meth. Enzym.* **88**: 74–81.

Bazhenov V Yu, Soskin MS, Taranenko VB (1985): Giant nonlinearity and optical bistability based on reversible transformations of bacteriorhodopsin in biochrome films. *Thesis at the XII All-Union Conference on Coherent and Nonlinear Optics*. Moscow Univ. **1**: 325.

Beece D, Brown SF, Czege J, et al. (1981): Effect of viscosity on the photocycle of bacteriorhodopsin. *Photochem. Photobiol.* **33**: 517–522.

Belchinsky G, Bogatov P, Gresko A (1988): Data storage light-sensitive matrix—uses p-n junction as memory cell for memory matrix, Patent 1,511,765 USSR.

Belchinsky G.Ya., Bogatov PN, Gresko AP, Kononenko AA, et al. (1991): Application of pigment-protein complexes as carriers of optic information. *Biophysics* **36**: 248–251.

Bennett N, Michel-Villaz M, Kuhn H (1982): Light-induced interaction between rhodopsin and the GTF-binding protein. Metharhodopsin II is the major photoproduct involved. *Eur. J. Biochem.* **127**: 97–103.

Bennett RR, Brown PK (1985): Properties of the visual pigments of the moth *Manduca sexta* and the effects of two digitonin and CHAPS. *Vision Res.* **25**: 1771–1781.

Birge R, Hubbard LM (1980): Molecular dynamics of *cis-trans* isomerization in rhodopsin. *J. Amer. Chem. Soc.* **102**: 2195–2204.

Birge RR, Hubbard LM (1981): Molecular dynamics of *trans-cis* isomerization in bacteriorhodopsin. *Biophys. J.* **34**: 517–534.

Birge RR, Pierce BM (1983): In: *Photochemistry and Photobiology* (Zewail AH, Ed.): New York: Harwood Press: 841–855.

Birge RR (1986): Two-photon spectroscopy of protein-band chromophores. *Acc. Chem. Res.* **19**: 138–146.

Birge RR, Murray LP, Zidovetsky R, Knapp HM (1987): Two-photon, 13C and two-dimensional 1N NMR spectroscopic studies of retinal Schiff's bases, protonated Schiff's bases, and Schiff's base salts: evidence for a protonation induced pp* excited state level ordering reversal. *J. Amer. Chem. Soc.* **109**: 2090–2101.

Birge RR, Einterz CM, Knapp HM, Murray LP (1988): The nature of the primary photochemical events in rhodopsin and isorhodopsin. *Biophys. J.* **53**: 367–385.

Birge R (1989): Optical random access memory based on bacteriorhodopsin. *Bull. Am. Phys. Soc.* **34**: 843.

Birge RR, Zhang CF, Lawrence AF (1989): Optical random access memory based on bacteriorhodopsin. In: *Molecular Electronics* (Hong F, Ed): Plenum: N.Y: 369–379.

Birge RR (1990): Nature of the primary photochemical events in rhodopsin and bacteriorhodopsin. *Biochem. et Biophys. Acta* **1016**: 293–327.

Birge RR (1990a): Photophysics and molecular electronic applications of rhodopsins. *Ann. Rev. Phys. Chem.* **41**: 683–733.

Birge R, Govender D (1991): Three-dimensional optical memory; Field-oriented bacteriorhodopsin. Patent 5,253,198 U.S..

Birge RR (1992): Protein-based optical computing and optical memories. *IEEE Computer* **25**: 56–67.

Birge RR, Lawrence AF (1993): Optical random access memory Patent 5.228.001 U.S..

Birge RR, ed. (1994): Molecular and Biomolecular Electronics. Advances in Chemistry, Series 240 Washington. DC: *American Chemical Society*.

Birge RR (1995): Protein-Based Computers. *Scientific American* March: 90–95.

Birge RR, Gross RB (1995): Biomolecular optoelectronics. In: *Introduction to Molecular Electronics*. (Petty MC, Bryce MR, Bloor D, Eds) London: Edward Arnold: **1**: 315–344.

Birge RR, Gross RB (1994): Biomolecular Optoelectronics. *In: Molecular and Biomolecular Electronics*. (Birge RR, Ed.) Advances in Chemistry series 240. Washington, DC. American Chemical Society.

Rosenfeld, Bivin DV, Stoeckenius W (1986): Photoactive retinal pigments in Halophylic bacteria. *J. Gen. Microbiol.* **132**: 2167–2177.

Bjorn LO (1979): Photoreversible photochromic pigments in organisms: properties and role in biological perception. *Quart. Rev. Biophys.* **12**: 1–23.

Blank A, Oesterhelt D (1987): The halo-opsin gene. II. Sequence, primary structure of halorhodopsin and comparison with bacteriorhodopsin. *EMBO J.* **6**: 265–273.

Blank A, Oesterhelt D, Fernando EE, Schek ES, Lotspeich F (1989): Primary structure of sensory rhodopsin I, a prokaryotic photoreceptor. *EMBO J.* **8**: 3963–3971.

Blaurock AE, Stoeckenius W (1971): Structure of the purple membrane. *Nature New Biol.* **233**: 152–155.

Bloor D (1991): Breathing new life into electronics. *Physics World* **11**: 36–40.

Blume H, Bader T, Luty F (1974): 13i—directional holographic information storage based on the optical reorientation of F centers in KCl:Na. *Opt. Commu.* **12**: 147–151.

Bochove AC, Griffiths WT (1984): The primary reaction in the photoreduction of protochlorophillide monitored by nanosecond fluorescence measuraments. *Photochem. Photobiol.* **39**: 101–106.

Bogomolni R, Taylor ME, Stoeckenius W (1984): Reconstitution of purified halorhodopsin. *Proc. Natl. Acad. Sci. U.S.* **81**: 5408–5411.

Bogomolni RA, Spudich JL (1992): Archaebacterial Rhodopsin: Sensory and energy transduction membrane proteins. Sensory receptors and signal transduction. Wiley-Liss, Inc.: 233–255.

Bolshakov BI, Drachev LA, Kalamkarov GR, Ostrovsky MA, Skulachev VP (1979): Cooperative processes in bacterial and visual rhodopsins: light energy conversion into electric potential difference. *Rep. USSR Acad. Sci.* **249**: 1462–1466.

Borisevitch GP, Lukashev EN, Kononenko AA, Rubin AB (1979): Bacteriorhodopsin BR570 bathochromic band shift in an external electric field. *Biochem. Biophys. Acta* **549**: 171–174.

Bortwick HA, Henrics SB, Toole EN, Toole VK (1954): Action of light on lettuce-seed germination. *Bot. Gaz.* **115**: 205–225.

Braiman MS, Mogi T, Marti T, Stern LJ, Khorana HG (1988): Vibration spectroscopy of BR mutants: light-driven proton transport involves protonation changes of aspartic acid residues 85, 96 and 212. *Biochemistry* (Washington) **27**: 8516–8520.

Bridges CDB, Yoshikami S (1970): The rhodopsin-porphyropsin system in freshwater fish. I. Effects of age and photic environment. *Vision Res.* **10**: 1315–1332.

Brith-Linder M, Avi-Dor Y (1979): Interaction of ionophore with bacteriorhodopsin: A flash photometric study. *FEBS Lett.* **101**: 113–115.

Brown, AD (1963): The peripheral structures of Gram-negative bacteria IV. The cation sensitive dissolution of the halophilic bacterium *Halobacterium halobium*. *Biochem. Biophys. Acta* **75**: 425–435.

Brown JK, Murakati M (1964): A new receptor potential of the monkey retina with no detectable latency. *Nature* **201**: 623–628.

Brown GH Ed. (1971): Photochromism. techn. chem. 111: N.Y: Wiley-Interscience.

Brown LS, Druzhko AB, Chamorovskii SK (1992): Light adaptation of a bacteriorhodopsin analog with 4-keto-retinal. *Biophysics* **37**: 66–71.

Bunkin FV, Vsevolodov NN, Druzhko AB, Mitsner BI, Prokhorov AM, Savranskii

VV, Tkachenko NV, Shevchenko TB (1981): Diffraction efficiency of bacteriorhodopsin and its analog. *Sov. Tech. Phys. Lett.* **7**: 630–631.

Burykin NM, Vsevolodov NN, Djukova TV, Korchemskaya E Ya, Soskin MS, Taranenko VB (1983): Photoinduced anisotropy of bacteriorhodopsin. *Ukrainsky Fizichesky Zhurnal* **28**: 1269–1270.

Burykin NM, Korchemskaya EY, Soskin MS, Taranenko VB, Djukova TV, Vsevolodov NN (1985): Photoinduced anisotropy on biochrome film. *Opt. Commun.* **54**: 68–71.

Byzov AL, Kuznetsova LP (1971): On the mechanisms of Visual adaptation. *Vision Res. Suppl.* **3**: 51–63.

Casadio R, Gutowitz H, Howery P et al. (1980): Light-dark adaptation of bacteriorhodopsin in triton-treated purple membrane. *Biochem. et Biophys. Acta.* **590**: 13–23.

Cazeca MJ, Kaplan DL, Kumar J, Marx KA, Samuelson LA, Sengupta SK, Tripathy SK, Wiley BG (1995): Dynamic adapted camouflage system with gps. of linked photodetector—having thin film layers of light-reactive proteins, conductive polymers and oligomers, Patent 5438192 U.S..

Chamorovskii SK, Lukashev EP, Kononenko AA, Rubin AB (1983): Effects of electric field on the photocycle of BR. *App. Opt.* **13**: 767–778.

Chamorovskii SK, Kononenko AA, Lukashev EP (1990): Photosensitive biosensors. In. Biosensors, Itogi Nauki, VINITI, *Biotechnology* **26**: 73–133. (In Russian).

Chekulaeva LN, Koryagin VV, Timuk OE (1978): New bacterial strain *Halobacterium halobium* 353 Pushchinsky—can be used in microbiology as a producer of bacteriorhodopsin used in studies of mechanism of sight. Patent 2,631,669 USSR.

Chen Z, Lewis A, Takei H, Nebezahl I (1991): Bacteriorhodopsin oriented in polyvinil alcohol films as an erasable optical storage medium. *Appl. optics* **30**: 5188–5196.

Chen Z, Birge RR (1993): Protein-based artificial retinas. *Trends Biotechnol.* **11**: 292–300.

Chen Z, Lewis A, Takei H (1994): Oriented biological material for optical information storage and processing. Patent 5,346,789 U.S..

Chen Z, Govender D, Gross R, Birge R (1995): Advance in protein-based three-dimentional optical memories. *Biosystems* **35**: 145–151.

Chernavsky DS, Khurgin YuI, Shnol SE (1987): Concept of "the protein-machine" and its implications. *Biophysics* **32**: 836–844.

Chibisov AK (1966): Photoisomerism of certain dyes during flash excitation. *Theor. Exp. Chemistry* **2**: 596–598.

Clark NA, Rotshild KJ, Luipold DA, Simon BA (1980): Surface induced lamellar orientation of multilayer membrane arrays. *Biophys. J.* **31**: 65–96.

Clayton RK, Sistrom WR, Eds (1978): *The Photosynthetic Bacteria.* N.Y. London: Plenum Press.

Collier RJ, Burckhardt CB. Lin LN (1971): *Optical Holography*. New York: Academic Press.

Cone RA (1967): Early receptor potential: Photoreversible charge displacement in rhodopsin. *Science* **155**: 1128–1131.

Conrad M (1972): Information processing in molecular systems. *Curr. Mod. Biol.* **5**: 1–14.

Conrad M (1973): Is the brain an effective computer? *Int. J. Neurosci.* **5**: 167–170.

Conrad M, Friedlander C, Acinbehin K (1982): Biochips: a feasibility study—a report submitted to National Geno Sciences, Inc. (Wayne State Univ.).
Conrad M (1984): Microscopic-macroscopic interface in biological information processing. *BioSystems* **16**: 345–363.
Conrad M (1985): On design principles for a molecular computer. *Commun. ACM* **28**: 464–480.
Conrad M (1990): Quantum mechanics and cellular information processing: The self-assembly paradigm. *Biomedica Biochimica Acta* **49**: 743–755.
Conrad M. (1990a): Molecular computing. In: *Advances in Computers* (Yovits M.C, Ed) San Diego: Academic Press **31**: 235–324.
Conrad M (1992): Quantum molecular computing: The self-assembly model. *Int. Quant. Chemistry: Quantum Biology Symposium* **19**: 125–143.
Cooper A (1979): Energy uptake in the first step of visual excitation. *Nature* **252**: 531–533.
Crescitelli F (1978): The chloride ionochromic response: an in situ effect. *Vision Res.* **18**: 1421–1422..
Crescitelli F, McFall-Ngai M, Horwitz J (1985): The visual pigments sensitivity hypothesis: further evidence from fish of varying habitats. *J. Compar. Physiol. A.* **157**: 323–333.
Crescitelli F (1991): The natural history of visual pigments: 1990. Progress in the retinal proteins **6**: 1: 32.
Crouch RK, Ebrey Th, Govinjee G (1981): A bacteriorhodopsin analog containing the retinal nitroxide free radical. *J. Am. Chem. Soc.* **103**: 7364–7366.
Cullin DW, Vsevolodov NN, Dyukova TV, Weetall HH (1995): Optical properties of Triton X-100-treated purple membrane embedded in gelatine film. *Supramolecular Science* **2**: 25–32.
Dameron CT, Winge DR (1991): Peptide-mediated formation of quantum semiconductors. *Trend Biotechnol.* **8**: 3–6.
Dancshazy Zs, Karvaly B (1976): Incorporation of bacteriorhodopsin into a bilpolar lipid membrane; a photoelectric-spectroscopic study. *FEBS Lett.* **72**: 136–138.
Danon A, Stoeckenius W. (1974): Photophosphorylation in *Halobacterium halobium. Proc. Natl. Acad. Sci. U.S.* **71**: 1234–1238.
Danon A (1977): Biogenesis of the purple membrane of *Halobacterium halobium. Biophys. Struct. and Mech.* **3**: 1–17.
Davson H, ed (1990): Physiology of the eye. N.Y: Pergamon Press: p. 280.
De-Grip WJ (1988): Recent chemical studies related to vision. *Photochem. and Photobiol.* **48**: 799–810.
Deininger W, Kroeger P, Hegemann U, Lottspeich F (1995): Chlamyrhodopsin represents a new type of sensory photoreceptor. *EMBO J.* **14**: 5849 .
Delaney JK, Schweiger U, Subramaniam S (1995): Molecular mechanism of protein-retinal coupling in bacteriorhodopsin. *Proc. Natl. Acad. Sci. U.S.* **92**: 11120–11124.
Dencher NA, Heyn MP (1978): Formation and properties of bacteriorhodopsin monomers in the non-ionic detergents octyl-b-D-glucoside, Triton X-100. *FEBS Lett.* **96**: 322–326.
Dencher NA, Hildebrand E (1979): Sensory transduction in *Halobacterium halobium:* retinal protein pigment controls UV-induced behavioral response. *Ztschr. Naturforsch. Bd.* **34**. S: 841–847.
Der A, Hargittai P, Simon J (1985): Time-resolved photoelectric and absorption

signals from oriented purple membranes immobilized in gel. *J. Biochem. Biophys. Methods* **10**: 259–300.

Derguini F, Caldwell CG, Motto MG, Balogh-Nair V, Nakanishi K (1983): Bacteriorhodopsin containing cyanin dye chromophores. Support for external point-charge model. *J. Amer. Chem.* **105**: 646–648.

Derguini F, Nakanishi K (1986): Synthetic rhodopsin analogs. *Photobiochem. Photobiophys.* **13**: 259–283.

Detwiler PB, Hodgkin AL, McNaughton PA (1980): Temporal and spatial characteristics of the voltage response of rod in the retina of the snapping turtle. *J. Physiol.* **300**: 213–250.

Djukova TV, Vsevolodov NN, Chekulaeva LN (1985): Dehydration-induced changes in photochemical activity of bacteriorhodopsin in polymeric matrix. *Biofizika* **30**: 613–616 (in Russian).

Dobrikov MI, Shishkin GV (1983): Photomaterial with enzymes. *Autometriya* **5**: 23–33.

Dorion GH, Wiebe AF (1970): Photochromism: Optical and photographic application. N.Y. Focal Press: 121–131.

Dorion GH, Weibe AF (1970a): *Photochromism in Living Systems.* Photochromism London: Focal Press: 240–250.

Dowling JE (1987): The Retina: *An Approachable Part of the Retina.* Cambridge, Ma: Harvard Univerity Press.

Downie JD (1994): Real-time holographic image correction using bacteriorhodopsin. *Appl. Opt.* **23**: 4353.

Drachev LA, Kaulen AD, Ostroumov SA, Skulachev VP (1974): Electrogenesis by bacteriorhodopsin incorporated in a planar phospholipid membrane. *FEBS Lett.* **39**: 43–45.

Drachev LA, Jasaitis AA, Kaulen AD, Kondrashin AA, Liberman EA, Nemechek LB, Ostroumov SA, Semenov AYu, Skulachev VP (1974a): Direct measurement of electric current generation by cytochromoxidase H-ATPase and bacteriorhodopsin. *Nature* **249**: 321–324.

Drachev LA, Florov VN, Kaulen AD et al (1976): Reconstitution of biological molecular generators of electric current. *J. Biol. Chem.* **251**: 7056–7065.

Drachev LA, Kaulen AD, Skulachev VP (1977): Time resolution of the intermediate steps in the bacteriorhodopsin-linked electrogenesis. *FEBS Lett.* **87**: 161–167.

Drachev L, Kaulen AD, Semenov AYu, Severina II, Skulachev VP (1979): Lipid-impregnated filters as a tool for studying the electric current-generating proteins. *Analy. Bioch. Dm.* **96**: 250–262.

Drachev LA, Kaulen AD, Khitrina LV, Skulachev VP (1981): Fast stages of photoelectric processes in biological membrane. I. Bacteriorhodopsin. *Eur. J. Biochem.* **117**: 461–470.

Drachev LA, Kalamkarov GR, Kaulen AD, Ostrovsky MA, Skulachev VP (1981a): Fast stages of photoelectric processes in biological membranes II: Visual rhodopsin. *Eur. J. Biochem.* **117**: 471–481.

Drachev LA, Kaulen AD, Skulachev VP (1984): Correlation of photochemical cycle, H^+ release and uptake, and electric events in bacteriorhodopsin. *FEBS Lett.* **178**: 331–335.

Druckmann S, Renthal K, Ottolenghi M, Stoeckenius W (1984): The radiolytic

reduction of the Schiff base in bacteriorhodopsin. *Photochem. and Photobiol.* **40**: 647–651.
Druzhko AB, Zharmukhamedov SK (1985): "Biochrome" films based on same analogs of bacteriohodopsin. In: *Light Sensitive Biological Complexes and Optical Information Recording*. Ivanitsky, Ed. Pushchino Press: 129–136.
Druzhko AB, Zharmukhamedov SK, Vsevolodov NN (1986): Photoinduced transformation of 4-keto bacteriorhodopsin in a polymeric matrix. *Biophysics*. **31**: 253–257.
Druzhko AB, Chamorovskii SK (1995): The cycle of photochromic reaction of a bacteriorhodopsin analog with 4-keto-retinal. *BioSystems* **35**: 133–136.
Druzhko AB, Chamorovskii SK, Lukashev EP, Kononenko AA, Vsevolodov NN (1995): 4-keto-bacteriorhodopsin films as a promising photochromic and electrochromic biological material. *BioSystems* **35**: 129–132.
Dujardin E, Kuiper Y, Cremer R, Sironval C (1976): Recording information by means of a photosensitive material of biological origin. Patent 1,423,991 Great Britain.
Dyukova TV, Vsevolodov NN, Chekulaeva LN (1985): Changing of bacteriorhodopsin photochemical activity in the polymeric matrix at dehydration. *Biophysics* **30**: 613–616.
Dyukova TV, Saphonova MV, Vsevolodov NN (1990): Optical parameters of purple membranes in polymeric films. Temperature and humidity effects. *Proc. 12th IEEE/EMBS Conference*, Philadelphia, U.S. **12**: 1716.
Dyukova TV, Vsevolodov NN (1996): Photochromic composition and material containing bacteriorhodopsin, Patent 5,518,858 U.S..
Eakin RM, Martin GG, Reed CT (1977): Evolution significance of fine structure of archiannelid eyes. *Zoomorphologie* **88**: 1–18.
Ebrey TG (1993): Light energy transduction in bacteriorhodopsin. In: *Thermodynamics of Membrane Receptors and Channels* (Jackson MB, Ed) CRC Press: 353–387.
Eguchi C, Waterman TH, Aciyama J (1973): Localization of the violet and yellow receptor cells in the crawfish *retinula*. *J. Jen. Physiol.* **62**: 355–374.
Eisenbach M, Caplan R (1979): Interaction of purple membrane with solvents: II. Mode in interaction. *Biochim. et Biophys. Acta* **554**: 281–292.
Eisenstein L. (1982): Effect of viscosity on the photocycle bacteriorhodopsin. *Meth. Enzymol.* **88**: 297–305.
Elkin RM (1982): Continuity and diversity of photoreceptors. *Visual Cell in Evolution*. N.Y. Raven Press: 91–105.
Engelmann DM, Handerson R, McLachlan ADF, Wallace BA (1980): Path of the polypeptide in bacteriorhodopsin. *Proc. Natl. Acad. Sci. U.S.* **77**: 2023–2927.
Erokhin VV, Feigin LA, Kayushina RL, Lvov YuM (1987): Langmuir films of photosynthetic reaction centers from purple bacteria. *Studia Biophys.* **122**: 231–236.
Fahr A, Lauger P, Bamberg E (1981): Photocurrent kinetics of purple membrane sheets bound to planar bilayer membranes *J. Membr. Biol.* **60**: 50–62.
Fang JM, Carriker JD, Balogh-Nair V, Nakanishi K (1983): Evidence for the necessity of double bond (13-ene) isomerization in the proton pumping of bacteriorhodopsin. *J. Am. Chem. Soc.* **105**: 5162–5164.
Feng Y, Menick DR, Katz B, Beischel CJ, Hazard ES, Misra S, Ebrey TG, Crouch

RK (1993): Probing of the retinal binding site of bacteriorhodopsin by affinity labeling. *Biochemistry* 33: **38**: 11624–11630.

Ferguson J, Mau AW-H (1974): *Molecul. Phys.* **27**: 377–387.

Fesenko EE, Kolesnikov SS, Lyubarsky AL (1985): Induction by cyclic GMP of cationic conduction on plasma membrane of retinal rod outer segment. *Nature* **313**: 310–313.

Fisher KA, Yanagimoto K, Stoekenius W (1978): Oriented absorption of purple membrane to cationic surfaces. *J. Cell. Biol.* **77**: 611–621.

Fisher VCh, Oesterhelt D. (1979): Chromophore equilibria in bacteriorhodopsin. *Biophys. J.* **28**: 311–230.

Foster KW, Smith RD (1980): Light antennas in phototactic algae. *Microbiol. Rev.* **44**: 572–630.

Foster KW, Saranak J, Patel N, Zarrillli G, Okabe M, Kline T, Nakanishi K (1984): A rhodopsin is the functional photoreceptor for phototaxis in the unicellular eukaryote *Chlamidomonas*. *Nature* (London) **311**: 756–759.

Foster KW, Saranak J, Derguini F, Rao VJ, Zarrilli GR, Okabe M, Fang G-M, Shimizu N, Nakanishi K (1988): Rhodopsin activation: A novel view suggestion by in vivid *Clamydomonas* experiments. *J. Am. Chem. Soc.* **110**: 6588–5689.

Foster KW, Saranak J (1989): The *Chlamidomonas* (chlorophyceae) eye as a model of cellular structure, intracellular signaling and rhodopsin activation. Algae as Experimental Systems. in: *Plant Biol.* **7**: 215–230.

Foster KW, Saranak J, Derguini F, Zarrilli GR, Johnson R, Okabe M, Nakanishi K (1989): Activation of *Chlamidomonas* rhodopsin *in vivo* does not require isomerization of retinal. *Biochemistry* **28**: 819–824.

Foster KW, Saranak J, Dowben PA (1991): Spectral sensitivity, structure and activation of eukaryotic rhodopsins: Activation spectroscopy of rhodopsin analogs in *Chlamidomonas*. *J. Photochem. Photobiol. B. Biol.* **8**: 835–408..

Frohlich R (1983): *Rev. Nuovo. Chimento.* **7**: 399–410.

Fukumaza K (1993): Position sensor apparatus. Patent 06,253,606 Japan.

Fuller BE, Okajima TL, Hong FT (1995): Analysis of the DC photoelectric signal from model BR membranes. *Bioelectochem. Bioenerg.*.

Fulton TA, Dolan J (1987): Observation of single-electron charging effects in small tunnel junctions. *Phys. Rev. Lett.* **59**: 109–112.

Gartner W, Oesterhelt D, Towner P, Hopf H, Ernst L (1981): 13-(Trifluoromethyl)-retinal forms an active and far-red-shifted chromophore in bacteriorhodopsin. *J. Am. Chem. Soc.* **103**: 7642–7643.

Gartner W, Oesterhelt D, Seifert-Shciller E, Towner P, Hopf H, Bohm I (1984): Acetylenic retinals form functional bacteriorhodopsin but do not form bovine rhodopsin. *J. Am. Chem. Soc.* **106**: 5654–5659.

Gat Y, Sheves M (1994): The origing of the red-shifted absorption maximum of the M412 intermediate in the BR photocycle. *Photochem. Photobiol.* **59**: 371–378.

Gatry H, Eisenbach M, Schuldman R, Caplan R (1979): Light induced pH changes in sub-bacterial particles of *Halobacterium halobium*. Effect of ionophores. *Biochim. et Biophys. Acta* **545**: 365–375.

Gawinowicz MA, Balogh-Nair V, Sabol JS, Nakanishi KJ (1977): A nonbleachable rhodopsin analog formed from 11,12-dihydroretinal. *J. Am. Chem. Soc.* **99**: 7720–7721.

Genter UM, Gartner W, Siebert F (1988): Rhodopsin-lumirhodopsin phototransition of bovine rhodopsin investigated by Fourier-transform infrared difference spectroscopy. *Biochemistry* (Washington) **27**: 7480–7488.

Gilmanshin RI, Lazarev PI (1988): Molecular Monoelectronics. *J. Mol. Electronics* **4**: 83–90.

Gilmanshin RI (1993): Proteins for molecular monoelctronics. *Mol. Electronics* **2**: 1–78.

Goldsmith TH (1958): The visual system of the honeybee. *Proc. Natl. Acad. Sci. U.S.* **44**: 123–126.

Govardovsky VI (1972): *Zhurnal Evolutsionnoii Biokhimii Fiziologii* **8**: 8–18 (in Russian).

Govindjee R, Dancshazy Z, Ebrey TG, Longstaff C, Rando RR (1988): Photochemistry of methilated rhodopsins. *Photochem. Photobiol.* **48**: 493–496.

Govindjee R, Balashov SP, Ebrey TG (1990): Quantum efficiency of the photochemical cycle of bacteriorhodopsin. *Biophys. J.* **58**: 597–608.

Green B, Monger T, Alfano B, Anton B, Callender RH (1977): *Cis-trans* isomerzation in rhodopsin occurs in picoseconds. *Nature* (London) **269**: 179–180.

Green RV, Lanyi JK (1979): Proton movement in response to a light-driven electrogenic pump for sodium in *Halobacterium halobium*. *J. Biol. Chem.* **254**: 10986–10984.

Greenhalgh DA, Farrens DL, Subramaniam S, Khorana HG (1993): Hydrophobic amino acids in the retinal-binding pocket of bacteriorhodopsin. *J. Biol. Chem.* **27**: 20305–20311.

Gribakin FG (1973): Perception of polarized light in insects by filter mechanism. *Nature* **246**: 357–358.

Gribakin FG (1987): Visual system: Parallelism and functional evolution. *Evolutionary Ideas in Histology and Embryology*: Leningrad University Press: 119–132 (in Russian).

Groma GI, Szabo G, Varo Gy (1984): Direct measurement of picosecond charge separation in bacteriorhodopsin. *Nature* **308**: 557–558.

Groma GI, Dancshazy (1986): How many M forms are there in the bacteriorhodopsin photocycle? *Biophys. J.* **50**: 357–366.

Grzesiek S, Dencher NA (1988): Monomeric and aggregated bacteriorhodopsin: Single-turnover proton transport stoichiometry and photochemistry. *Proc. Natl. Acad. Sci. U.S.* **85**: 5901–5913.

Gudkov ND, Stolovitsi Yu M, Yevstigneyev VB (1975): Flash photoconductivity of solution of chlorophyll and its analogs-II. Influence of the polarity of the medium on the yield of ion-radicals on photo-oxidation of chlorophyll. *Biophysics* **20**: 215–221.

Gulbik WL (1971): Hearing. N.Y. Oxford University Press: 258 pp..

Gunter P, Huignard J-P (Eds.) (1988, 1989): Photorefractive materials and their application I and II. Springer-Verlag: Berlin .

Haarer D (1989): How to tailor molecular electronics or why is nature taking the "soft" approach? *Adv. Mater.* **11**: 362–365.

Hackett NH, Stern LJ, Chao BH, Kronis KA, Khorana HG (1987): Structure-function studies on bacteriorhodopsin, V. Effects of amino acid substitutions in the putative helix *F*. *J. Biol. Chem.* **262**: 9277–9284.

Hameroff SR (1987): Ultimate computing. *Biomolecular Consciousness and Nanotechnology*. Pg 357: North-Holland: Amsterdam, N.Y. Oxford, Tokyo:.

Hamdorf K, Paulsen R, Schwemer J (1973): In: *Biochemistry and Physiology of Visual Pigments* (Langer H, ed.). New-York: Springer-Verlag: 155–166.

Hampp N, Brauchle, Oesterhelt D (1990): Bacteriorhodopsin as a reversible holographic medium in optical processing. *Ann. Int. Conf. of the IEEE Engineering in Medicine and Biology Soc.* **12**: 1719–1720.

Hampp N, Brauchle C, Oesterhelt D (1990a): Bacteriorhodopsin wildtype and aspargate-96—aspargine as reversible holographic media. *Biophys. J.* **58**: 83–93.

Hampp N, Popp A, Oesterhelt D, Brauchle Ch (1992): Preparation containing BR variants with increased memory times and their use. Patent 4,226,868 Germany.

Hampp N, Thoma R, Oesterhelt D, Brauchle C (1992a): Biological photochrome bacteriorhodopsin and its genetic variants Asp96—Asn as media for optical pattern recognition. *Applied Optics* **31**: 1834–1841.

Hara T, Hara R (1965): New photosensitive pigment found in the retina of the squid *Ommastrephes. Nature* **206**: 1331–1334.

Hara T, Hara R (1972): Cephalopod retinochrome. In: *Handbook of Sensory Physiology*. 7/1 (Dartnall HJA, Ed.): Berlin, Springer-Verlag: 720–746.

Hara T, Hara R (1973): Isomerization of the retinal catalyzed by retinochrome in the light. *Nature* **242**: 39–43.

Hara T, Hara R, Tokunaga F, Yoshizawa T (1981): Photochemistry of retinochrome. *Photochem. Photobiol.* **33**: 883–891.

Hara T, Hara R (1987): Retinochrome and its function. Retinal protein. Utrecht: VNU Science. press: 457–466.

Hargrave PA, McDovell JH, Curtis DR, Wang JK, Juszczak E, Fong SL, Rao JK, Argos P (1983): The structure of bovine rhodopsin. *Biophys. Struct. Mech.* **9**: 235–244.

Hariharan P (1984): *Optical Holography: Principles, Techniques and Applications.* London, N.Y: Cambridge University Press: 319 pp.

Haronian D, Lewis A (1991): Element of a unique bacteriorhodopsin neural network architecture. *Appl. Optics* **30**: 597–608.

Haronian D, Lewis A (1991a): Microfabricating bacteriorhodopsin films for imaging and computing. *Appl. Phys. Lett.* **61**: 2237–2239.

Harurkar V, Spudich J. (1981): Evidence that the light-driven sodium ion pump is a phototaxis receptor in *Halobacterium halobium. Biophys. J.* **33** No 2: 218a.

Hassal M, Mogi T, Karnik SS, Khorana HG (1987): *J. Biol. Chem* **262**: 9264–9270.

Havelka WA, Henderson R, Oesterhelt D (1995): Three-dimensional structure of halorhodopsin at 7 A resolution. *J. Mol. Biol.* **247**: 726–738.

Hegelman R, Oesterhelt D, Bamgerg E. (1985): The transport activity of the light-driven chloride pump halorhodopsin is regulated by green and blue light. *Biochim. et Biophys. Acta* **819**: 195–205.

Hegemann P, Gartner F, Uhi R (1991): All-*trans* retinal constituted the functional chromophore in Chlamydomonas rhodopsin. *Biophys. J.* **60**: 1477–1489.

Hegemann P, Kroger P, Zhang Y, Holland EM and Braun FJ (1995): Chlamydorhodopsin—mediator of signalling. *Abstr. of the 23th Ann. Meet. of the Amer. Soc. for Photobiol.* 33S, MAM B4.

Henderson R, Unwin PNT (1975) Three-dimensional model of the purple membrane obtained by electron microscopy. *Nature* **257**: 28–32.

Henderson R, Baldwin JM, Ceska TA, Zemlin F, Beckmann E, Downing KH (1990):

Model for the structure of bacteriorhodopsin based on high-resolution electron crio-microscopy. *J. Mol. Biol.* **213**: 899–929.
Herbert J, Riesle J, Thiedemann G, Oesterhelt D, Dencher NA (1994): Proton migration along the purple membrane surface and retarded surface/bulk transfer indicate localised proton circuits. *Abstr. of Inter. Conf. of Retinal Protein* 19–24 June, 1994, Leiden, NL: 104.
Hess B, Kuschmitz D (1977): The photochemical reaction of the 412 nm chromophore of bacteriorhodopsin. *Nature* **258**: 766–768.
Heyman PM (1969): Photoconductivity of photochromic Gd:CaF2 . *Appl. Phys. Lett.* **14**: 81–84.
Heyn MP, Braun D, Dencher NA, Fahr A, Holz M, Lindau M, Seiff F, Wallat I, Westerhausen J (1988): Chromophore location and charge displacement in bacteriorhodopsin. *Ber. Bunsenges. Phyz. Chem.* **92**: 1045–1050.
Hibino J, Hashida T, Suzuki M (1994): Multiple memory using aggregated photochromic compounds. In: *Photo-reactive materials for Ultrahigh Density Optical Memory* (Irie M, Ed.): N.Y. Elsevier: 25–53.
Hightower KR, McCready JP (1992): Mechanisms involved in cataract development following near-UV radiation of lenses. *Curr. Eye Res.* **11**: 579–689.
Hildebrand E, Dencher NA (1975): Two photosystems controling behavioral responses of *Halobacterium halobium. Nature* **257**: 46–48.
Hiraki K, Hamanaka T, Mitsuo T, Kito Y (1981): Structural studies of brown membrane by X-ray diffraction, circular dichroism and absorption spectra. *Photochem. Photobiol.* **33**: 419–427.
Hirayama J, Imamoto Y, Shichida Y, Kamo N, Tomioka H, Yoshizawa T (1992): Photocycle of phoborhodopsin from haloalkaliphilic bacterium (*Natronobacterium pharaonis*) studied by low-temperature spectrophotometry. *Biochemistry* **31**: 2093–2098.
Hirshberg Y (1956): Reversible formation and iradiation of color by irradiation at low temperatures. A photochromical memory model. *J. Am. Chem. Soc.* **78**: 2304–2312.
Hoff WD, Dux P, Hard K, Devrees B, Nugteren-Roodzant IM, Crealaard W, Boelens R, Kaptein R, Beeumen JV, Hellingwert KJ (1994): Thiol ester-linked p-Coumaric acid as a new photoactive prosthetic group in a protein with rhodopsin-like photochemistry. *Biochemistry* **33**: 13959–13962.
Hoff WD, van Stokkum IHM, van Ramesdonk HJ, van Brederode ME, Brouver AM, Fitch JC, Meyer TE, van Grondelle R, Hellingwert KJ (1994a): Measurament and global analysis of the absorption changes in the photocycle of the photoactive yellow protein from *Ectothiorhodospira halophila. Biophys. J.* **67**: 1691–1705.
Hoff WD, Kort R, Crielaard W, Hellingwert KJ (1995): The photoactive yellow protein. *Abstr. of the 23rd Ann. Meet. of the Amer. Soc. Photobiol.*: 33S.
Holz M, Drachev LA, Mogi T, Otto H, Kaulen AD, Heyn MP, Skulachev VP, Khorana HG (1989): Replacement of aspartic acid -96 by sparagin in bacteriorhodopsin slows both the decay of the intermediate and the associated proton movement. *Proc. Natl. Acad. Sci. U.S.* **86**: 2167–2171.
Hong FT (1986): The bacteriorhodopsin model membrane system as a prototype molecular computing element. *BioSystems* **19**: 223–236.
Hong FT, Okajima TL (1987): Rapid light-induced charge displacements in bacteriorhodopsin membranes: An electrochemical and electrophysiological study. *In:*

Biophysical Studies of Retinal Protein. (Ebrey TG, Fraunfelder H, Honig B, Nakanishi K, Eds.), Urbana-Champaign IL: University of Illinois Press. 188–198.

Hong FT (1994): Molecular electronics: science and technology for the future. *IEEE Eng. Med. Biol.* **13**: 24–32.

Hong FT (1994a): Photovoltaic effects in biomembranes: Reverse-engineering naturally occurring molecular optoelectronic devices. *IEEE Eng. Med. Biology.* **13**: 75–85.

Hong FT (1994b): Retinal proteins in photovoltaic devices. In: *Molecular and Biomolecular Electronics. Advances in Chemistry, Ser. 240* (Birge RR, Ed.) Am. Chem. Soc.: Washington, DC: 527–559.

Hong FT (1995): Biomolecular electronics. In: *Handbook of Chemical and Biological Sensors* (R.F Taylor and J.S. Schuts, Eds.): IOP Publishing, Inc.: Philadelphia.

Hong FT (1995a): Fundamentals of photoelectronic effects in molecular electronic thin film devices: applications to bacteriorhodopsin-based. *BioSystems* **35**: 117–121.

Hong FT (1997): Interface photochemistry of retinal proteins. Progress in Surface Science (to appear).

Honig B, Callender RH, Dinur H, Ottolenghi M (1979): Photoisomerisation, energy storage and charge separation. *Proc. Natl. Acad. Sci. U.S.* **76**: 2503–2507.

Honig B (1982): Special added commentary: Visual pigments: A new type of molecular switching device. in: *Molecular Electronic Devices* (Carter FL, Ed.).

Hoshino M, Koizumi M (1972): Order of quencher participation in photochemistry: I. Proton transfer from the exited p-hydroxybenzophenon in mixed solvents of cyclohexene and alcohols. *Bull. Chem. Soc. Japan.* **45**: 2731–2735.

Hristova SG, Der A, Varo G, Kesthelyi L (1986): Effect of pH on photocycle and electric signal kinetics in the purple membrane subjected to digestion with proteolytic enzymes. *Photobiochem. Photobiophys.* **12**: 231–241.

Huang Y, Chen Z, Lewis A (1989): Second-harmonic generation in purple membrane-poly(vinyl alcohol) films: probing the dipolar characteristics of the bacteriorhodopsin chromophore in bR570 and M412. *J. Phys. Chem.* **93**: 3314–3320.

Hubel DH (1988): Eye, brain and vision. Scientific American Library series: #22..

Hurley JB, Ebrey TG, Honig B, Ottolenghi M (1977): Temperature and wavelength effects on the photochemistry of rhodopsin, isorhodopsin, bacteriorhodopsin and their photoprodacts. *Nature* (London) **270**: 540–542.

Hurley JB, Becher B, Ebrey TG (1978): More evidence that light isomerizes the chromophore of purple membrane protein. *Nature* **272**: 87–88.

Hurley JB, Ebrey TG (1978): Energy transfer in the purple membrane of *Halobacterium halobium. Biophys. J.* **22**: 49–66.

Hwang S-B, Korenbrot JI, Stoeckenius W (1977): Structural and spectroscopic characteristics of BR in air-water interface films. *J. Membrane Biol.* **36**: 115–135.

Hwang S-B, Korenbrot JI, Stoeckenius W (1977a): Proton transport by bacteriorhodopsin through an interface film. *J. Membr. Biol.* **36**: 137–158.

Hwang SB, Stoeckenius W (1977): Purple membrane vesicles: morphology and proton translocation. *Membrane Biol.* **33**: 325–350.

Ikonen M, Peltonen J, Vuorima E, Lemmetyinen H (1992): Study of photocycle and spectral properties of bacteriorhodopsin in Langmuir-Blodgett films. *Thin Solid Films* **213**: 277–284.

Imamoto Y, Yoshizava T, Takahashi T, Tomioka H, Kamo N, Kobatake Y (1988):

Low temperature spectrophotometric study on photoreaction cycle of phoborhodopsin. In: *Molecular Physiology of Retinal Proteins.* (Hara T, Ed.) Osaka, Japan: Yamada Science Foumdation: 361–362.
Imamoto Y, Shichida Y, Yoshizawa T, Tomioka H, Takanashi T, Fujikawa K, Kamo N, Kobatake Y (1991): Photoreaction cycle of phoborhodopsin studied by low-temperature spectrophotometry. *Biochemistry* **30**: 7416–7424.
Imamoto Y, Shichida Y, Hirayama J, Tomioka H, Kamo N, Yoshizava T (1992): Chromophore configuration of pharaonis rhodopsin and its isomerization on photon absorption. *Biochemistry* **31**: 2523–2528.
Inatomi K (1986): ATP-regeneration bioreactor equipped with immobilized ATPase. Patent 61124384 Japan.
Irie M (1994): Design and synthesis of photochromic memory media. In: *Photoreactive Materials for Ultrahigh density Optical Memory* (Irie M, Ed.): N.Y. Elsevier: 1–12.
Ivanitsky GR, Vsevolodov NN (Eds.) (1985): Photosensitive biological complexes and optical recording of information. Biol Sci. Res. Center: Pushchino, USSR: 209 pp (in Russian).
Iwasa T, Takao M, Ymada M, et al. (1984): Properties of an analog pigment of bacteriorhodopsin synthesized with naphthyl retinal. *Biochemistry* **23**: 838–843.
Iwasa T, Tokunaga F, Yoshizawa T (1979): Photoreaction of *trans*-bacteriorhodopsin at liquid helium temperature. *FEBS Lett.* **101**: 121–124.
Iwasa T, Tokunaga F, Ebrey T, Yoshizawa T. (1981): The photoreactions and photosensitivity of 3,4-dihydro-bacteriorhodopsin at low temperatures. *Photochem. Photobiol.* **33**: 547–557.
Iwashita H, Kato K, Yamamoto N (1987): Optical recording medium and color image-formation method, Patent 01,116,536 Japan.
Jackson G (1969): The properties of photochromic materials. *Optical Acta* **16**: 1–16.
Jackson MB, Sturtevant JM (1979): Phase transition of the purple membranes of *Halobacterium halobium. Biochemistry* **17**: 911–915.
James TN (1967): The theory of the photograph. N.Y. Macmillan.
Jung K-H, Spudich JL (1996): Protonatable residues at the cytoplasmic end of transmembrane helix-2 in the signal transducer Htr-I. Control photochemistry and function of sensory rhodopsin I. *Proc. Natl. Acad. Sci. U.S.* **93**: 6557.
Kabayashi T, Ohiani H, Iwai J et al. (1983): Effect of pH on the photoreaction cycle of bacteriorhodopsin. *FEBS Lett.* **162**: 197–200.
Kakichashvili SD (1972): Polarization recording of holograms. *Opt. Spectrosc.* **33**: 171–179 (in Russia).
Kakichashvili SD (1982): Polarizational holography. *Vestnik USSR Acad. Sci.* **7**: 51–61.
Kalisky O, Goldschmidt CR, Ottolehghi M (1977): On the photocycle and light adaptation of dark-adapted bacteriorhodopsin. *Biophys. J.* **19**: 185–189.
Kaminuma T, Matsumoto G (Eds.) (1991): Biocomputers. London-New York-Tokyo-Melburne-Madras: Chapman and Hall.
Kantcheva MP, Popdimitrova N, Stoilov S (1982): Electrophoretic mobility of purple membrane from *Halobacterium halobium. Studia Biophys.* **90**: 125–129.
Kaplan E, Barlow RB (1984): Circadian clock in *Limulus* brain increases response and decreases noise of retinal photoreceptors. *Nature* **286**: 393–395.

Karplus M, McCammon JA (1983): Dynamics of proteins: elements and function. *Ann. Rev. Biochem.* **53**: 263–300.

Kaufman H, Vratsanos SM, Erlanger BF (1968): Photoregulation of an enzyme by process by means of a light-sensitive ligand. *Science* **162**: 1487–1489.

Kaushin LP, Sibeldina LA, Lasareva AV, Vsevolodov NN, Kostikov AS, Richireva GT, Chekulaeva LN (1974): Membrane protein bacteriorhodopsin from halophilic bacteria. *Studia Biophys.* **42**: 71–74.

Keszthelyi L (1980): Orientation of membrane fragment by electric field. *Biochim. Biophys. Acta.* **598**: 429–436.

Keszthelyi L (1982): Orientation of purple membrane by electric field. *Meth. Enzymol.* **88**: 287–197.

Ketis BP (1981): Piezoelectric effects of bacteriorhodopsin. *Biol. Membr.* **1**: 1307–1315.

Khan Sh, Amoyaw K, Spudich JL, Reid JP, Trentham DR (1992): Bacterial chemoreceptor signaling probed by flash photoreliase of a *Caged serine*. *Biophys. J.* **62**: 67–68.

Khitrina LV, Drachev LA, Kaulen AD, Skulayeva LN (1976): BR inhibition by formalin and lanthanum ions. *Biochemistry* **47**: 1763–1762.

Khitrina LV, Lazarova TSR (1989): Study of 13-*cis* and all *trans*-isomers of 4-keto-bacteriorhodopsin. *Biokhimya.* **54**: 136–139 (in Russian).

Khorana HG, Gerber GE, Herlihy WC, Gray CP (1979): Amino acid sequence of bacteriorhodopsin. *Proc. Natl. Acad. Sci. U.S.* **76**: 5046–5052.

Khorana HG (1988): Bacteriorhodopsin, a membrane protein that uses light to translocate protons. *J. Biol. Chem.* **263**: 7439–7442.

Khorana HG (1992): Rhodopsin, photoreceptor of the rod cell. *J. Biol. Chem.* **267**: 1–4.

Kimura Y, Ikegami A, Stoeckenius W (1984): Salt and pH-dependent changes of the purple membrane absorption spectrum. *Photochem. Photobiol.* **40**: 641–646..

Kinumi T, Tsujimoto K, Ohashi M, Hara R, Hara T, Ozaki K, Sakai M, Katsuta Y, Wada A, Ito M (1993): The conformation analysis and photoisomerization of retinochrome analogs with polyenals. *Photochem. Photobiol.* **58**: 409–412.

Kirschfeld K, Franceschini N, Minke B (1977): Evidence for a sensitising pigment in fly photoreceptors. *Nature* **269**: 386–390.

Kiselev BA, Kozlov Yu N (1980): Photo/electrochemistry of chlorophyll monolayers. *Bioelectrochem. Bioenergetics* **7**: 247–254 and references therein.

Knowles A, Dartnall HJA (1977): The photobiology of vision. In: *The Eye 2B* (Davson H Ed.) N.Y. Academic Press.

Kobayashi T (1979): Hypsorhodopsin: the first intermediate of the photochemical process in vision. *FEBS Lett.* **106**: 313–316.

Kononenko AA, Lukashev Ye.P., Broun DS, Chamorovskii SK, Kruming BA, Yakushev SA (1993): Model device based on bacteriorhodopsin prototype of position-sensitive detectors. *Biophysics* **38**: 1025–1030.

Kononenko AA, Lukashev (1995): Light biosensors based on bacteriorhodopsin and photosynthetic reaction centers. *Adv. Biosensors* **3**: 191–211: London, JAI Press Inc..

Korchemskaya E Ya, Soskin MS, Taranenko V (1990): Enhancement of the contrast of low-noise optical signals in the course of nonlinear absorption in media based on bacteriorhodopsin. *Sov. J. Quantum Electron.* **20**: 381–382.

Korchemskaya E, Soskin M, Dyukova T, Vsevolodov N (1993): Contrast enhansement and phase conjugation of low-power optical signals in dynamic recording material based on bacteriorhodopsin. *SPIE* **2083**: 217–224.

Korchemskaya E Ya, Soskin MY, Vsevolodov NN (1994): Photoinduced anisotropy and dynamic polarization holography recording in polymer films with bacteriorhodopsin for biomedical applications. BIOS Europe 94, International Symposium on Biomedical Optics (Sept. 6–10, 1994) Universite de Lille, Lille, France 2329: 36.

Korenbrot JJ, Hwang S-B (1980): Proton transport by bacteriorhodopsin in purple membranes assembled from air-water interface film. *J. Gen. Physiol.* **76**: 649–659.

Korenbrot JJ (1982): The assembly of bacteriorhodopsin-containing planar membranes by the sequential transfer of air-water interface films. *Meth. Enzymol.* **88**: 45–55.

Korenstein R, Hess B (1977): Hydration effect on the photocycle bacteriorhodopsin in thin layers of purple membrane. *Nature* **270**: 184–186.

Korenstein R, Hess B (1982): Analysis of photocycle and orientation in thin layers. *FEBS Lett.* **81**: 180–193.

Kostilev GD, Vsevolodov NN (1985): On the possibility of biochrome films application for polychromatic hologram recording. In: *Photosensitive Biological Complexes and Optical Recording of Information.* Sci. Biol. Res. Center: Pushchino: 142–144 (in Russian) Ivanitsky GR, Vsevolodov NN Eds..

Kouyama T, Kinosita K Jr, Ikegama A (1988): Structure and function of bacteriorhodopsin. *Adv. Biophys.* **24**: 123–175..

Kozhevnikov NM, Lipovskaya M Yu (1990): Adaptive holographic fiber-optic interferometer. *Proc. SRIE* **1121**: 293–297.

Krah M, Marvan W, Vermeglio A, Oesterhelt D (1994): Phototaxis of *Halobacterium salinarium* requires a signalling complex of sensory rhodopsin I and its methilaccepting tranduser Htr-I. *EMBO. J.* **13**: 2150–2155.

Krasnovsky AA (1948): Reversible photochemical reduction of chlorophill by ascorbic acid. *Dokl. Acad. Nauk* USSR **60**: 421–424 (in Russian).

Krasnovsky AA (1966): In: *Elementary Processes in Molecules.* Moscow-Leningrad: "Nauka": 213–242 (in Russian).

Krasnovsky AA, Nikandrov VV, Nikiforova SA (1985): Interfacing of non-organic catalyzers-semiconductors with *clastridia* cells: photoformation of hydrogene. *DAN SSSR* **285**: 1467–1471 (in Russian).

Krongauz VA, Shifrina RR, Fedorovich IB, Ostrovsky MA (1975): Photochromism on visual pigments. III. Comparative study of phototransformations of bovine and frog rhodopsin. *Biophysics* **20**: 427–431.

Kropf A, Hubbard R (1958): The mechanism of bleaching rhodopsin. *Ann. N.Y. Acad. Sci.* **74**: 266–280.

Kuan KN, Lee YY, Melius P (1980): Preparetion of tripsin-like enzyme from pronase by carbobenzoxy-l-phenile lange-triethylene tetraminyl sepharose. *Am. Chem. Soc.* Abstracts: BIOL. 224.

Kuhn H (1986): Electron transfer mechanism in the reaction center of photosynthetic bacteria. *Phys. Rev. A.* **34**: 3409–3425.

Kuhn H (1994): Organized monolayer assemblies: Their role in constracting supramolecular devices and in modeling of early life. *IEEE Eng. Med. Biol.* **13**: 33–44.

Kuzmin LS, Likharev KK (1987): Direct experimental observation of discrete correlated single-elution tunneling. *JETP Lett.* **45**: 495–497.

Langer H, Schmeinck G, Anton-Erxleben F (1986): Identification and localization of visual pigments in the retina of the moth, *Antheraea polyphemus* (Insecta, Saturnidae). *Cell and Tissue Res.* **245**: 81–89.

Lanyi JK, Oesterhelt D (1982): Identification retinal-binding protein in halorhodopsin. *J. Biol. Chem.* **257**: 2674–2677 .

Lanyi J (1984): Nature of the principal photointermediate of halorhodopsin. *Biochem. and Biophys. Res. Commun.* **122**: 91–96.

Lanyi JK, Duschl A, Varo G, Zimany L (1990): Anion binding to the chloride pump, halorhodopsin, and its implications for the transport mechanism. *FEBS Lett.* **265**: 1–6.

Lanyi JK (1993): Proton translocation mechanism and energetics on the light-driven pump bacteriorhodopsin. *Biochem. Biophys. Acta* **1183**: 241–261.

Lanyi JK, Varo G (1995): The photocycles of bacteriorhodopsin. *Isr. J. Chem.* **35**: 365–385.

Law WC, Kim S, Rando RR (1989): Stereochemistry of the visual cycle. *J. Am. Chem. Soc.* **111**: 793–795.

Lawrence AF, Birge RR (1984): Communication with submicron structures. Perspective in the application of biomolecules to computer technology. In: *Nonlinear Electrodynamics in Biological Systems* (Adey HR, Lawrewnce AF, Eds.): Plenum: New York: 207–218.

Lawson MA, Zacks DN, Derguini F, Nakanishi K, Spudich JL (1991): Retinal analog restoration of photophobic responce in a blind *Clamidomonas Reinhartii* mutant. Evidence for an archaebacterialike chromophore in a eukaryotic rhodopsin. *Biophys. J.* **60**: 1490–1498.

Lazarev Yu A, Shnyrov BL (1979): The study of thermal denaturation of bacteriorhodopsin of *Halobacterium halobium. Bioorganich. Khimiya* **5**: 105–112 (in Russian).

Lazarev Yu A, Terpugov E (1980): Effect of water on the structure of bacteriorhodopsin and photochemical processes in purple membrane. *Biochem. Biophys. Acta* **590**: 324–328.

Lear JD, Wasserman ZR, DeGrado WF (1988): Synthetic amphiphylic peptide models for protein ion channels. *Science* **240**: 1177–1181.

LeGrance J, Cahen D, Caplan SR (1982): Photoacoustic calorimetry of purple membranes. *Biochem. Biophys. Acta* **710**: 4–6.

Lenci F, Gretti F, Colombetti G, Hader D-P, Song P-S, Eds. (1991): Biophysics of photoreceptors and photomovements on microorganisms. NATO ASI: Series A: Life sciences V.211. N.Y. and London: Plenum Press.

Levitt M, Sander S, Stern PS (1985): Protein normal-mode dynamics: trypsin inhibitor, crambin, ribonuclease and lisozyme. *J. Mol. Biol.* **181**: 423–447.

Lewis A, Spoonhower J, Bogomolni R, Lozier RH, Stoeckenius W (1974): Tunable laser resonance Raman spectroscopy of bacteriorhodopsin. *Proc. Natl. Acad. Sci. U.S.* **71**: 4462–4466.

Li J-R, Wang J-P, Jiang L (1994): Biphasic photocurrent of a photocell containing retinal Langmuir-Blodgett film. *Biosensors and Bioelectronics* **9**: 147–150.

Liberman EA (1978): Molecular computer. Biological physics and physics of the real world. *Biofizika* **23**: 1118–1121 (in Russian).

Liebman PA, Parker KR, Dratz EA (1987): The molecular mechanism of visual excitation and its relation to the structure and composition of the rod outer segment. *Annu. Rev. Physiol.* **49**: 765–791.

Lisman J (1985): The role of metarhodopsin in the generation of spontaneous quantum bump in the ultraviolet receptor of *Limulus* median eye. *Gen. Physiol.* **85**: 171–187.

Lisyutenko VN, Barachevskii VA, Pankratov AA. Konoplev GG (1990): Extraction-energy relaxation in photochromic indoline spiropiranes. *Theoretical and Experimental Chemistry* **25**: 390, and references therein.

Litvin FF, Balashov SP, Sineshchekov VA (1975): The investigation of the primary photochemical conversions of bacteriorhodopsin in purple membranes and cells of *Halobacterium halobium* by the low temperature spectrophotometry method. *Bio-organic Chem.* **1**: 1767–1777.

Litvin FF, Balashov SP (1977): New intermediates in the photochemical conversions of bacteriorhodopsin. *Biofizika* **22**: 1111–1114 (in Russian).

Litvin FF, Sineshchekov OA, Sineshchekov VA (1978): Photoreceptor electric potential in the phototaxis of the alga *Haematococcus pluvialis*. *Nature* **271**: 476–478.

Liu RSH, Mead D, Asato AE (1985): Application of the H.T.-n mechanism of photoisomerization to the photocycle of bacteriorhodopsin. A model study. *J. Am. Chem. Soc.* **107**: 6609–6614.

Liu RSH, Krogh E, Li X-Y, Mead D, Colmenares LU, Thiel JR, Ellis J, Wong D, Asato AE (1993): Analyzing the red-shift characteristics of azulenic, naphthyl, other ring-fused and retinal pigment analogs of bacteriorhodopsin. *Photochem. Photobiol.* **58**: 701–705.

Liu SY, Ebrey TG (1988): Photocurrent measurements of the purple membrane oriented in a polyacrylamide gel. *Biophys. J.* **54**: 321–329.

London E, Korana H, (1982): Denaturation and renaturation of bacteriorhodopsin in detergents and lipid-detergent mixtures. *J. Biol. Chem.* **257**: 7003–7011.

Lozier R, Bogomolni R, Stoeckenius W. (1975): Bacteriorhodopsin: a light driven proton pump in *Halobacterium halobium*. *Biophys. J.* **15**: 955–962.

Lukashev EP, Vozary E, Kononenko AA, Rubin AB (1980): Electric field promotion of the bacteriorhodopsin BR570 to BR412 photoconversion in film of *Halobacterium halobium* purple membranes. *Biochem. Biophys. Acta* **592**: 258–266.

Lukashev EP, Druzhko AB, Kononenko NG (1992): Electrically induced bathochromic shift of the absorption band of 4-keto-bacteriorhodopsin in gelatine-based films. *Biophysics* **37**: 72–75.

Lukashev EP, Kononenko NG, Abdulaev NG, Ebrey TG (1993): Retardation of the photocycle in bacteriorhodopsin by cyclohexanedione. *Bio-organic Chem.* **19**: 395–405.

Lukashev EP, Robertson B (1995): Bacteriorhodopsin retains its light-induced proton-pumping function after being heated to 140°C. *Bioelectrochem. and Bioenergetics* **37**: 157–160.

Macnab RM (1984): The bacterial flagellar motor. *TIBS* **9**: 185–188.

Maillart N (1978): Photographic elements containing vesicles of rhodopsin and lipids, Patent 4,084,967 U.S..

Maksimichev AB, Lukashev EP, Kononenko AA, Chekulaeva LN, Timashev SF (1984): Photopotentials and BR photocycle regulation by an electricfield in high-oriented purple membrane films. *Biologicheskiye Membrany* **1**: 294–304 (in Russian).

Maksimichev AB, Chamorovskii SK (1988): Bacteriorhodopsin as possible element of membrane bioreactors. *Russ. Chem. Rev.* **57**: 592–604.

Mao BR, Govinjee T, Ebrey M, Arnaboldi V, Balogh-Nair, K Nakanishi, R Crouch, (1981): Photochemical and functional properties of BR analogs formed 5,6-dihydroretinals and 5,6-dihydrodesmethylretinals. *Biochemistry* **20**: 428–435..

Margus J, Eisenstein L (1984): Pressure effect on the photocycle of PM. *Biochemistry* **23**: 5556–5563.

Marti T, Otto H, Mogi T, Rosselet SJ, Heyn MP, Khorana HG (1991): *J. Biol. Chem.* **266**: 6919–6927.

Martinek K, Berezin IV (1979): Artificial light-sensitive enzymatic systems as chemical amplifiers of weak light signals. *Photochem. Photobiol.* **29**: 637–649.

Mathewson PR, Finley JW (Eds.) (1992): Biosensor design and application. Washington, DC: American Chemistry Soc..

Mathies RA (1991): *Rev. Biophys. Chem.* **20**: 491–518.

Matsui S, Seidon M, Uchiyama N, Seniya N, Hiraki K, Yoshihara K, Kito Y (1988): 4-hydroxy-retinal, a new visual pigments chromophore found in the bioluminiscent squid *Watasenia scinillans*. *Biochem. et Biophys. Acta* **966**: 370–374.

McCain DA, Amici LA, Spudich JL (1987): Kinetically resolved states of the *Halobacterium halobium* flagellar motor switch and modulation of the switch by sensory rhodopsin I. *J. Bacteriol.* **169**: 4750–4758.

McConell DG, Rafferty CN, Dilley RA (1968): The light-induced proton uptake in bovine retinal outer segment fragments. *J. Biol. Chem.* **243**: 5820–5826.

McRee DE, Meyer TE, CU.S.novich MA, Parge HE, Getzoff ED (1986): Crystallographic characterization of a photoactive yellow protein with photochemistry similar to sensory rhodopsin. *J. Biol. Chem.* **261**: 13850–13851.

McReynolds JS, Gorman ALF (1970): Photoreceptor potentials of opposite polarity in the eye of the scallop, *Pecten irradians*.? *J. Gen. Physiol.* **56**: 376–391.

Meyer TE (1985): Isolation and characterization of soluble cytochromes, ferredoxins and other chromophoric proteins from the halophylic phototrophic bacterium *Catothiorhodospira halophila*. *Biochem. Biophys. Acta* **806**: 175–183.

Michaile S, Duschl A, Lanyi JK, Hong FT (1990): Chloride ion modulation of the fast photoelectric signal in halorhodopsin thin films: *Proc. 12th Ann. Int. Conf. IEEE Engin. Med. and Biol. Soc.* **12**: 1721–1723.

Michel H, Oesterhelt D, Henderson K (1980): Orthorombic two-dimensional crystal form of purple membrane. *Proc. Natl. Acad. Sci. U.S.* **77**: 338–342.

Michel H (1982): Characterization and crystal packing of three-dimensional bacteriorhodopsin crystal. *EMBO J.* **1**: 435–451.

Michel H, Oesterhelt D (1982): Preparation of new 2-dimensional and 3-dimensional crystal forms of bacteriorhodopsin. *Methods Enzymol.* **88**: 111–126.

Michel-Wolvertz M-R, Duhardin E, Sironval C (1976): Photosensitive material of biological origin, Patent 1,423,992 Great Britain..

Mikaelian AL, Bobrynev BI (1974): *Radiotekhnika i Elektronika* **19**: 898–926 (in Russian).

Mikaelian AL, Salakhutdinov VK, Vsevolodov NN, Dyakova TV (1991): High capacity optical spatial switch based on reversible holograms. In: *Optical Memory and Neural Networks* (Mikaelian AL, Ed.) Proc. SPIE 1621, 148–157.

Mikaelian AL, Salakhutdinov VK (1992): Use of dynamic holograms for information channel switching. *J. Optical Memory and Neural Networks* **1**: 315–324.

Miller A, Oesterhelt D (1990): Kinetic optimization of bacteriorhodopsin by aspartic acid 96 as an internal proton donor. *Biochem. Biopys. Acta* **1020**: 57–64.

Mirzadega T, Liu RSH (1990): Probing the visual pigment rhodopsin and its analogs by molecular modeling and computer graphic analysis. *Progress in the Retinal Proteins*. **6**: 57–74.

Miyasaka T (1990): Purple membrane photovoltaic cell, Patent 04,006,420 Japan.

Miyasaka T, Koyama K, Iton I (1992): Quantum conversion and image detection by a bacteriorhodopsin-based artificial photoreceptor. *Science* **255**: 342–344.

Mogi T, Stern LJ, Hackett NR, Khorana HG (1987): Bacteriorhodopsin mutants containing single tyrosine to phenylalanine substitutions are all active in proton translocation. *Proc. Natl. Acad. Sci. U.S.* **84**: 5595–5599.

Mogi T, Stern LJ, Marti T, Chao BH, Khorana HG (1988): Aspartic acid substitution affects proton translocation by bacteriorhodopsin M. *Proc. Natl. Acad. Sci. U.S.* **85**: 4148–4152.

Montal M, Mueller P (1972): Formation of biomolecular membranes from lipid monolayers and a study of their electrical properties. *Proc. Natl. Acad. Sci. U.S.* **69**: 3561–3566.

Mote MI (1974): Polarization sensitivity. *J. Comp. Physiol.* **90**: 389–403.

Mukohata Y, Sugiyama Y, Kaji Y, et al. (1981): The white membrane of crystalline bacteriorhodopsin in *Halobacterium halobium*. Strain R Lin3 1MW and its conversion into purple membrane by exogenous retinal. *Photochem. and Photobiol.* **33**: 593–600.

Mukohata Y, Kaji Y(1981): Light-induced membrane-potential increase ATP synthesis and proton uptake in *Halobacterium halobium* RmR catalysed by halorhodopsin. *Arch. Biochem. Biophys.* **206**: 72–76.

Mukohata Y, Sugiyama Y (1982): Isolation of white membrane of crystalline bacterio-opsin from *Halobacterium halobium* R1mR lacking carotinoid. *Meth. Enzymol.* **88**: 407–411.

Munz FW, Beatty DD (1965): A critical analysis of the visual pigments of salmon and trout. *Vision Res.* **5**: 1–17.

Muradin-Szweykowska M, Pardoen JA, Dobelstein D, Amsterdam LGP, Lugtenburg J (1981): Bacteriorhodopsin with chromophores modified at the β-ionone site. *J. Biochem.* **140**: 173–176.

Myasaka T (1990): Manufacturing of protein laminate membranes for use as photosensitive membranes in biosensors. Patent 04,101,837 Japan.

MyBsaka T (1992): Quantum conversion and image detection by a bacteriorhodopsin-based artifitial photoreceptor. *Science* **255**: 342–344.

Nagayama K (1992): *Nanobiology* **1**: 25–37.

Nagy K (1978): Photoelectric activity of dried oriented layers of purple membrane from *Halobacterium halobium*. *Biochim. Biophys. Res. Commun.* **85**: 383–390.

Nakanishi K, Balogh-Nair, Arnaboldi M, Tsujimoto K, Honig B (1980): An external point-charge model got bacteriorhodopsin to account for its purple color. *Amer. Chem. Soc.* **102**: 7945–7947.

Nakanishi K, Derguini F, Rao VJ, Zarrilli G, Okabe M, Lien T, Johnson R, Foster KW, Saranak J (1989): Theory of rhodopsin activation: Probably charge redistribution at excited state chromophore. *Pure Appl. Chem.* **61**: 361–364.

Nakanishi K (1991): A wandering natural products chemist. Washington, DC: *American Chemical Soc*: 230 pg..

Nakanishi K, Crouch R (1995): Application of artificial pigments to structure determination and study of photoinduced transformation of retinal proteins. *Isr. J. Chem.* **35**: 253–272.

Nashima K, Mitsudo M, Kito Y (1979): Molecular weight and structural studies of cephalopod rhodopsin. *Biochem et Biophys. Acta* **579**: 155–167.

Nathans J (1987): Molecular biology of the visual pigments. *Annu. Rev. Neurosci.* **10**: 163–194.

Neugebauer DCh, Blaurock AE, Worcester DL (1977): Magnetic orientation of purple mebrane demonstration by optical measurements and neutron scattering. *FEBS Lett.* **78**: 31–35.

Neugebauer DCh, Zingsheim HP, Oesterhelt D (1978): Recrystallization of the purple membrane *in vivo* and *in vitro*. *J. Mol. Biol.* **123**: 247–257.

Neuman S, Leigeber H (1989): Verfahren zur Herstellung von Purpurmembran enthaltend Bacteriorhodopsin, German patent Application DE 3922133AI, 3 pp.

Ninneman H (1980): Flavoproteins are the same as photoreceptors of blue light. *Bioscience* **30**: 166–170..

Novokhvatsky AS, Alyi AB (1983): Does human retina contain iodopsin? *Voprosy Klinicheskoi Oftalmologii* **5**: 279–281 (in Russian).

O'Brien DF, Tyminski PN (1983): Photographic composition using rhodopsin and light-activatable enzymes, Patent 0,073,628 U.S..

Oesterhelt D, Stoeckenius W (1971): Rhodopsin-like protein from the purple membrane of *Halobacterium halobium*. *Nature New Biol.* **233**: 149–152.

Oesterhelt D, Stoeckenius W (1973): Function of a new photoreceptor membrane. *Proc. Natl. Acad. Sci. U.S.* **70**: 2838–2857..

Oesterhelt D, Schulmannn L (1974): Reconstitution of bacteriorhodopsin. *FEBS Lett.* **44**: 262–265.

Oesterhelt D, Schulmannn L, Gruber H (1974): Light-dependent reaction of bacteriorhodopsin with hydroxylamine in cell suspension of *Halobacterium halobium*: demonstration of an apo-membrane. *FEBS Lett.* **44**: 257–161.

Oesterhelt D (1982): Anaerobic growth of halobacteria. *Meth. Enzymol.* **88**: 417–420.

Oesterhelt D, Krippahl G (1983): Phototrophic growth of halobacteria and its use for isolation of photosynthetically deficient mutants. *Ann. Microbiol.* (Inst. Pasteur) **134B**: 137–150.

Oesterhelt D, Hegemann P, Tavan P, Schulten K (1986): *Trans-cis* isomerization of retinal and a mechanism for ion translocation in halorhodopsin. *Eur. Biophys. J.* **14**: 123–129.

Oesterhelt D, Soppa J, Krippahl G (1987): Method for producing strains of *Halobacterium halobium* with a modified bacteriorhodopsin, Patent 3,730,424 Germany.

Oesterhelt D, Tittor J (1989): Two pumps, one principle: Light-driven ion transport in halobacteria. *Trends Biochem. Sci.* **14**: 57–61.

Oesterhelt D (1989): Photosynthetic systems in procariotes. The retinal proteins of halobacteria and the reaction center of purple bacteria. *Biochemistry Intern.* **18**: 673–694.

Oesterhelt D, Brauchle Ch, Hampp N (1991): Bacteriorhodopsin: a biological material for information processing. *Quart. Rev. Biophys.* **24**: 425–478.

Ogama (1987 1990): Color image sensor obtained from visual photosensitive materials derived from biological substances, Patents 05,135,013 Japan, 4,896,049 U.S..
Ohno K, Govindjee R, Ebrey TG (1983): Blue light effect of proton pumping by bacteriorhodopsin. *Biophys. J.* **43**: 251–254.
Okon EB, Vsevolodov NN (1987): Does bacteriorhodopsin energize the membranes of animal mitochondria under light? *FEBS Lett.* **216**: 241–244.
Olson KD, Deval P, Spudich JL (1992): Absorption and photochemistry of sensory rhodopsin I: pH effects. *Photochem. Photobiol.* **56**: 1181–1187.
Oren A, Shilo M (1981): Bacteriorhodopsin in a bloom of Halobacteria in the Dead Sea. *Arch. Mikrobiol. Bd.* **130**: S. 185–187.
Ormos P, Dancshazy Z, Karvaly B (1978): Mechanism of generation and regulation of photopotential by bacteriorhodopsin in biomolecular lipid membrane. The quenching effect of blue light. *Biochem. Biophys. Acta* **503**: 304–315.
Ormos P, Reinisch L, Kesztheyi L (1983): Fast electric response signals in the bacteriorhodopsin photocycle. *Biochim. Biophys. Acta* **722**: 471–479.
Oseroff AR, Callender RH (1974): Resonanse Raman spectroscopy of rhodopsin in retinal disk membrane. *Biochemistry* **13**: 4243–4248.
Ostrovsky MA (1978): Light-induced changes in the photoreceptor membrane. In: Membrane Transport Processes (Tosteson DC, Ovchinnikov YuA, Lattorre R, Eds) N.Y: Raven Press: **2**: 217–243.
Ostrovsky MA (1989): Animal rhodopsin as a photoelectric generator.In: *Molecular Electronics: Biosensors and Biocomputors* (Hong FT, Ed.) N.Y: Plenum Press: 187–201.
Ostroy SE (1977): Phodopsin and the visual process. *Biochem. Biophys. Acta* **463**: 91–125.
Otto H, Marti T, Holz M, Stern LJ, Engel F, Khorana HG, Heyn MP (1990): Substitution of amino acids Asp85, Asp212, and Arg82 in bacteriorhodopsin affects the proton release phase of the pump and PK of the Schiff base. *Proc. Natl. Acad. Sci. U.S.* **87**: 1018–1022.
Ovchinnikov Yu A, Abdulaev NG, Feigina MY, Kiselev AV, Lobanov NA (1979): The structural basis of the functioning of bacteriorhodopsin: an overview. *FEBS Lett.* **100**: 219–224.
Ovchinnikov Yu A, Abdulaev NG, Modyanov NN (1982): Structural basis of proton translocating protein function. *Annu. Rev. Biophys.* **11**: 445–363.
Ovchinnikov Yu A (1987): Rhodopsin and bacteriorhodopsin: Structure-function relationships. *FEBS Lett.* **148**: 179–191..
Ovchinnikov Yu A, Abdulaev NG, Zolotarev AS, Artamonov ID, Bespalov IA, Dergachev AE, Tsuda M (1988): Octopus rhodopsin: Amino acid sequence deduced from cDNA. *FEBS Lett.* **232**: 69–72.
Oyama J, Tomita Y, Sakuranaga M (1990): Photoelectrical device, Patent 03,295,278 Japan.
Ozaki K, Terakita A, Hara R, Hara T (1987): Isolation and characterization of a retinal-binding protein from the squid retina. *Vision Res.* **26**: 691–705.
Palings I, Pardoen JA, van der Berg E, Winkel C, Lugtenburg J, Mathies RA (1987): Assignment of fingerprint vibrations in the Raman resonance. *Biochemistry* **26**: 2544–2556.
Pande AJ, Callender RH, Ebrey TG, Tsuda M (1984): Resonance Raman study of the primary photochemistry of visual pigments. *Biophys. J.* **45**: 573–576.

Padros E, Dumach M, Sabes M (1984): Induction of the blue form of bacteriorhodopsin by low concentrations of sodium dodecylsulfate. *Biochem. et Biophys. Acta* **769**: 1–7.

Passchier MF, Bonjakovic BFM (Eds.) (1987): *Human Exposure to Ultarviolet Radiation*. Elsevier Sci. Publishers B.V.: Biomedical Division..

Patent 2,149,714 Germany (1974).

Patent 05,227,765 Japan (1974).

Patent 4,029,677 U.S.(1977).

Patent 3,609,093 U.S. (1971).

Patent 3,869,292 U.S. (1977).

Patent. 3,597,054 U.S. (1971).

Pepe IM, Dchwemer J, Pauslen R (1984): Characterization of retinal-binding proteins from the honeybee retina. *Vision Res.* **22**: 475–481.

Perking G, Liu Ed, Burkard F, Berry EA, Glaeser RM (1992): Characterization of the Conformation Change in the M1 an M2 substates of bacteriorhodopsin by the combined use of visible and infrared spectroscopy. *J. Struct. Biol.* **109**: 142–151.

Pettei MJ, Yudd AP, Nakanishi K, Henselman R, Stoeckenius W (1977): Identification of retinal isomers isolated from bacteriorhodopsin. *Biochemistry* **16**: 1955–1959.

Petty MC, Bryce MR, Bloor D (Eds.) (1995): *Introduction to Molecular Electronics*. London: Edward Arnold,.

Polyak S (1941): *The Retina*. Chicago: University. Press..

Popp A, Wolperdinger M, Hampp N, Bruchle C, Oesterhelt D (1993): Photochemical conversion of the O-intermediate to 9-*cis*-retinal-containing products in bacteriorhodopsin film. *Biophys. J.* **65**: 1449–1459.

Powell RJ, Stetson KA (1965): Photoinduced potentials across a polymer stabilized planar membrane in the present at bacteriorhodopsin. *J. Opt. Soc. Am.* **55**: 1593.

Protasova TB, Fedorovich IB, Borisevich GP, Ostrovsky MA, Rubin AB (1987): Photo and electro-induced spectral changes of rhodopsin in dried photoreceptor membrane films. *Biologicheskie Membrany* **4**: 695–701 (in Russian).

Protasova TB, Fedorovich IB, Ostrovsky MA (1991): Retinal isomers in photo- and electroinduced rhodopsin intermediates. In: *Light in Biology and Medicine* (Douglas RH, Moan J, Ronto G, Eds.) **2**: 545–555: N.Y and London: Plenum Press.

Racker E, Stoekenius W (1974): Reconstitution of purple membrane vesicles catalizing light-driven proton uptake and ATP formation. *J. Biol. Chem.* **249**: 662–663.

Rafferty CN, Shichi H (1981): The involvement of water at the retinal binding site in rhodopsin and early light-induced intramolecular proton transfer. *Photochem. Photobiol.* **33**: 229–234.

Ramon Y, Cajal S (1892): La retine des vertebres. *Cellule* **9**: 119–275.

Rao V, Derguini F, Nakanishi K (1986): 5(trifluorometyl) bacteriorhodopsin does not translocate protons. *J. Am. Chem. Soc.* **108**: 6077–6078.

Rayfield GW (1988): Ultra-high speed bacteriorhodopsin photodetector. Symposium on molecular electronics-Biosensors and biocomputers. Santa Clara, California: **27**.

Rayfield G (1989): Bacteriorhodopsin as an ultrafast electrooptic material. *Phys. Bull.* **34**: 483.

Rayfield GW (1994): Photodiode based on bacteriorhodopsin. In: *Molecular and*

Biomolecular Electronics. Advances in Chemistry ser. (Birge R.R, Ed.): Washington, DC: Amer. Chem. Soc: No. 240: 561–575 .

Rebrik TI, Kalamkarov GR, Ostrovsky MA.(1986): The absence of channel structures in the photoreseptor membrane of the rod disk. *Biofizika* **3**: 985–989 (in Russian).

Regan I, DeGrado WF (1988): Characterization of a helical protein design from first principles. *Science* **241**: 976–978.

Renner T, Hampp N (1993): Bacteriorhodopsin-films for dynamic time average interferometry. *Optics Commun.* **96**: 142–149.

Rental R, Perez MN (1982): Bleaching of purple membrane with O-substituted hydroxylamines. *Photochem. Photobiol.* **36**: 345–348.

Richardson JS, Richardson DC (1989): The de novo design of protein structures. *TIBS* **14**: 304–309.

Robinson ID, Gerlach JC (1969): Photochromic aminotriarylmethane solutions for flash blindness protection. *Appl. Opt.* **8**: 2285–2292.

Rodionov, AB, Shkrob AM (1979): Aldimine retinal hydrolysis in silver ions-induced bacteriorhodopsin. *Bioorgan. Khim.* **5**: 376–394 (in Russian).

Rogers PJ, Morris CA (1979): Regulation of bacteriorhodopsin synthesis by growth rate in continuous cultures of *Halobacterium halobium*. *Arch. Microbiol. Bd.* **119**: S 323–325.

Rosenfeld T, Honig B, Ottolenghi M, Hurley J, Ebrey TG (1977): Cis-*trans* isomerization in the photochemistry of vision. *Pure Appl. Chem.* **49**: 341–351.

Rosenheek K, Brith-Linder M, Linder P, Zakaria A, Caplan S (1978): Proteolysis and flash photolysis of bacteriorhodopsin in purple membranes fragments. *Biophys. Struct. Mech.* **4**: 301–313.

Rothschild KJ, Roepe P, Ahl RI, Earnest TN, Bogomolni RA, Das Gupta SK, Muliken CM, Herzfeld J (1986): Evidence for a tyrosine protonation change during the primary phototransition of bacteriorhodopsin at low temperature. *Proc. Natl. Acad. Sci. U.S.* **83**: 347–351.

Rothschild KJ, He Y-W, Sonar S, Marti T, Khorana HG (1992): Vibration spectroscopy of bacteriorhodopsin mutants. *J. Biol. Chem.* **267**: 1615–1622.

Rothschild KJ, Warti T, Sonar S, He Y-W, Parshuram R, Fischer W, Khorana HG (1993): Asp96 deprotonation and transmembrane α-helical structural changes in bacteriorhodopsin. *J. Biol. Chem.* **268**: 27046–27052.

Saito M, Koyano T, Myamoto H, Umibe K, Kato M (1993): Photoactive artificial membrane and manufactory thereof, Patent 05,048,176 Japan.

Sakmar TP, Franke RR, Khorana HG (1989): Glutamic acid-113 serves as the retilydene Schiff base counterion in bovine rhodopsin. *Proc. Natl. Acad. Sci. U.S.* **86**: 8309–8313.

Salvini-Plawen L, van Mayer E (1977): On the evolution of photoreceptors and eyes. *Evol. Biol.* **10**: 207–263.

Salvini-Plawen L, van Mayer E (1982): *On the Polyphyletic Origin of Photoreceptors.* Visual cell in Evolution: N.Y: Raven Press: 137–154.

Sarma SD, Rajbhandary UL, Khorana HG (1984): Bacterio-opsin mRNA in wild type and bacterio-opsin deficient mutation in *Halobacterium halobium* strains. *Proc. Natl. Acad. Sci. U.S.* **81**: 125–129.

Savransky VV, Tkachenko NV, Chucharev VI (1987): Refraction index changes in bacteriorhodopsin photocycle. *Biologicheskie Membrani* **4**: 479–485.

Scherrer P, McGinnis K, Bogomolni RA (1986): Biochemical and spectoscopic characterization of the blue-green photoreceptor in *Halobacterium halobium*. *Proc. Natl. Acad. Sci. U.S.* **84**: 402–406.

Schimz A, Hinsch K-D, Hildebrand E (1989): Enzymatic and immunological detection of a G-protein in *Halobacterium halobium*. *FEBS Lett.* **249**: 59–61.

Schobert B, Lanyi JK, (1982): Halorhodopsin is a light-driven chloride pump. *J. Biol. Chem.* **257**: 10306–10313.

Schorbert B, Lanyi J, Gragoe E (1983): Evidence for a halide-binding site in halorhodopsin. *J. Biol. Chem.* **258**: 15158–15164.

Schobert B, Lanyi JK (1986): Electrostatic interaction between anions bound to site I and the retinal Schiff of halorhodopsin. *Biochemistry* **25**: 4163–4167.

Schroder DR, Nakanishi K (1987): A simplified isolation procedure for azadirachtin. *J. Natl. Prod.* **50**: 241–244.

Schulz R (1992): Molecular computers based on atomic chips in combination with optical transistors and associative memory storage, Patent 4,241,871 Germany.

Schweiger U, Tittor J, Oesterhelt D (1994): Bacteriorhodopsin can function without a covalent linkage between retinal and protein. *Biochemistry* **33**: 535–541.

Schwemer J. (1983): Pathway of visual pigment regeneration in fly photoreceptor cells. *Biophys. Struct. Mech.* **9**: 287–298.

Sharkov AV, Pakulev AV, Chekalin SV, Matveetz YuA (1985): Primary events in bacteriorhodopsin probed by subpicosecond spectroscopy. *Biochem. Biophys. Acta* **808**: 94–102.

Sherman WV, Slifkin MA, Caplan SR (1976): Kinetic studies of phototransients in bacteriorhodopsin. *Biochem. Biophys. Acta* **423**: 238–248.

Sherman WV, Caplan RS (1979): Influence of membrane lipids on the photochemistry of bacteriorhodopsin in the purple membrane of *Halobacterium halobium*. *Biochem. Biophys. Acta* **502**: 222–231.

Sheves M. Friedman N, Albeck A, Ottolenghi M (1985): Primary photochemical event in bacteriorhodopsin: Study with artificial pigments. *Biochemistry* **24**: 1260–1265.

Shichida Y (1986): Primary intermediates of photobleaching of rhodopsin *Photobiochem. Photobiophys.* **13**: 287–307.

Shieh P, Packer L (1976): Photoinduced potential across a polymer stabilized planar membrane, in the present bacteriorhodopsin. *Biochem. Biophys. Res. Commun.* **71**: 603–609.

Shinar K, Drukmann S, Ottolenghi, Korenstein K (1977): Electric field effect in bacteriorhodopsin. *Biophys. J.* **19**: 1–6.

Shnyrov VL, Tarakhovsky Yu S, Borovyagin VL (1981): The study of structural changes in the purple membranes of *Halobacterium halobium* at thermal and alkaline denaturation. *Bioorgan. Khim.* **7**: 1054–1059 (in Russian).

Shuvalov VA, Parson WW (1981): Energies and kinetics of radical pairs involving bacteriochlorophill and bacteriopheophytin in bacterial reaction centers. *Proc. Natl. Acad. Sci. U.S.* **78**: 957–961.

Shuvalov VA, Klevanik AV (1983): The study of the state [P870+ B800–] in bacteria reaction centers by selective picosecond and low-temperature spectroscopies. *FEBS Lett.* **160**: 51–55.

Shuvalov VA (1984): In. *Advances in Photosyntesis Research* (Sybesma C, Ed.) The Hague, Boston, Lancaster: M.Nijhaff/Dr.W.Link: 93–98.

Skulachev VP (1976): Conversion of light-energy into electric energy by bacteriorhodopsin. *FEBS Lett.* **64**: 23–25.
Skulachev VP (1982): A single turnover study of photoelectric current generating proteins. *Meth. Enzymol.* **88**: 35–45.
Skulachev VP (1984): Membrane bioenergetics—should we build the bridge across the river or alongside of it? *TIBS*: 184–185.
Slobodyanskaya EA, Abrastchin EB, Ostrovsky MA (1980): The study of ionochromic properties of chicken visual pigments. *Bioorgan. Khim.* **6**: 223–229 (in Russian).
Smets G (1983): *Adv. Polymer* **50**: 17–38.
Sokolov NA, Mitsner BI, Zakis VI (1979): The synthesis of 4-keto and 4-oxyderivatives all-E- and 13-z-retinals and their interaction with bacterioopsin. *Bioorgan. Khim.* **5**: 1053–1058 (in Russian).
Soppa J, Oesterhelt D (1989): Bacteriorhodopsin mutants of *Halobacterium* spec. GRB. I. The 5-bromo-2-deoxyuridin-selection as a method to isolate point mutants in Halobacteria. *J. Biol. Chem.* **264**: 13043–13048.
Soppa J, Otomo J, Straub J, Tittor J, Meessen S, Oesterhelt D (1989): Bacteriorhodopsin mutants of *Halobacterium* spec. GRB. II. Characterization of mutants. *J. Biol. Chem.* **264**: 13049–13056.
Sperling L, Hubbard R (1975): Squid retinochrome. *J. Gen. Physiol.* **65**: 235–251.
Sperling W, Carl P, Rafferty ChN, Dencker NA (1977): Photochemistry and dark equilibrium of retinal isomers and bacteriorhodopsin isomers. *Biophys. Struct. and Mech.* **3**: 79–94.
Spudich JL, Bogomolni RA (1983): Spectroscopic discrimination of the three rhodopsin-like pigments in *H. halobium* membranes. *Biophys. J.* **43**: 243–246.
Spudich JL, Bogomolni RA (1984): The mechanism of color discrimination by a bacterial sensory rhodopsin. *Nature* (London) **312**: 509–513.
Spudich JL, Stoeckenius W (1979): Photosensory and chemosensory behavior of *Halobacterium halobium. Photochem. and Photobiophys.* **1**: 43–53.
Spudich EN, Spudich JL (1982): Control of transmembrane ion fluxes to select halorhodopsin-deficient and other energy transduction mutant of *Halobacterium halobium. Proc. Natl. Acad. U.S.* **79**: 4308–4312.
Spudich EN, Sundberg SA, Manor D, Spudich JL (1986): Properties of a second sensory receptor protein in *Halobacterium. Proteins* **1**: 239–246.
Spudich JL, McCain DA, Nakanishi K, Okabe M, Shimizu, N, Rodman H, Honig B, Bogomolni RA (1986a): Chromophor/protein interaction in bacterial sensory rhodopsin and bacteriorhodopsin. *Biophys J.* **49**: 479–483.
Spudich JL, Bogomolni RA (1988): Sensory rhodopsin in halobacteria. *Annu. Rev. Biophys. Chem.* **17**: 193–215.
Spudich JL, Bogomolni RA (1992): Sensory rhodopsin I: Reception activation and signal relay. *J. Bioenerg. Biomem.* **24**: 193–200.
Spudich JL (1994): Protein-protein interaction converts a proton pump into a sensory receptor. *Cell* **79**: 747–750.
Spudich JL, Lanyi JK (1996): Shutting between two protein conformation: The common mechanism for sensory transduction and ion transport. *Current Opinion in Cell Biology* **8**: 452.
Steinberg G, Friedman N. Sheves M, Ottolenghi M (1991): Isomer composition and spectra at the dark and light adapted forms at artifitial bacteriorhodopsin. *Photochem. Photobiol.* **54**: 969–976.

Stern LJ, Khorana HG (1989): Structure-function studies on bacteriorhodopsin. X. Individual substitution of arginine residues by glutamine affect chromophore formation, photocycle, and proton translocation. *J. Biol. Chem.* **264**: 14202–14208.

Stoeckenius W, Kunau WH (1968): Further characterization of particulate fractions from lysed cell envelopes of *Halobacterium halobium* and isolation of gas vacuole membranes. *J. Cell Biol.* **38**: 337–357.

Stoeckenius W, Lozier RH (1974): Light energy conversion in *Halobacterium halobium. J. Supramol. Struct.* **2**: 769–774.

Stoeckenius W (1976): Structure of biological membranes. Bacteriorhodopsin and purple membrane. In: *Biological and Artificial Membranes and Desalinization of Water*. Rome: Pontificae Academiae Scientiarum Scripta Varia **40**: 65–84.

Stoeckenius W (1976a): The purple membrane of salt-loving bacteria. *Sci. Am.* **234**: 38–46.

Stoeckenius W (1976b): Light energy transduction in halobacteria. *Bioelectrochem. Bioenerg.* **3**: 371–372.

Stoeckenius W (1979): A model for the function of bacteriorhodopsin. In. Membrane Transduction Mechanisms (R.A. Cone, J.E. Dowling, Eds.) Raven, New York, 39–47.

Stoeckenius W, Lozier RH, Bogomolni (1979): Bacteriorhodopsin and purple membrane of halobacteria. *Biochem. Biophys. Acta* **505**: 215–278.

Stoeckenius W, Bogomolni RA (1982): Bacteriorhodopsin and related pigments of halobacteria. *Annu. Rev. Biochem.* **52**: 587–616..

Stoeckenius W (1994): From membrane structure to bacteriorhodopsin. *J. Membr. Biol.* **139**: 139–148 .

Stryer L (1991): Visual excitation and recovery. *J. Biol. Chem.* **226**: 10711–10714.

Subramaniam S, Marti T, Khorana HG (1990): Protonation state of Asp (Glu)-85 regulates the purple-to-blue transition in bacteriorhodopsin mutants Arg82–Ala and Asp85–Glu: The blue form is inactive in proton translocation. *Proc. Natl. Acad. Sci. U.S.* **87**: 1013–1017.

Subramaniam S, Greenhalgh DA, Rath P, Rothschild KJ, Khorana HG (1991): Replacement of leucine-93 by alanine or threonine slows down the decay of the N and O intermediates in the photocycle of bacteriorhodopsin: Implications for proton uptake and 13-*cis*-retinal—all-*trans*-retinal reisomerization. *Proc. Natl. Acad. Sci. U.S.* **88**: 6873–6877.

Subramaniam S, Gerstein M, Oesterhelt D, Henderson R (1993): Electron diffraction analysis of structural changes in the photocycle of bacteriorhodopsin. *EMBO J.* **12**: 1–8.

Sumper M, Reitmeier H, Oesterhelt D (1976): Biosynthesis of the purple membrane of *Halobacterium halobium. Angew. Chem. Intern. Ed.* **15**: 187–194.

Sumper M, Hermann G (1978): Studies of the biosynthesis of bacteriorhodopsin. *Eur. J. Biochem.* **89**: 229–235.

Sumper M (1982): The brown membrane of *Halobacterium halobium*: The biosynthetic precursor of the purple membrane. *Meth. Enzymol.* **88**: 391–399.

Sunberg SA, Alam N, Lebert M, Spudich JL, Oesterhelt D, Hazelbauer GL (1990): Characterization *Halobacterium halobium* mutant defective in taxis. *J. Bacteriol.* **172**: 2328.

Suzuki T, Makino-Tasaka M, Eguchi E (1984): 3-dehydroretinal (vitamin A2 aldehyde) in crayfish eye. *Vision Res.* **24**: 783–787.

Sybesma C (Ed.) (1984): Advances in photosynthesis research. The Hague, Boston, Lancaster: M.Nijhaff/Dr. W. Junk.

Takanashi T, Mochizuki Y, Kamo N, Kobatake Y. (1985): Evidence that the long-lifetime photointermediate of s-rhodopsine is a receptor for negative phototaxis in *Halobacterium halobium*. *Biochem. and Biophys. Res. Commun.* **127**: 99–105.

Takahashi T, Yan B, Mazur P, Derguini F, Nakanishi K, Spudich JL (1990): Color regulation in the archaebacterial phototaxis receptor. *Biochemistry* **29**: 8467–8474.

Takei H, Lewis A, Chen Ah, Nebenzahl I (1991): Implementing receptive fields with excitatory and inhibitory optoelectrical responses of bacteriorhodopsin films. *Appl. Opt.* **30**: 500–509.

Tan EH, Govender DSK, Birge RR (1996): Large organic cations can replace Mg^{2+} and Ca^{2+} ions in bacteriorhodopsin and maintain proton pumping ability. *J. Am. Chem. Soc.* **118**: 2752–2753.

Tanny G, Caplan SP, Eisenbach M (1979): Interaction of purple membrane with solvents. I. Applicability of solubility parameter mapping. *Biochem. Biophys. Acta* **554**: 269–278.

Thiedemann G, Dencher N (1994): The activity of bacteriorhodopsin at reduced hydration. Abstr. of VI Inter. Conf. of Retinal Protein 19–24 June, 1994, Leiden, NL: 156.

Tiede DM, Mueller P, Dutton PL (1982): Spectrophotometric and voltage clamp characterization of monolayers of bacterial photosynthetic reaction centers. *Biochem. Biophys. Acta* **681**: 191–210.

Tierno ME, Mead D, Asato AE, Liu SRH, Sekiya N, Yishihara K, Chang CW, Nakanishi K, Govindjee R (1990): 14-Fluorobacteriorhodopsin and other fluorinated and 14-substituted analogs. An extra, unusually red-shifted pigment formed during dark adaptation. *Biochemistry* **29**: 5948–5953.

Todorov T, Nikolova L, Tomova N (1984): Polarization holography. 1: A new high-efficiency organic material with reversible photoindused birefringence. *Appl. Opt.* **23**: 4309.

Todorov T, Nikolova L, Tomova N (1984a): Polarization holography. 2: Polarization holographic gratings in photoanisotropic materials with and without intrinsic birefringence. *Appl. Opt.* **23**: 4588–4591.

Tokunaga F, Govindjee R, Ebrey TG (1977): Synthetic pigment analogs of the purple membrane protein. *Biophys. J.* **19**: 191–198.

Tokunaga F, Ebrey TG, Crouch R (1981): Purple membrane analogs synthesized from C17 aldehyde. *Photochem. Photobiol.* **33**: 495–499.

Tokunaga F, Iwasa T (1982): The photoreaction cycle of bacteriorhodopsin: Low-temperature spectrophotometry. *Meth. Enzymol.* **88**: 163–167.

Tokunaga F, Watanabe T, Uematsu J, Hara R, Hara T (1990): Photoreaction of retinochrome at very low temperature. *FEBS Lett.* **262**: 266–268.

Tomioka H, Takahashi T, Kamo N, Kobatake Y (1986): Flash spectrophotometric identification of a fourth rhodopsin-like pigment in *Halobacterium halobium*. *Biochem. Biophys. Res. Commun.* **139**: 389–395.

Tomioka H, Otomo J, Hirayama J, Kamo N, Sasabe H (1990): IV International Conference on Retinal Proteins, Santa Cruz.

Tomioka H, Sasabe H (1995): Isolation of photocemical active archebacterial photoreceptor, pharaonis phoborhodopsin from *Natrobacterium pharaonis*. *Biochem. Biophys. Acta* **1234**: 261–267.

Tomita T, Kaneko A, Murakami M, Pautler EL (1967): Spectral response curves of single cones in the carp. *Vision Res.* **7**: 519–531.

Torrealba F, Guillerty RW, Eysel U, Polley EH, Masson CA (1982): Studies of retinal representations within the cat's optic tract. *J. Comp. Neurol.* **211**: 377–396.

Towner P, Gaertner W, Wackhoff B, Oesterhelt D, Hopf H (1980): α-Retinal as a prostetic group in bacteriorhodopsin. *FEBS Lett.* **117**: 363–367.

Trissl H-W, Montal M (1977): Electrical demonstration of rapid light-induced conformational changes in bacteriorhodopsin. *Nature* (London) **266**: 655–657.

Trissl H-W (1979): Light-induced conformational changes of cattel rhodopsin as probed by measurements of the interfacial potential. *Photochem. and Photobiol.* **29**: 579–588.

Trissl H-W (1985): I. Primary electrogenic processes in bacteriorhodopsin probed by photoelectric measurement with capacitative metal electrodes. *Biochem. Biophys. Acta* **806**: 124–135.

Trissl H-W, Kunze U (1985): II. Primary electrogenic reaction in chloroplast probed by picosecond flash-induced dielectric polarization. *Biochim. Biophys. Acta* **806**: 136–144.

Trissl H-W (1987): Eine biologische photodiode mit hochster Zeitauflosung. *Optoelektronik Magazin* **3**: 105–107 (in German).

Trissl H-W, Gartner W (1987): Rapid charge separation and bathochromic absorption shift of flash-excited bacteriorhodopsin containing 13-*cis* or all-*trans* forms of substituted retinals. *Biochem.* **26**: 751–758.

Trissl H-W (1990): Photoelectric measurement of purple membrane. *Photochem. Photobiol.* **51**: 793–818.

Tsuda M (1978): Kinetic study of photoregeneration process of digitonin-solubilized squid rhodopsin. *Biochem. Biophys. Acta* **502**: 495–506.

Tsuda M (1979): Transient spectra of intermediates in the photolytic sequence of octopus rhodopsin. *Biochem. Biophys. Acta* **545**: 537–546.

Tsuda M, Hazemoto N, Kondo M, Kamo N, Kobatake Y, Terayama Y (1982): Two photocycles in *Halobacterium halobium* that lacks bacteriorhodopsin. *Biochem. Biophys. Res. Commun.* **108**: 970–976.

Tsuda M, Govindjee R, Ebrey TG (1983): Effect of pressure and temperature on the M412 intermediate of the bacteriorhodopsin photocycle. *Biophys. J.* **44**: 249–254.

Tsuji K, Rosenheck K (1979): The low pH species of bacteriorhodopsin. *FEBS Lett.* **98**: 368–372.

Tsuji K, Neumann E (1981): Structural changes in bacteriorhodopsin induced by electric impulses. *Int. J. Biol. Macromol.* **3**: 231–242.

Tsuji K, Neumann E (1981a): Structural changes induced by electric fields in membrane bound bacteriorhodopsin. *Biophys. Struct. Mech.* **7**: 284–298.

Tucker JB (1984): Biochip: can molecules compute? *High Technol.* **4**: 37–47.

TuShu-T, Hutchinson H, Cavanaugh JR (1981): Interaction between gramicidin-a and bacteriorhodopsin in reconstituted purple membrane. *Biochem. Biophys. Res. Commun.* **106**: 23–29.

Vanfleteren JR (1982): A monophyletic line of evolution: Ciliary induced photoreseptor membrane. Visual cell in evolution. N.Y.: Raven Press: 107–136.

Varo G (1981): Dried oriented purple membrane samples. *Acta Biol. Acad. Sci. Hung.* **32**: 301–310.

Varo G, Keszthelyi L (1983): Photoelectric signals from dried oriented purple membranes of *Halobacterium halobium*. *Biophys. J.* **43**: 47–51.
Varo G, Lanyi JK (1991): Distortion in the photocycle of bacteriorhodopsin at moderate dehydration. *Biophys. J.* **59**: 313–322.
Varo G, Lanyi JK (1991a): Thermodynamics and energy coupling in the bacteriorhodopsin photocycle. *Biochemistry* **30**: 5016–5022.
Vekker L, Allen J (1997): Mental representation of physical reality. Mechanism and processes (in preparation).
Vernon LP, Ke B (1971): *Photochromism in Living Systems. Technical Chemical Photochromism.* Pergamon Press **3**: 687–732: N.Y.
Vodt K (1987): Chromophores of insect visual pigments. *Photobiochem. Photobiophys.* Suppl.: 273–296.
Vsevolodov NN, Kostikov A.P, Rikhireva GT (1974): Study of photoinduced transformations in bacteriorhodopsin membrane complex. *Biofizika* **19**: 942–946 (in Russian).
Vsevolodov NN, Kayushin LP (1976): The spectral transformation in the purple membranes of *Halobacterium halobium*. *Stud. Biophys.* **59**: 81–87.
Vsevolodov NN, Chekulaeva LN (1977): Spectral transfomation in purple membranes by *Halobacterium halobium*. *Biofizika* **32**: 1019–1023 (in Russian).
Vsevolodov NN, Chekulaeva L (1979): Spectral transition in purple membranes from *Halobacterium halobium*. I. Effect of preliminary illumination on photochemical processes. *J. Bioenerg. Biomembranes* **10**: 13–22.
Vsevolodov NN, Druzhko AB, Evstigneeva RP, Mitsner BI, Chekulaeva LN (1983): Photochromic material, Patent No. 1,032,912 USSR.
Vsevolodov NN (1984): Biochrom application for the creation of a Biocomputer. *Mikroprotses. Sredstva i Sistemi* **3**: 13–17 (in Russian).
Vsevolodov NN, Djukova TV, Korchemskaya EYA, Taranenko VB (1984): Nonlinear Veiger-effect in film from bacteriorhodopsin. *Ukrainsky Fizicheskiy Zhurnal* **29**: 1120–1122.
Vsevolodov NN, Dyukova TV, Chekulaeva LN (1985): Photochromic material, Patent No.1,194,177 USSR.
Vsevolodov NN, Poltoratsky VA (1985): The holograms on biological photochromic materials—biochrome. *Zhurn. Sov. Tech. Fiziki* **55**: 2093–2094.
Vsevolodov NN, Gaynullina SM, Chekulaeva LN (1986): Phototrophic properties of heterotrophic microorganisms in purple membrane-containing media. *Biofizika* **31**: 437–439 (in Russian).
Vsevolodov NN, Ivanitski GR, Soskin MS, Taranenko VB (1986a): Biochrome film is a reversive medium for optical recording. *Avtometriya* **2**: 43–48 (in Russian).
Vsevolodov NN (1988): Biopigments-photoregistrators: Photomaterial based on bacteriorhodopsin. *Nauch. izd. Moskva*: Nauka (in Russian).
Vsevolodov NN, Poltoratsky VA, Razumov LA (1989): Photoplate Biochrome in the holographic interferometer. *Zurn. Tchn. Fiziki* **59**: 176–177.
Vsevolodov NN, Dyakova TV, Druzhko AB, Shakhbazyan VYu (1991): Optical recording material bazed on bacteriorhodopsin modified with hydroxylamine. *Pros. SPIE* **1621**: 11–20.
Vsevolodov NN, Kotov VB, Salakhutdiniv VK (1992): Interaction of a light wave with a volume diffraction grating in a dynamic photosensitive medium. *Soviet Physics. Technical Physics* **37**: 792.

Vsevolodov NN, Dyukova TV, Druzhko AB, Shakhbazyan VY (1995): Optical recording material based on bacteriorhodopsin modified with hydroxylamine. *SPIE Milestone Ser.* **114**: 511.

Vsevolodov NN, Dyukova TV (1995): Retinal-protein complexes as optoelectronic components. *Trends in Biotechnology* **12**: 81–88..

Vsevolodov NN, Korchemskaya YeA, Soskin MS (1996): Holographic investigation of the process of light adaptation of the molecules of bacteriorhodopsin. *Biophysics* **41**: 315–328.

Vsevolodov NN, Korchemskaya YeA, Soskin MS (1996a): Study of the adaptation of the bacteriorhodopsin molecules by means of the holographic method. *Biophysics* **41**: 329–333.

Wagner G, Oesterhelt D, Krippahl G, Lanyi JK (1981): Bioenergetic role of halorhodopsin in *Halobacterium Halobium* cells. *FEBS Lett.* **131**: 341–345.

Wald G (1937): Visual purple system in fresh-water fish. *Nature* **139**: 1017–1018.

Wald G (1968): The molecular basis of visual excitation. *Nature* **219**: 800–810.

Warshel A (1976): Bicycle-pedal model for the first step in the vision process. *Nature* **260**: 679–683.

Wassle H (1982): Morphological types and central projection ganglion cells in the cat retina. In: *Progress in Retinal Research* Oxford: 125–152.

Weber HJ, Bogomolni RA (1981): P588 a second retinal-containing pigment in *Halobacterium halobium. Photochem. Photobiol.* **33**: 601–698.

Weber HJ, Sarma S, Leighton T (1982): The Halobacterium group—microbiological methods. *Methods. Enzymol.* **88**: 369–173.

Weetall HH, Robertson B, Cullin D, Brown J, Walch M (1993): Bacteriorhodopsin immobilized in sol-gel glass. *Biochim. Biophys. Acta* **1142**: 211–213.

Werber M (1980): Halophilism. *Biochim.* **62**: 411–422.

Werner O, Fisher B, Lewis A (1992): Strong self-defocusing effect and four-wave mixing in bacteriorhodopsin films. *Opt. Lett.* **17**: 241–243.

Wilden U, Hall SV, Kuhn H (1986): Phosphodiesterase activation by photoexcited rhodopsin is quenched when rhodopsin is phosphorilated and binds the intrinsic 48-kDa protein of rod outer segments. *Proc. Natl. Acad. Sci. U.S.* **83**: 1174–1178.

Wilkens LA (1984): Ultraviolet sensitivity in hyperpolarizing photoreceptors of the giant clam *Tridacna. Nature* **309**: 446–448.

Yan B, Takahashi T, McCain DA, Rao J, Nakanishi K, Spudich JL. (1990): Effects of modifications of the retinal beta-ion ring on archaebacterial sensory rhodopsin I. *Biophys. J.* **57**: 477–483.

Yan B, Takanashi T, Johnson R, Derguini F, Nakanishi K, Spudich JL (1990a): All-*trans*/13-*cis* isomerization of retinal is required for phototaxis signaling by sensory rhodopsins in *Halobacterium halobium. Biophys. J.* **57**: 807–814.

Yan B, Spudich JL (1991): Evidence of the repellent receptor form of sensory rhodopsin-I in an attractant signalling state. *Photochem. Photobiol.* **54**: 1023–1026.

Yan B, Nakanishi K, Spudich JL (1991): Mechanism of activation of sensory rhodopsin I: Evidence for a steric trigger. *Proc. Natl. Acad. Sci. U.S.* **88**: 9412–9416.

Yao VJ, Spudich JL (1992): Primary structure of an archaebacterial transducer, a methyl-accepting protein associated with sensory rhodopsin I. *Proc. Natl. Acad. Sci. U.S.* **89**: 11915–11919.

Yoshizawa T, Wald G (1963): Pre-lumirhodopsin and the bleaching of visual pigment. *Nature* **197**: 1279–1286.

Yoshizawa T (1972): The behavior of visual pigment at low temperature. In: *Handbook of Sensory Physiology*. B. Springer 7/1: 145–179.

Yoshizawa T, Shichida Y, Matuoka S (1984): Primary intermediates of rhodopsin studied by low temperature spectrophotometry and laser photolysis. Bathorhodopsin, hypsorhodopsin and photorhodopsin. *Vision Res.* **24**: 1455–1463.

Zaitsev SYu, Kalabina NA, Zubov VP, Lukashev EP (1992): Monolayers of photosynthetic reaction centers of green and purple bacteria. *Thin Solid Films* **210/211**: 723–735.

Zaitsev SYu, Kozhevnikov NM, Barmenkov Yu O, Lipovskaya M Yu (1992a): Kinetics of dynamic hologram recording in polymer films with immobilized bacteriorhodopsin. *Photochem. Photobiol.* **55**: 851–856.

Zavarzin AA (1941): *Essays on Evolutionary Histology of the Nervous System.* Moscow-Leningrad: Acad. Sci. Press: 419 Zbx. (in Russian).

Zeldovich B YA, Shkunov VV (1979): Spatial polarizational reversal of wave front at four-photon interaction. *Sov. J. Quantum. Electron* **6**: 629.

Zhang CF, Birge RR (1990): *J. Chem. Phys.*.

Zhykovsky EA, Oprian DD (1989): Effect of carboxylic acid side chains on the absorption maximum of visual pigments. Science **249**: 928–930.

Zimanyi L, Keszhelyi L, Lanyi JK (1989): Transient spectroscopy of bacterial rhodopsins with an optical multichannel analyzer. 1. Comparison of the photocycles of bacteriorhodopsin and halorhodopsin. *Biochemistry* **28**: 5165–5172.

Zingoni J, Or YS, Crouch R, Chang C-H, Govindjee R, Ebrey TG (1986): Effect of variation of polien side-chain length on formation and function of bacteriorhodopsin analog pigments. *Biochemistry* **25**: 2022–2017.

Zubov VP, Zaytsev S Yu, Novikova MB, Vsevolodov NN, Zharmukhamedov SK (1987): Photochromic material, Patent No. 1,389,491 USSR.

9

Glossary and Abbreviations

Glossary

Apoproteins	See opsins
Bacterioopsin	Protein of bacteriorhodopsin
Bacteriorhodopsin	The first retinal-containing protein of the plasma membrane of the halophilic bacteria *Halobacterium halobium*. Can conduct light-dependent proton transfer across the bacterial membrane
BR Analogs	Bacteriorhodopsins from wild strains with substituted artificial retinal
BR Variants	Bacteriorhodopsin analogs synthesized by replacement of natural individual amino acids in bacterioopsin
Ciliaric	Tubular
Cyclicity	The number of record-erase cycles after which irreversible destruction of the photochrome occurs
D (optical absorption)	Sometimes the figures are written with "D" and sometimes with "Absorption." The author apologizes for this discrepancy
Davson-Danielli model	Model postulates the membrane as a three-layer structure—a lipid layer included between two layers of proteins
Dichroic (or optical anisotropy)	Dichroic materials polarized light in one direction more strongly than polarized light at right angles to that direction. Such materials are different from birefringent materials which may have different refractive indices for electrical vectors vibrating at right angles to each other and often have negligible absorption coefficients
Dichroic analyzer	An instrument for measuring the rate of dichroism

Dichroism	The property of exhibiting two colors. Under reflection of light, one color is seen, and another when viewed under transmitted light. A solution of chlorophyll is an example. Substances which have this property are called dichroic
Dynamic hologram	An image recorded in a material with laser beams which is not constant, but persists for a defined period of time and then fades spontaneously
Electrochromic effect	The capacity of a material (including orientated purple-membrane layers) to change reversibly its absorption maximum when an electric field is applied
Halobacterium halobium	Latin name of one of the species of extreme halophilic bacteria capable of synthesizing bacteriorhodopsin.
Halobacterium salinarium	New name recommended to use since 1990 instead of *Halobacterium halobium*
Halophilic bacteria	Bacterias dwelling in naturally saline reservoirs, containing up to 35% salt (for example the Dead Sea in Israel; lakes near Krasnovodsk, Russia; Great Salt Lakes in Utah, U.S.)
Halorhodopsin	The second retinal-protein complex of *Halobacterium halobium*. Accomplishes light-dependent chloride ion translocation across the bacterial membrane
Holography	The process of three-dimensional image recording with laser beams by creating a regular pattern of bright and dark lines produced by the interference of optical waves
Josephson effect	Spontaneous tunneling of current through an insulating barrier that separates two super-conducting materials
Mitchel's theory	Hemoosmotic theory suggested by British biochemist Peter Mitchell in 1961. In his theory, cell respiration involves the creation of a proton concentration (hydrogen ions) difference on both sides of the membrane. The osmotic energy of this concentration difference supports the chemical synthesis of ATP from ADP.
Opsin	Protein of rhodopsin
Opsin shift	This shift represents the influence of the opsin binding site on the absorption spectrum of the chromophore. It is expressed by the λ_{max} of the protonated Schiff base (in cm^{-1}) minus the λ_{max} of the pigment (in cm^{-1}) [see Balogh-Nair et al. 1981]

Phoborhodopsin	Same as Sensory rhodopsin II
Photochromy	The capacity of a material to reversibly change its optical density or color (absorption spectrum shape and/or position) when irradiated with light
Phototaxis	The response of an organism to light. A negative phototaxis is motion of the organism away from light while a positive phototaxis is in the reverse direction
Purple membrane	Membrane fragments of a lipid-protein monolayer of hexagonally packed bacteriorhodopsin molecules
Schiff base	Chemical linkage between protein (for example opsin) and chromophore (for example retinal)
Sensory rhodopsins	The third and the fourth retinal–proteins from *Halobacterium halobium* (see Sensory rhodopsin I and II)
Sensory rhodopsin I	Retinal–protein complex controlling the reorientation of cell movement in a light gradient
Sensory rhodopsin II	Retinal–protein complex controlling the repulsion of cell movement (negative phototaxis) in blue-green light

Abbreviations

ADP	adenosinediphosphate
ATP	adenosinetriphosphate
STM	scanning tunneling microscope
PM	purple membranes
BR	bacteriorhodopsin from wildtype
BR-326	different from BR by the exchange of Asp-96 for Asn
VR	visual rhodopsin
ABR	bacteriorhodopsin analogs
AR	rhodopsin analogs
WT	wild type
LA	light-adapted
DA	dark-adapted
ERP	early receptor potential
PD	photovoltaic device
SSM	site-specific mutagenesis
RM	random mutagenesis
SB	Schiff base
SBH+	protonated Schiff base
RP	retinal–protein
FTIR	Furie transform infra-red method
RBP	retinal-binding protein
DD model	The Davson–Danielli model

eBR	prepared from the expression of a synthetic BR wild gene in *Esch. Coli*
deH	dehydro (for example: 3,4-deH = 3,4-dehydro-retinal)
diH	dihydro
Chl	chlorophyll

Names of amino acids

Amino acid	Triple letter code	Single letter code
Alanine	Ala	A
Arginine	Arg	R
Asparagine	Asn	N
Aspartic Acid	Asp	D
Cysteine	Cys	C
Glutamine	Gln	Q
Glutamic Acid	Glu	E
Glycine	Gly	G
Histidine	His	H
Isoleucine	Ile	I
Leucine	Leu	L
Lysine	Lys	K
Methionine	Met	M
Phenylalanine	Phe	F
Proline	Pro	P
Serine	Ser	S
Threonine	Thr	T
Tryptophan	Trp	W
Tyrosine	Tyr	Y
Valine	Val	V

Example of abbreviations for BR mutants

Name	Substituted
D85A	Asp85—Ala
D85N	Asp85—Asn
D96A	Asp96—Ala
D96G	Asp96—Gly
D96N	Asp96—Asn
R82Q	Arg82—Gln

10

Appendix

CHEMICAL STRUCTURES AND FORMULAS
(NAME OF RETINAL ANALOGS IS SHORTENED)

Number and name		Number and name	
1. All-*trans*-retinal	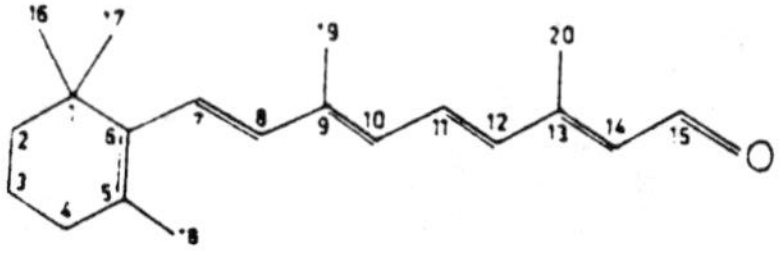	2. 11-cis-locked retinal	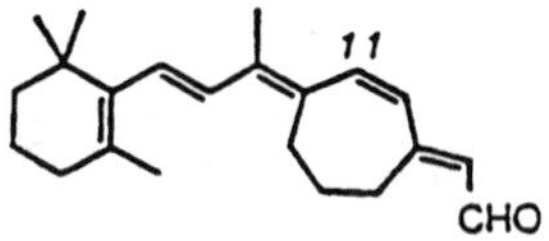
3. All-*trans*-locked-retinal	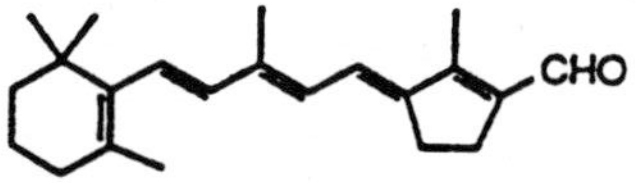	4. Dienal	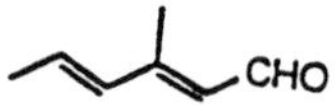
5. Hexanal		6. 11,12-dihydro-retinal	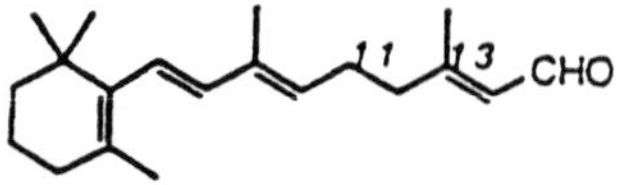
7. 11-cis-retinal	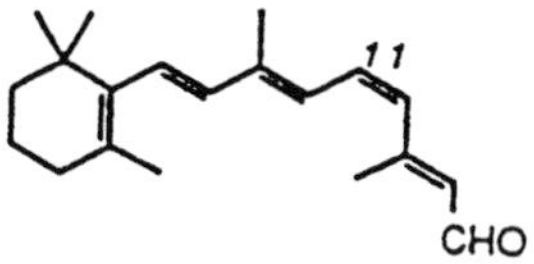	8. Aromatic 9,11-bridged	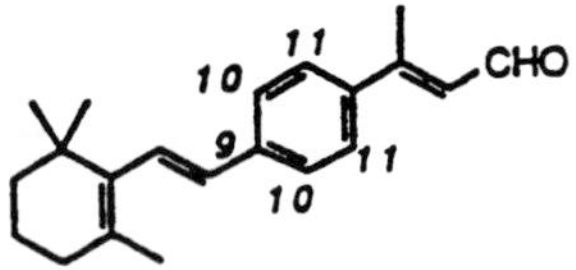

Number and name	Number and name
9. n-hexanal	 10. locked 11-cis 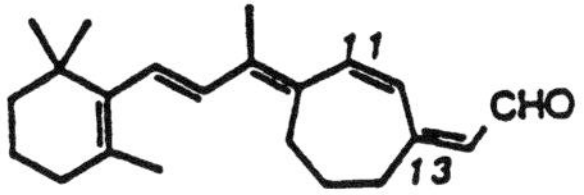
11. azulene analog	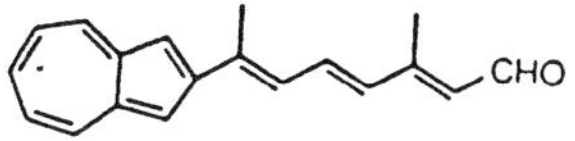 12. naphtyl-retinal 11-cis 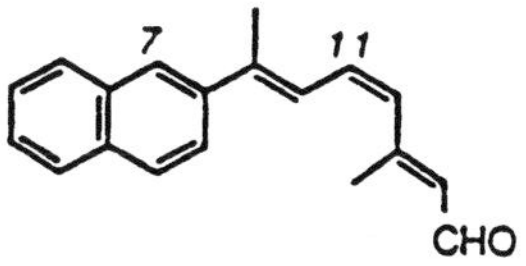
13. Tetraenal	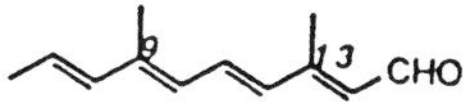 14. Trienal 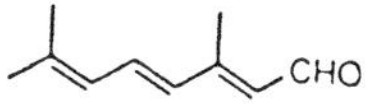
15. Locked 13-cis- analog	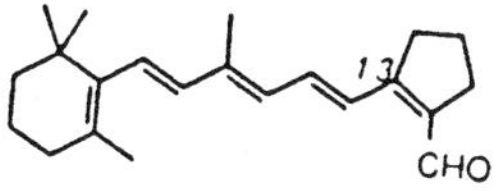 16. 9-cis 12, 14-briged-retinal 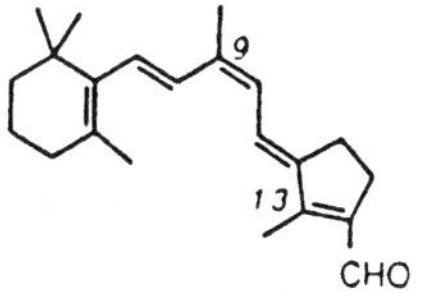

Number and name		Number and name	
17. 9,11-dicis-retinal. Locked 13-trans	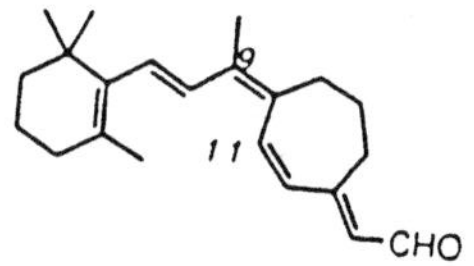 11 CHO	18. 9,11-dicis-retinal. Locked 13-cis	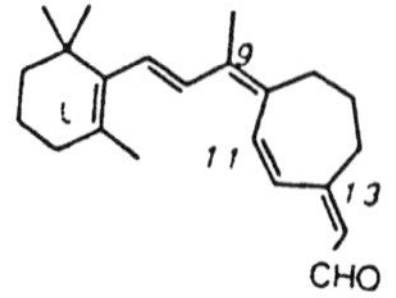 11 13 CHO
19. Aromatic-locked	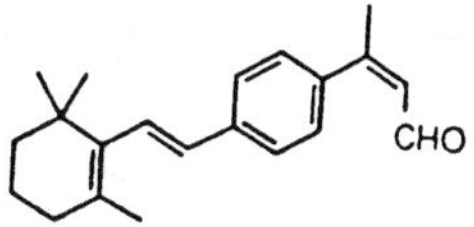 CHO	20. Naphtyl-retinal	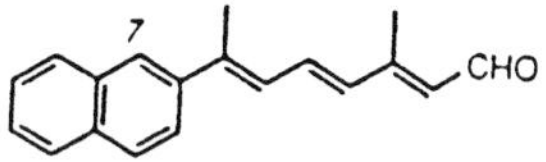 7 CHO
21. 3,4-deH- retina	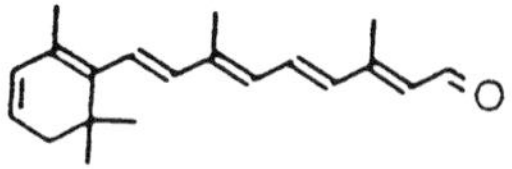 O	22. bicyclic-retinal	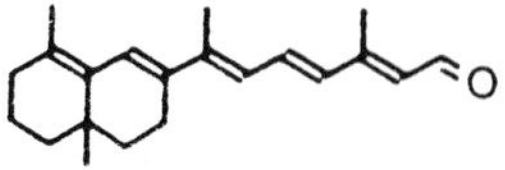 O
23. 3-hydroxy- retinal	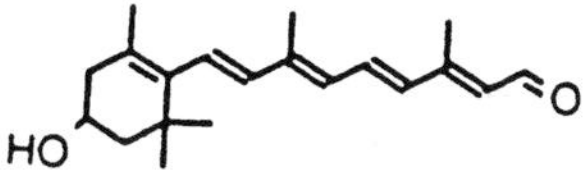 HO O	24 13-des-methiy- retinal	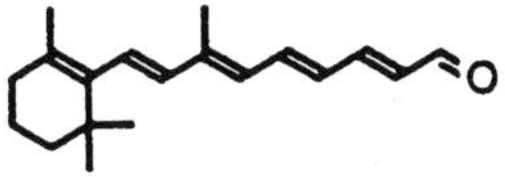 O
25. Pentaenal-retinal	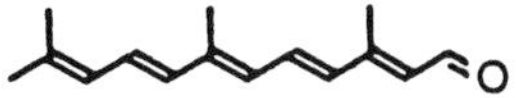 O	26. 1,1,5-demethyl retinal	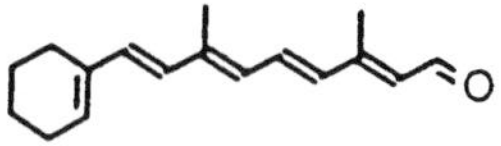 O

Number and name		Number and name	
27. Aromatic-retinal	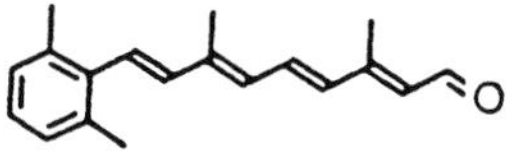 O	28. Phenyl-retinal	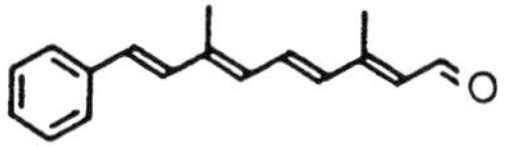 O
29. Naphtyl-retinal	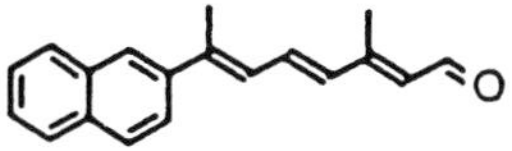 O	30. 9-desmethyl-retinal	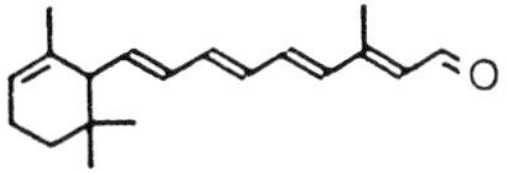 O
31. 5,6,7,8-tetraH-retinal	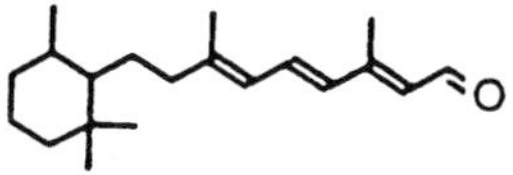 O	32. 9,10-diH-retinal	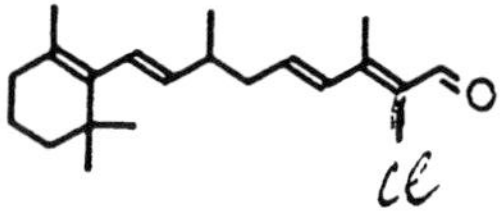 O
33. Didesmethyl	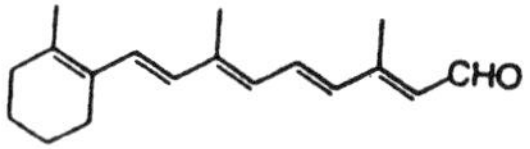 CHO	34. 5-desmethyl-retinal	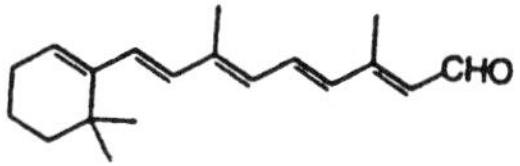 CHO
35. All-desmethyl-retinal	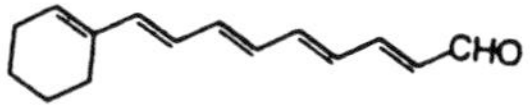 CHO	36. Acyclic-retinal	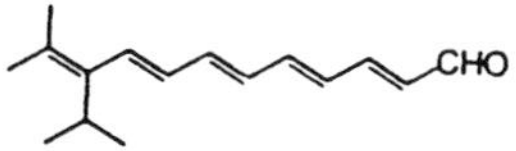 CHO

Number and name		Number and name	
37. 5,6-diH- retinal	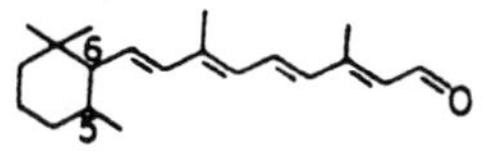	38. 7,8-diH- retinal	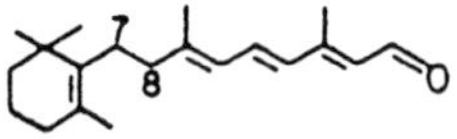
39. 9,10-diH- retinal	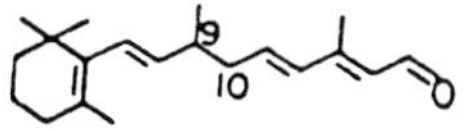	40. 11,12-diH- retinal	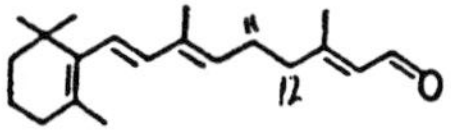
41. N-oxide- retinal	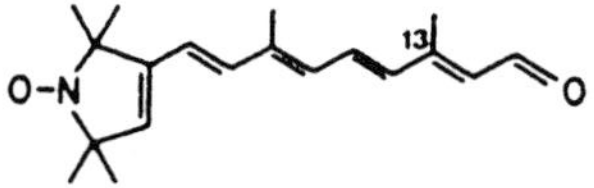	42. 4-keto- retinal	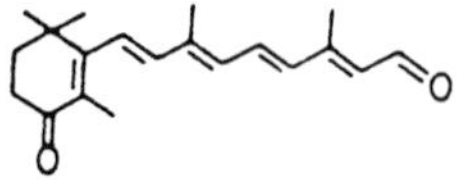
43. 5,6-epoxy- retinal	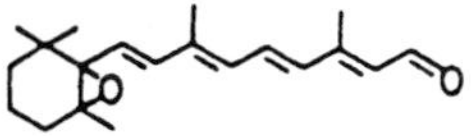	44. 3-methoxy- retinal	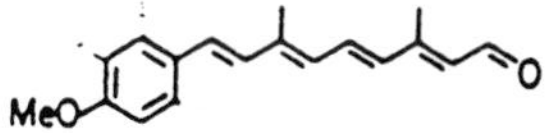
45. 3,4-deH- retinal (Retinal-2 or Vitamin A2)	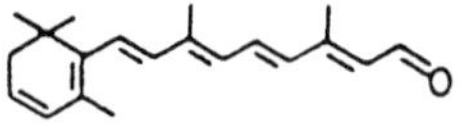	46. 13-CF_3- retinal	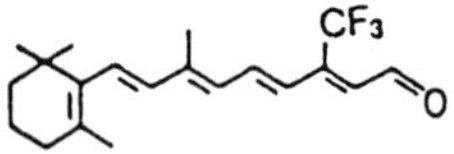

Number and name		Number and name	
47. 11-methyl-merocyanine-retinal	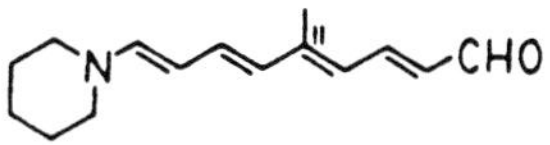 N, CHO	48. C-22-retinal	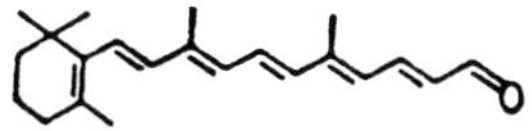 O
49. 5,6-diH-desmetyl-retinal	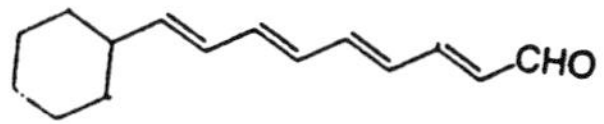 CHO	50. 4-hydroxy-retinal	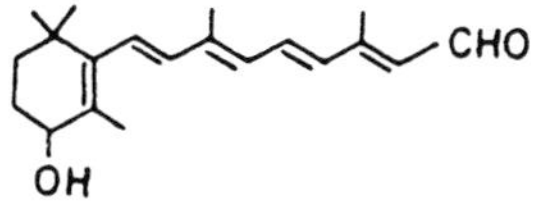 CHO, OH
51. 4-dimethyl-amino-retinal	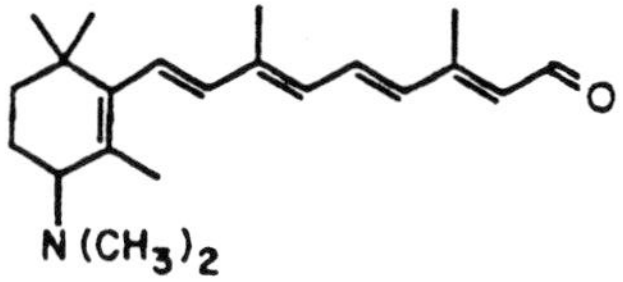 O, $N(CH_3)_2$	52. α-retinal	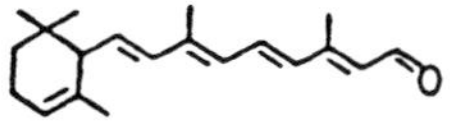 O

Structures of some retinal isomers

Number and name	Structure	Number and name	Structure
53. All-*trans*	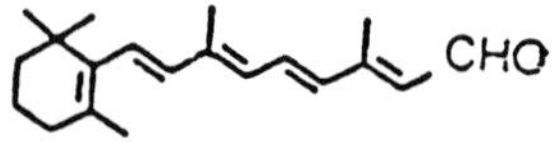 CHO	54. *13-cis*	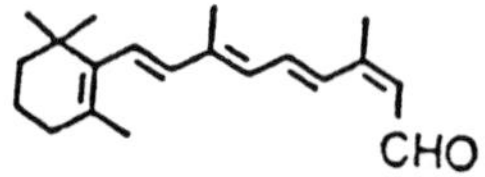 CHO
55. 11-*cis*	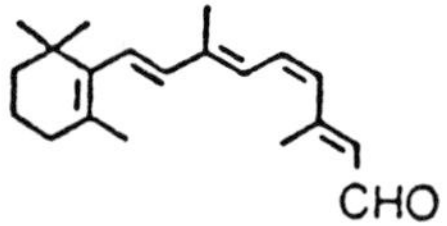 CHO	56. *9-cis*	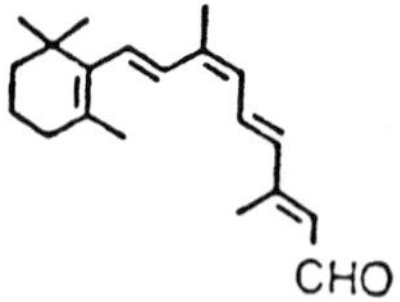 CHO
57. 7-*cis*	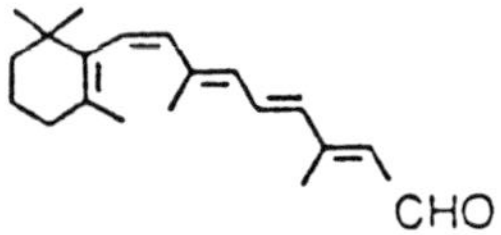 CHO	58. *7,9-dicis*	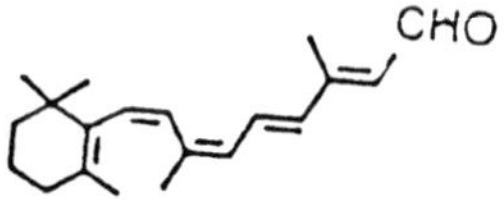 CHO
59. 7,9,11-tricis	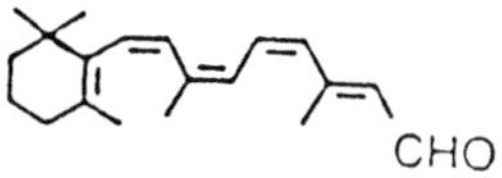 CHO	60. all-*cis*	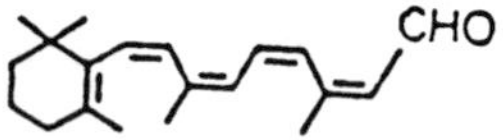 CHO

Number and name	Number and name
61. Chlorophyll	62. Pheophytine

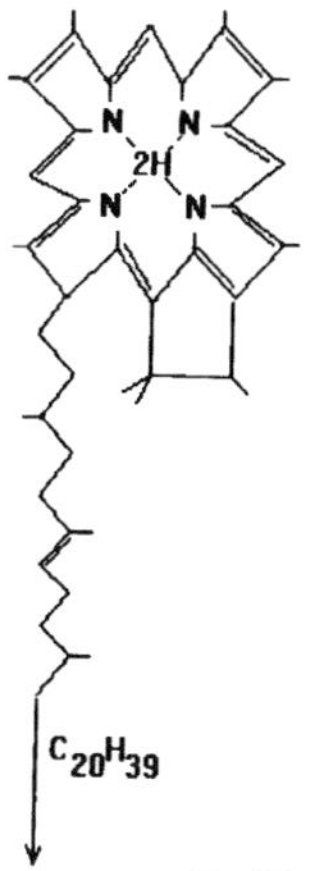

Chlorophyll

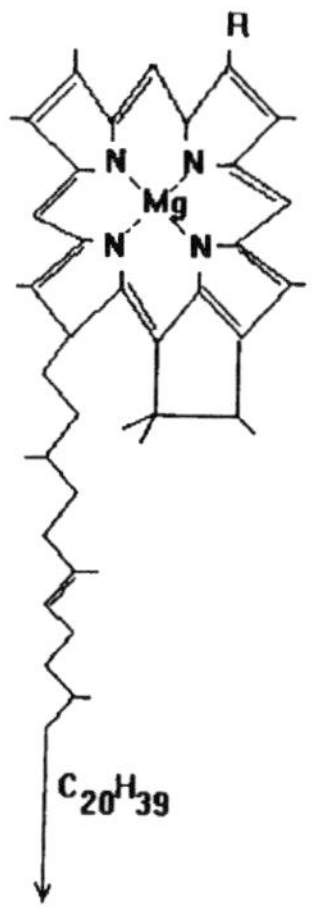

R

N N Mg N N

$C_{20}H_{39}$

Pheophytine

N N 2H N N

$C_{20}H_{39}$

R = CH_3 (chlorophyll a)

R = CHO (chlorophyll b)

Index